A Specialist Periodical Report

Chemical Thermodynamics

Volume 1

A Review of the Recent Literature Published up to December 1971

Senior Reporter

M. L. McGlashan, *Department of Chemistry, University of Exeter*

Reporters

D. Ambrose	**S. G. Frankiss**	**O. Kubaschewski**
J. F. Counsell	**J. H. S. Green**	**I. J. Lawrenson**
J. D. Cox	**A. J. Head**	**J. F. Martin**
W. A. Dench	**E. F. G. Herington**	**P. J. Spencer**

All of: Division of Chemical Standards, National Physical Laboratory, Teddington, Middlesex

The Chemical Society
Burlington House, London, W1V 0BN

ISBN: 0 85186 253 5

Library of Congress Catalog Card No. 72-83456

Organic formulae composed by Wright's Symbolset method

PRINTED IN GREAT BRITAIN BY JOHN WRIGHT AND SONS LTD., AT THE STONEBRIDGE PRESS, BRISTOL BS4 5NU

Foreword

In this first volume of Specialist Periodical Reports on Chemical Thermodynamics I have deliberately made no attempt to cover the whole of the field, but rather have chosen to concentrate on what I have called in Chapter 1 'primary thermodynamic tables' and on the several kinds of measurement that contribute more or less directly to them. These are tables, such as the U.S. National Bureau of Standards 'Technical Note 270' which superseded 'Circular 500', from which the standard equilibrium constants of chemical reactions can be obtained. Except for Chapter 1, which is intended to be an introduction not only to this volume but also to subsequent ones, all of this volume has been written by members of the thermodynamics group of the Division of Chemical Standards of the National Physical Laboratory at Teddington.

I plan in the second volume to concentrate especially on the thermodynamic properties of liquid mixtures of non-electrolytes, and in subsequent volumes to deal with other aspects of chemical thermodynamics (see pp. 29 and 30) as well as bringing up to date topics previously covered.

I regret that I am largely responsible for this volume appearing about a year later than had originally been planned, and I am most grateful to the team of Reporters from the N.P.L. for their patient forbearance during a most trying period.

M. L. McG.

Note: In this Report the word stoichiometric has been so spelled in accord with the IUPAC 'Manual of Symbols and Terminology for Physicochemical Quantities and Units' and the Report 'Quantities, Units, and Symbols' of the Symbols Committee of the Royal Society representing the Royal Society, the Chemical Society, the Faraday Society, and the Institute of Physics, but at variance with the spelling stoicheiometric which is usual Chemical Society practice.

M. L. McG.

Contents

Chapter 8 Statistical Methods for Calculating Thermodynamic Functions

By S. G. Frankiss and J. H. S. Green

1

The Scope of Chemical Thermodynamics

BY M. L. McGLASHAN

1 Introduction

The nature and size of Chemical Thermodynamics make it impracticable to deal with the whole subject in any one volume of Specialist Periodical Reports. Except for the present introductory Chapter, the whole of this first volume has been written by members of the thermodynamics group of the Division of Chemical Standards of the National Physical Laboratory at Teddington, which under Dr Herington's distinguished leadership has made an enormous contribution to the measurement of reliable values of useful chemical-thermodynamic quantities. Such measurements can seldom be made on any scale except in great national laboratories. Until fairly recently the Laboratories of the U.S. Bureau of Standards and of the U.S. Bureau of Mines had produced a quite disproportionate share of the values tabulated, especially in what are called below 'primary thermodynamic tables'. With the passage of the years the N.P.L. group has grown steadily in scientific stature, through the number, the reliability, and the comprehensiveness of the chemical-thermodynamic measurements, and the invention of apparatus, of measuring equipment, and of theoretical ideas, for which its members have been responsible. It is a fitting tribute to Dr Herington and his colleagues that they should have been largely responsible for this first volume.

Thermodynamics consists of a collection of mathematical equations (and also some inequalities) which inter-relate the equilibrium properties of macroscopic systems. Every quantity which occurs in a thermodynamic equation is independently measurable. What does such an equation 'tell one' about one's system? Or, in other words, what can we learn from thermodynamic equations about the microscopic or molecular explanation of macroscopic changes? Nothing whatever. What is a 'thermodynamic theory'? (The phrase is used in the titles of many papers published in reputable chemical journals.) There is no such thing. What then is the use of thermodynamic equations to the chemist? They are indeed useful, but only by virtue of their use for the calculation of some desired quantity which has not been measured, or which is difficult to measure, from others which have been measured, or which are easier to measure.

It will be the aim of successive volumes of these Specialist Periodical Reports on Chemical Thermodynamics to present critical reviews of the experimental methods which have been used for thermodynamic measurements of interest to chemists, to discuss the analysis and accuracy of the results of such measurements, and to review the results in the light of statistical-mechanical theories, where these exist.

From a severely chemical point of view, by far the most important thermodynamic equations are those that relate the yield of a chemical reaction to other thermodynamic quantities. The yield of a chemical reaction, defined for our purpose as the mole fraction of a desired product present at equilibrium at a specified temperature and a specified pressure and for specified initial amounts of all the substances (those that do not appear in the equation for the chemical reaction as well as those that do), can of course be measured directly. To measure the yield of every conceivable chemical reaction under every conceivable condition of temperature, pressure, and initial amounts of all the substances would, however, be a most daunting task to say the least. Happily, it is often unnecessary. Thermodynamic equations can be used to calculate the yield of a given reaction under given conditions, with varying degrees of accuracy, from such measured quantities as a value of the yield of the same reaction under some different conditions, or from the measured values of the yields of a set of reactions for which the equations sum to that of the reaction in question, or from the measured values of quantities, for example calorimetrically determined enthalpy changes and heat capacities, which do not include the yield of any reaction, or from combinations of these.

For some reactions use can also be made of statistical-mechanical equations either (rarely) alone or in combination with some of the quantities alluded to above. These are reactions taking place in systems for which we have a model which is at once realistic enough and mathematically tractable enough to be useful. An example is the calculation of the standard equilibrium constant of a gas reaction (and thence of the yield, but only if the gas mixture is nearly enough perfect) from spectroscopically determined molecular properties. Another example is the use of the Debye–Hückel theory or its extensions to improve the calculation of the yield of a reaction in a dilute electrolyte solution from the standard equilibrium constant of the reaction when, as is usually so, it is not accurate enough to assume that the solution is ideal-dilute.

The results of measurements and calculations of these kinds are made available in *thermodynamic tables.* Such thermodynamic tables can conveniently be divided into two classes. *Primary thermodynamic tables* contain values of the standard equilibrium constant $K^{\ominus}(T)$, which depends only on thermodynamic temperature T, or of other standard thermodynamic functions from which these can be calculated, and when $K^{\ominus}$ is given for only one value of T (commonly 298.15 K), of standard thermodynamic functions which can be used to estimate it at other values of T. The yield of a reaction under specified conditions can be calculated from the standard equilibrium constant $K^{\ominus}$ only when the specified conditions are such that the reacting mixture is nearly enough 'ideal'; in other cases further information is needed, for example the virial coefficients or fugacity coefficients in real gas mixtures, or the activity or osmotic coefficients of substances in real solutions. *Secondary thermodynamic tables* contain such information.

As the result of prodigious efforts, primary thermodynamic tables are now sufficiently extensive to be most useful. The U.S. Bureau of Standards' famous Circular 500, now superseded by the parts of Technical Note 270, forms the best-known example.[1] The primary tables at present available are listed and described by Dr Herington in Chapter 2.

Secondary thermodynamic tables are unfortunately much more sparse. For example, while the deviations from perfect-gas behaviour have been studied for many *pure* gases, relatively little information is yet available even for binary gas mixtures, let alone for the usually multi-component mixtures relevant to chemical reactions in gases. Again, while the activity coefficient of the solute or the osmotic coefficient of the solvent has been measured over useful molality ranges for many solutions of a *single* electrolyte, little such information is yet available for the mixed electrolyte solutions relevant to chemical reactions in solutions.

2 Standard Thermodynamic Functions of Substances

The quantities listed in primary thermodynamic tables are the standard thermodynamic functions of substances, and sums and differences of these, leading to the standard equilibrium constants $K^{\ominus}$ of chemical reactions.

The definition of a standard ($^{\ominus}$) thermodynamic function of a substance B often involves an arbitrarily chosen standard value $p^{\ominus}$ of pressure and an arbitrarily chosen standard value $m^{\ominus}$ of molality.† The value assigned to $p^{\ominus}$ (usually, though not necessarily, 101.325 kPa) and the value assigned to $m^{\ominus}$ (usually, though not necessarily, 1 mol kg^{-1}) must be specified.‡

[1] 'Technical Note 270-3', 1968, and 'Technical Note 270-4', 1969, U.S. National Bureau of Standards, Washington.

† The superscript ° has also been used with the meaning 'standard'; unfortunately it has also been used with other meanings.

‡ I have often been asked recently whether it is not implied by the adoption of the SI, or whether it is not desirable in view of the SI, to change from the generally adopted value of $p^{\ominus}$ to a round number of Pa such as 100 kPa. It is certainly not implied. Whether it is desirable is at least arguable. Were we starting from scratch to compile primary thermodynamic tables it would no doubt be more convenient to adopt the value $p^{\ominus} = 100$ kPa, just as it would then be more convenient to adopt 300 K instead of 298.15 K as the standard temperature for tabulation. In the event, the long-term advantages of a more easily remembered and arithmetically more convenient standard value of pressure, as of temperature, must be weighed against the probable, albeit short-term, confusion which the change would cause. In practice, the change from 101.325 kPa to 100 kPa as the standard value of pressure would cause a value of $K^{\ominus}$ to change by a factor of only 1.013 raised to a power equal to the stoichiometric sum Σ (see Section 3) of the reaction, which for almost all reactions would leave the value of $K^{\ominus}$ effectively unchanged (within experimental error). To the best of my knowledge, none of the responsible international organizations, such as IUPAC, has suggested that this change be made, or has even received a suggestion that this change be made. 'If it were done, when 'tis done, then 'twere well it were done quickly'.

Values having been specified where necessary for $p^\ominus$ and $m^\ominus$, the standard chemical potential $\mu_B^\ominus$ of a substance B, and hence any other standard thermodynamic function of the substance B, depends only on the thermodynamic temperature T. The definition of the standard chemical potential $\mu_B^\ominus$ of a substance B is given: for a component of a gas mixture in Section 4, for a component of a liquid or solid (one-phase) mixture in Section 5, and for a solute in a solution and for the solvent (which may be a mixed one) in Section 6.†

Once the standard chemical potential $\mu_B^\ominus$ has been defined, formulae follow for all the other standard thermodynamic functions obtained by differentiation of $\mu_B^\ominus$ with respect to temperature. In particular,

$$S_B^\ominus = -\,\mathrm{d}\mu_B^\ominus/\mathrm{d}T, \tag{1}$$

$$H_B^\ominus = \mu_B^\ominus - T\,\mathrm{d}\mu_B^\ominus/\mathrm{d}T, \tag{2}$$

$$C_{p,B}^\ominus = -\,T\,\mathrm{d}^2\mu_B^\ominus/\mathrm{d}T^2. \tag{3}$$

The standard thermodynamic functions of a substance are often described as relating to the substance 'in its standard state'. Any difficulties associated with a verbal description of such a 'standard state' disappear when an algebraic definition of the standard chemical potential is given.

3 Standard Equilibrium Constants of Reactions: Part I

We write the general equation for a chemical reaction in the form:

$$0 = \Sigma_B \nu_B B, \tag{4}$$

where B is the symbol of a chemical substance and ν_B is its stoichiometric number, positive for 'products' and negative for 'reactants'. The sum $\Sigma_B\nu_B$ will be denoted by Σ and called the stoichiometric sum of the reaction. The extent of reaction ξ is defined in terms of the initial ($\xi = 0$) amounts of substance n_B^i and the amounts n_B when the extent of reaction is ξ by the relation:

$$n_B = n_B^i + \nu_B\xi. \tag{5}$$

The standard equilibrium constant $K^\ominus$ of reaction (4) is *defined* by the relation:

$$-\,RT\ln K^\ominus = \Sigma_B\nu_B\mu_B^\ominus. \tag{6}$$

Like the $\mu_B^\ominus$'s, $K^\ominus$ therefore depends only on temperature T. It follows immediately that

$$\Delta S^\ominus = \Sigma_B\nu_B S_B^\ominus = R\ln K^\ominus + RT\,\mathrm{d}\ln K^\ominus/\mathrm{d}T, \tag{7}$$

[2] 'Manual of Symbols and Terminology for Physicochemical Quantities and Units', Butterworths, London, 1970; *Pure Appl. Chem.*, 1970, **21**, 1.

† The word 'mixture' is used here to describe a gas or liquid or solid phase containing more than one substance, when the substances are all treated in the same way. The word 'solution' is used to describe a liquid (or rarely solid) phase containing more than one substance, when for convenience one of the substances, called the 'solvent', is treated differently from the other substances, called 'solutes'.[2]

and

$$\Delta H^{\ominus} = \Sigma_B \nu_B H_B^{\ominus} = RT^2 \, \mathrm{d} \ln K^{\ominus}/\mathrm{d}T. \tag{8}$$

The notation $\Delta S^{\ominus}$ for $\Sigma_B \nu_B S_B^{\ominus}$ and $\Delta H^{\ominus}$ for $\Sigma_B \nu_B H_B^{\ominus}$ is a curious one in view of the usual reservation of 'ΔX' to mean

$$\Delta X = X(\text{final state of system}) - X(\text{initial state of system}). \tag{9}$$

This notation is nevertheless widely used, and similarly $\Sigma_B \nu_B \mu_B^{\ominus}$ is often denoted by the symbol $\Delta G^{\ominus}$ and called the standard change of Gibbs function or standard (Gibbs) free energy change for the reaction. Equation (6) may then be rewritten in the more usual form:

$$\Delta G^{\ominus} = \Sigma_B \nu_B \mu_B^{\ominus} = - RT \ln K^{\ominus}. \tag{10}$$

The increasingly common omission of the word 'standard' from the names of $\Delta G^{\ominus}$ and $\Delta S^{\ominus}$ is most deplorable. It results inevitably in confusion of the standard change of Gibbs function $\Delta G^{\ominus}$ with the usually quite different actual change of Gibbs function ΔG, and of the standard entropy change $\Delta S^{\ominus}$ with the usually quite different actual entropy change ΔS, which result when a chemical reaction actually takes place under specified conditions. This confusion is a convenience to those who do not wish to say too clearly what they mean, and especially to those who wish to 'interpret' or 'explain' their observations in terms of vague would-be-thermodynamic arguments about 'order' and 'randomness' or 'mixed-upness' or 'structure-breaking' and 'structure-making', based neither on thermodynamics itself nor on any testable statistical-mechanical model. It will not be among the aims of these Specialist Periodical Reports to deal with such undisciplined 'hand-waving'.

If the value of $K^{\ominus}$ for a reaction is known at one temperature T_1 (often 298.15 K), its value at another temperature T_2 can be obtained by integration of equation (8):

$$\ln K^{\ominus}(T_2) = \ln K^{\ominus}(T_1) + \int_{T_1}^{T_2} \{\Delta H^{\ominus}(T)/RT^2\} \, \mathrm{d}T, \tag{11}$$

provided that $\Delta H^{\ominus}(T)$ is known over the range T_1 to T_2. It is often sufficient to assume that $\Delta H^{\ominus}$ is independent of temperature or to use an average value $\langle \Delta H^{\ominus} \rangle$, in which case

$$\ln K^{\ominus}(T_2) = \ln K^{\ominus}(T_1) + \langle \Delta H^{\ominus} \rangle (T_2 - T_1)/RT_1T_2. \tag{12}$$

Alternatively, $\Delta H^{\ominus}$ can be expressed as a function of temperature, and the integration in equation (11) can be carried out accurately, if, as is often the case, the standard heat capacities $C_{p,B}^{\ominus}$ have been measured and tabulated (either as such or in the form of coefficients of a series in powers of T) at temperatures over the required range. We then use the relation:

$$\Delta H^{\ominus}(T) = \Delta H^{\ominus}(T_0) + \int_{T_0}^{T} \Sigma_B \nu_B C_{p,B}^{\ominus}(T) \, \mathrm{d}T, \tag{13}$$

where T_0 is some temperature which can be outside the range T_1 to T_2, to rewrite equation (13) in the form:

$$\ln K^{\ominus}(T_2) = \ln K^{\ominus}(T_1) + \Delta H^{\ominus}(T_0)(T_2 - T_1)/RT_1T_2 + \int_{T_1}^{T_2}\int_{T_0}^{T} \Sigma_B \nu_B C^{\ominus}_{p,B}(T)\,dT(1/RT^2)\,dT. \quad (14)$$

An alternative route to $K^{\ominus}$, entirely calorimetric except for small corrections, exists by way of the relation:

$$-RT \ln K^{\ominus} = \Sigma_B \nu_B H^{\ominus}_B - T\Sigma_B \nu_B S^{\ominus}_B, \quad (15)$$

whenever $\Sigma_B \nu_B H^{\ominus}_B = \Delta H^{\ominus}$ and $\Sigma_B \nu_B S^{\ominus}_B = \Delta S^{\ominus}$ can each be measured calorimetrically.

The standard enthalpy change $\Delta H^{\ominus}$ can be determined, apart from corrections which are usually small and easily made, from calorimetric measurements of energies or enthalpies of combustion or of other reactions which can be carried out quantitatively and quickly enough in a calorimeter, or by combining these according to Hess's law. The determination of enthalpies of reaction from calorimetric results is briefly introduced in Section 8. The subject of combustion and reaction calorimetry is reviewed by Dr Head in Chapter 3, and as seen from a high-temperature and metallurgical point of view by Drs Kubaschewski, Spencer, and Dench in Chapter 9.

The standard entropy $S^{\ominus}_B$(g or l or s, T) of a substance B at temperature T may be divided into two terms:

$$S^{\ominus}_B(\text{g or l or s}, T) = \{S^{\ominus}_B(\text{g or l or s}, T) - S_B(\text{s}, T \to 0)\} + S_B(\text{s}, T \to 0), \quad (16)$$

of which the first, {...}, can be determined, apart from corrections which are usually small and easily made, from calorimetric measurements of heat capacities and of enthalpies of phase transition. The determination of the term {...} from calorimetric results is briefly introduced in Section 9. The subject of heat-capacity calorimetry is reviewed, with particular reference to organic substances and to modern high-precision methods, by Dr Martin in Chapter 4, and with particular reference to high-temperature and metallurgical applications by Drs Kubaschewski, Spencer, and Dench in Chapter 9. Enthalpies of phase transition will be dealt with explicitly in a subsequent volume. They are, however, frequently mentioned in this volume, especially in Chapter 9. Enthalpies of evaporation (or 'vaporization') $\Delta_l^g H$ and enthalpies of sublimation $\Delta_s^g H$ can be calculated from measurements of the temperature-dependence of the vapour pressure p_σ together with measured values of the molar volumes of the two orthobaric phases according to Clapeyron's equation:

$$\Delta^g_{\text{l or s}} H = T\Delta^g_{\text{l or s}} V\,dp_\sigma/dT. \quad (17)$$

Methods of measurement of vapour pressures and of analysis of the results are reviewed by Dr Ambrose in Chapter 7, and from the point of view of high-temperature and metallurgical applications by Drs Kubaschewski, Spencer, and Dench in Chapter 9.

When $\{S_B^\ominus(\text{g or l or s}, T) - S_B(\text{s}, T \to 0)\}$ has been determined for each substance, $\Delta S^\ominus = \Sigma_B \nu_B S_B^\ominus$ can be calculated except for the quantity $\Sigma_B \nu_B S_B(\text{s}, T \to 0)$, which according to the third law of thermodynamics may nearly always be set equal to zero.

The calculation of standard entropies of *gaseous* substances from spectroscopically determined molecular properties according to statistical-mechanical formulae is mentioned again briefly in Section 4 and is extensively reviewed by Drs Frankiss and Green in Chapter 8.

For some *gas* reactions it is possible to calculate $K^\ominus(T)$ either completely, or more usually with the help of a calorimetric value of $\Delta H^\ominus(298.15\ \text{K})$, by substitution of spectroscopically determined values of molecular properties into a statistical-mechanical formula. Such calculations are briefly introduced in Section 7 and are extensively reviewed by Drs Frankiss and Green in Chapter 8.

For some reactions in *electrolyte solutions* it is possible to calculate $K^\ominus(T)$ from the results of e.m.f. measurements on galvanic cells. This subject is mentioned again in Section 7, is reviewed from a high-temperature and metallurgical point of view by Drs Kubaschewski, Spencer, and Dench in Chapter 9, and will be dealt with thoroughly for ordinary-temperature liquid electrolyte solutions in a subsequent volume.

For some reactions in electrolyte solutions it is also possible to calculate $K^\ominus(T)$ from the results of measurements of electric conductivity; this subject will also be dealt with in a later volume.

Apart from standard molar enthalpies of formation $\Delta_f H_B^\ominus(T)$ of substances B (see Section 8), most commonly given at $T = 298.15$ K and at $T \to 0$, and standard molar entropies $S_B^\ominus(T)$ (see Section 9) and standard molar heat capacities $C_{p,B}^\ominus(T)$, each most commonly given at $T = 298.15$ K, other quantities found in thermodynamic tables include values of the increments $\{H_B^\ominus(T_2) - H_B^\ominus(T_1)\}$ in the standard molar enthalpies, especially for $T_2 = 298.15$ K, $T_1 \to 0$, and also values of the quantity:

$$\begin{aligned}\Phi_B^\ominus(T) &= \{H_B^\ominus(T) - H_B^\ominus(0)\}/T - \{S_B^\ominus(T) - S_B(0)\} \\ &= \{\mu_B^\ominus(T) - H_B^\ominus(0)\}/T + S_B(0),\end{aligned} \qquad (18)$$

which is sometimes known by the curious name (standard molar) 'free energy function'. The use of the word 'function' to distinguish the (standard molar) 'free energy' (standard molar Gibbs function or standard chemical potential) $G_B^\ominus$ or $\mu_B^\ominus$, from the (standard molar) 'free energy function' ('standard molar Gibbs function function'?) defined by (18) is indefensible. The quantity (18), here denoted by $\Phi_B^\ominus(T)$, is, however, a most useful one, being given (for a monocrystalline solid, or, with appropriate allowance for phase changes, for other states of matter) by the formula:

$$\Phi_B^\ominus(T) = T^{-1}\int_0^T C_{p,B}^\ominus\, \mathrm{d}T - \int_0^T C_{p,B}^\ominus\, \mathrm{d}\ln T, \qquad (19)$$

and being related to the standard equilibrium constant $K^{\ominus}(T)$ by the formula:

$$-R\ln K^{\ominus}(T) = \Delta\Phi^{\ominus}(T) + \Delta H^{\ominus}(0)/T - \Delta S(0). \qquad (20)$$

4 Standard Thermodynamic Functions of a Substance in a Gas Mixture

The standard chemical potential $\mu_{\mathrm{B}}^{\ominus}(\mathrm{g}, T)$ of a gaseous substance B at temperature T is defined by the relation:

$$\mu_{\mathrm{B}}^{\ominus}(\mathrm{g}, T) = \mu_{\mathrm{B}} - RT\ln(x_{\mathrm{B}}p/p^{\ominus}) - \int_0^p (V_{\mathrm{B}} - RT/p)\,\mathrm{d}p, \qquad (21)$$

where $\mu_{\mathrm{B}} = \mu_{\mathrm{B}}(\mathrm{g}, T, p, x_{\mathrm{B}}, x_{\mathrm{C}}, \ldots)$ and $V_{\mathrm{B}} = V_{\mathrm{B}}(\mathrm{g}, T, p, x_{\mathrm{B}}, x_{\mathrm{C}}, \ldots)$ are the chemical potential and partial molar volume, respectively, of the substance B in a gas mixture containing mole fractions x_{B} of B, x_{C} of C, ..., at temperature T and pressure p, and $p^{\ominus}$ is a standard value of pressure.

Formulae for all the other standard functions of a gaseous substance B follow from equation (21). In particular,

$$S_{\mathrm{B}}^{\ominus}(\mathrm{g}, T) = S_{\mathrm{B}} + R\ln(x_{\mathrm{B}}p/p^{\ominus}) + \int_0^p \{(\partial V_{\mathrm{B}}/\partial T)_p - R/p\}\,\mathrm{d}p, \qquad (22)$$

$$H_{\mathrm{B}}^{\ominus}(\mathrm{g}, T) = H_{\mathrm{B}} - \int_0^p \{V_{\mathrm{B}} - T(\partial V_{\mathrm{B}}/\partial T)_p\}\,\mathrm{d}p, \qquad (23)$$

$$C_{p,\mathrm{B}}^{\ominus}(\mathrm{g}, T) = C_{p,\mathrm{B}} + \int_0^p T(\partial^2 V_{\mathrm{B}}/\partial T^2)_p\,\mathrm{d}p. \qquad (24)$$

The fugacity f_{B} (or p_{B}^{*}, but here we shall reserve the superscript * for the meaning 'pure') [2] of a substance in a gas mixture is a quantity, having the same dimension as a pressure, defined by the relation:

$$f_{\mathrm{B}} = (x_{\mathrm{B}}p)\exp\left\{\int_0^p (V_{\mathrm{B}}/RT - 1/p)\,\mathrm{d}p\right\}, \qquad (25)$$

so that equation (21) may be rewritten in the abbreviated form:

$$\mu_{\mathrm{B}}^{\ominus}(g, T) = \mu_{\mathrm{B}} - RT\ln(f_{\mathrm{B}}/p^{\ominus}). \qquad (26)$$

For a *pure* gaseous substance, equations (21)—(26) still apply, but with $x_{\mathrm{B}} = 1$ so that $x_{\mathrm{C}} = x_{\mathrm{D}} = \ldots = 0$.

A gas mixture of substances B, C, ..., for which $V_{\mathrm{B}} = V_{\mathrm{C}} = \ldots = RT/p$, or equivalently $f_{\mathrm{B}} = x_{\mathrm{B}}p$, $f_{\mathrm{C}} = x_{\mathrm{C}}p$, ..., is called a *perfect gas mixture.* Any real gas mixture behaves more and more nearly as a perfect gas mixture the lower the pressure p.

According to Technical Note 270:[1] 'For a gas the standard state is the hypothetical ideal gas at unit fugacity, in which state the enthalpy is that of the real gas at the same temperature and at zero pressure.' This complicated verbal description of the 'standard state' is useful only insofar as it enables one correctly to construct equation (21). Happily, all that is ever needed is the definition of the standard chemical potential given in equation (21), together of course with a declaration of the value arbitrarily assigned to the standard pressure $p^{\ominus}$.

It might, however, be helpful to show where equation (21) comes from, or alternatively to show how we know that the right-hand side of equation (21) depends for a given substance only on temperature. We make use of the identity:

$$\mu_{\mathrm{B}}(\mathrm{g}, T, p \to 0, x_{\mathrm{B}}, x_{\mathrm{C}}, \ldots) = \mu_{\mathrm{B}}(\mathrm{pg}, T, p \to 0, x_{\mathrm{B}}, x_{\mathrm{C}}, \ldots), \qquad (27)$$

where pg denotes a perfect gas, which expresses the fact that any gas becomes perfect in the limit as the pressure tends to zero. We use equation (27) to write the further identity:

$$\begin{aligned}\mu_{\mathrm{B}}(\mathrm{g}, T, p, x_{\mathrm{B}}, x_{\mathrm{C}}, \ldots) &= \mu_{\mathrm{B}}(\mathrm{pg}, T, p, x_{\mathrm{B}}, x_{\mathrm{C}}, \ldots) \\ &\quad + \{\mu_{\mathrm{B}}(\mathrm{g}, T, p, x_{\mathrm{B}}, x_{\mathrm{C}}, \ldots) - \mu_{\mathrm{B}}(\mathrm{g}, T, p \to 0, x_{\mathrm{B}}, x_{\mathrm{C}}, \ldots)\} \\ &\quad - \{\mu_{\mathrm{B}}(\mathrm{pg}, T, p, x_{\mathrm{B}}, x_{\mathrm{C}}, \ldots) - \mu_{\mathrm{B}}(\mathrm{pg}, T, p \to 0, x_{\mathrm{B}}, x_{\mathrm{C}}, \ldots)\}. \qquad (28)\end{aligned}$$

In view of the relations $(\partial\mu_{\mathrm{B}}/\partial p) = V_{\mathrm{B}}$ and $V_{\mathrm{B}}(\mathrm{pg}) = RT/p$ we can rewrite equation (28) in the form:

$$\begin{aligned}\mu_{\mathrm{B}}(\mathrm{g}, T, p, x_{\mathrm{B}}, x_{\mathrm{C}}, \ldots) &= \mu_{\mathrm{B}}(\mathrm{pg}, T, p, x_{\mathrm{B}}, x_{\mathrm{C}}, \ldots) \\ &\quad + \int_0^p (V_{\mathrm{B}} - RT/p)\,\mathrm{d}p \\ &= \mu_{\mathrm{B}}(\mathrm{pg}, T, p^{\ominus}, x_{\mathrm{B}}, x_{\mathrm{C}}, \ldots) \\ &\quad + RT\ln(p/p^{\ominus}) + \int_0^p (V_{\mathrm{B}} - RT/p)\,\mathrm{d}p. \qquad (29)\end{aligned}$$

We know from experiment that

$$\mu_{\mathrm{B}}(\mathrm{pg}, T, p, x_{\mathrm{B}}, x_{\mathrm{C}}, \ldots) = \mu_{\mathrm{B}}^{*}(\mathrm{pg}, T, p) + RT\ln x_{\mathrm{B}}, \qquad (30)$$

where μ_{B}^{*} denotes the chemical potential of *pure* substance B. It is to be emphasized that equation (30) is extra-thermodynamic, however familiar it might be in the light of its successful application in the formula for the equilibrium constant of a reaction in a perfect gas mixture, and however plausible it might be in the light of the statistical-mechanical theory of an assembly of non-interacting non-localized particles of two or more kinds. Substituting equation (30) into equation (29) we obtain

$$\begin{aligned}\mu_{\mathrm{B}}(\mathrm{g}, T, p, x_{\mathrm{B}}, x_{\mathrm{C}}, \ldots) &= \mu_{\mathrm{B}}^{*}(\mathrm{pg}, T, p^{\ominus}) + RT\ln(x_{\mathrm{B}}\,p/p^{\ominus}) \\ &\quad + \int_0^p (V_{\mathrm{B}} - RT/p)\,\mathrm{d}p. \qquad (31)\end{aligned}$$

We now recover equation (21) by identifying the standard chemical potential $\mu_{\mathrm{B}}^{\ominus}(T)$ with the chemical potential $\mu_{\mathrm{B}}^{*}(\mathrm{pg}, T, p^{\ominus})$ of the pure substance B in the hypothetically perfect gaseous state at the same temperature T and at the standard pressure $p^\ominus$.

The evaluation of the integral in equation (21) demands measurements of the molar volume $V_m(g, T, p, x_B, x_C, \ldots)$ of the gas mixture as a function of $p, x_B, x_C, \ldots$. Extensive measurements of $V_m(g, T, p)$ have been made for many *pure* gases. Unfortunately, however, relatively few gas mixtures have been extensively studied, and this is especially true of the usually multi-component mixtures relevant to chemical reactions.

The experimental methods which have been used and the results obtained for *pure* gases are reviewed by Drs Cox and Lawrenson in Chapter 5. Gas mixtures will be dealt with in a subsequent volume.

Measurements of quantities closely related to the integrals in equations (23) and (24) can also be made by flow calorimetry. If power P is supplied (usually electrically) to a gas flowing in an adiabatic calorimeter from temperature T_1 and pressure p_1 to temperature T_2 and pressure p_2 with flow rate f, then

$$P/f = H(T_2, p_2) - H(T_1, p_1)$$
$$= \int_{T_1}^{T_2} C_p(T, p_2)\,dT + \int_{p_1}^{p_2} \{V - T(\partial V/\partial T)_p\}(T_1, p)\,dp. \qquad (32)$$

Such measurements are usually made with no throttle ($p_1 = p_2$; heat capacity), or with no power ($P = 0$; Joule–Thomson), or with the power adjusted so as to restore the temperature of the gas after it has passed through the throttle ($T_1 = T_2$; isothermal Joule–Thomson). The equations for these special cases can be readily derived from equation (32).

Flow-calorimetric measurements on pure gases are reviewed by Dr Counsell in Chapter 6.

The calorimetric determination of standard entropies of gaseous substances is briefly introduced in Section 9.

The calculation of standard entropies of gaseous substances from spectroscopically determined molecular properties according to statistical-mechanical formulae is reviewed by Drs Frankiss and Green in Chapter 8.

The standard entropies of gaseous substances found in primary thermodynamic tables are either measured calorimetric values of $\{S^\ominus(g, T) - S(s, T \to 0)\}$ or 'spectroscopic' values of $S^\ominus(g, T)$ calculated from the appropriate formula ignoring the effects of nuclear spins and of the existence of isotopes. The values of these two quantities are known to be identical for almost all gases. For a few gases, notably CO, N_2O, NO, and H_2O, however, these two quantities are not quite the same, the latter being the greater. It is usually the latter which is tabulated in these cases,[1] and this is the value which should be used in calculations of equilibrium constants. For H_2 (and D_2) the former value agrees with the latter only if the extrapolation of the heat capacity of the solid to $T \to 0$ is made from $T < 1.5$ K, so as to include the 'heat capacity anomaly' found below 12 K.

5 Standard Thermodynamic Functions of a Substance in a Liquid or Solid Mixture

The standard chemical potential $\mu_B^\ominus(\text{l or s}, T)$ of a liquid or solid substance B at temperature T is defined by the relation:

$$\mu_B^\ominus(\text{l or s}, T) = \mu_B^* + \int_p^{p^\ominus} V_B^* \,\mathrm{d}p, \tag{33}$$

where $\mu_B^* = \mu_B(\text{l or s}, T, p, x_B = 1)$ and $V_B^* = V_B(\text{l or s}, T, p, x_B = 1)$ are the chemical potential and molar volume respectively of the *pure* substance B at temperature T and pressure p, and where $p^\ominus$ is a standard value of pressure.

Formulae for all the other standard thermodynamic functions of a liquid or solid substance B follow from equation (33). In particular,

$$S_B^\ominus(\text{l or s}, T) = S_B^* - \int_p^{p^\ominus} (\partial V_B^*/\partial T)_p \,\mathrm{d}p, \tag{34}$$

$$H_B^\ominus(\text{l or s}, T) = H_B^* + \int_p^{p^\ominus} \{V_B^* - T(\partial V_B^*/\partial T)_p\} \,\mathrm{d}p, \tag{35}$$

$$C_{p,B}^\ominus(\text{l or s}, T) = C_{p,B}^* - \int_p^{p^\ominus} T(\partial^2 V_B^*/\partial T^2)_p \,\mathrm{d}p. \tag{36}$$

When $p^\ominus$ is chosen, as it usually is, to be 101.325 kPa, and when p is close to atmospheric pressure, the integrals in equations (33)—(36) make only small contributions to the standard functions and are often neglected.

According to Technical Note 270:[1] 'For a pure solid or liquid, the standard state is the substance in the condensed phase under a pressure of one atmosphere.' For each substance in a liquid or solid mixture the 'standard state' is the same as for the pure substance.

The activity coefficient $f_B(\text{l or s}, T, p, x_B, x_C, \ldots)$ of a substance B in a liquid or solid mixture containing mole fractions $x_B, x_C, \ldots$, of substances B, C, ..., is a number defined by the relation:

$$\mu_B(\text{l or s}, T, p, x_B, x_C, \ldots) = \mu_B^*(\text{l or s}, T, p) + RT\ln(x_B f_B). \tag{37}$$

It follows from this definition that

$$\lim_{x_B \to 1} f_B = 1, \quad (\text{all B}). \tag{38}$$

A liquid mixture of substances B, C, ..., for which $f_B = f_C = \ldots = 1$ is called an *ideal mixture*.[2] Few real mixtures, however, behave with useful accuracy as though they were ideal.

Formulae for all other partial molar thermodynamic functions of B expressed in terms of the activity coefficient f_B follow from (37). In particular,

$$S_B(\text{l or s}, T, p, x_B, x_C, \ldots) = S_B^*(\text{l or s}, T, p) - R\ln(x_B f_B) - RT(\partial \ln f_B/\partial T)_p, \tag{39}$$

$$H_B(\text{l or s}, T, p, x_B, x_C, \ldots) = H_B^*(\text{l or s}, T, p) - RT^2(\partial \ln f_B/\partial T)_p, \tag{40}$$

$$V_B(\text{l or s}, T, p, x_B, x_C, \ldots) = V_B^*(\text{l or s}, T, p) + RT(\partial \ln f_B/\partial p)_T. \tag{41}$$

Eliminating μ_B^* between (33) and (37), we obtain

$$\mu_B^{\ominus}(\text{l or s}, T) = \mu_B - RT\ln(x_B f_B) + \int_p^{p^{\ominus}} V_B^* \,dp, \tag{42}$$

which is the most useful form for insertion into equation (6) to obtain an equilibrium constant.

Formulae for all the other standard thermodynamic functions follow from equation (42). In particular,

$$S_B^{\ominus}(\text{l or s}, T) = S_B + R\ln(x_B f_B) + RT(\partial \ln f_B/\partial T)_p - \int_p^{p^{\ominus}} (\partial V_B^*/\partial T)_p \,dp, \tag{43}$$

$$H_B^{\ominus}(\text{l or s}, T) = H_B + RT^2(\partial \ln f_B/\partial T)_p + \int_p^{p^{\ominus}} \{V_B^* - T(\partial V_B^*/\partial T)_p\} \,dp. \tag{44}$$

The molar change of Gibbs function on mixing, $\Delta_M G_m$, is defined, for a binary mixture containing mole fractions $(1 - x)$ of A and x of B, by the relation:

$$\begin{aligned}\Delta_M G_m(T, p, x) &= G_m(T, p, x) - (1 - x)G_m(T, p, 0) - xG_m(T, p, 1) \\ &= (1 - x)\{\mu_A(T, p, x) - \mu_A(T, p, 0)\} \\ &\quad + x\{\mu_B(T, p, x) - \mu_B(T, p, 1)\},\end{aligned} \tag{45}$$

or, in view of equation (37),

$$\begin{aligned}\Delta_M G_m(T, p, x) &= RT\{(1 - x)\ln(1 - x) + x\ln x\} \\ &\quad + RT\{(1 - x)\ln f_A(T, p, x) + x\ln f_B(T, p, x)\}.\end{aligned} \tag{46}$$

Formulae for all the other mixing functions follow from equations (45) or (46).

The molar excess functions X_m^E are defined by the relation:

$$X_m^E = X_m - X_m^{id}, \tag{47}$$

where the superscript ^{id} refers to an ideal mixture at the same temperature, pressure, and composition. Thus

$$G_m^E = \Delta_M G_m - \Delta_M G_m^{id} = RT\{(1 - x)\ln f_A + x\ln f_B\}, \tag{48}$$

$$\mu_A^E = \mu_A - \mu_A^{id} = RT\ln f_A, \tag{49}$$

$$\mu_B^E = \mu_B - \mu_B^{id} = RT\ln f_B, \tag{50}$$

$$\begin{aligned}S_m^E &= \Delta_M S_m - \Delta_M S_m^{id} \\ &= -R[(1 - x)\{\ln f_A + T(\partial \ln f_A/\partial T)_p\} \\ &\quad + x\{\ln f_B + T(\partial \ln f_B/\partial T)_p\}],\end{aligned} \tag{51}$$

$$\begin{aligned}H_m^E &= \Delta_M H_m - \Delta_M H_m^{id} = \Delta_M H_m \\ &= -RT^2\{(1 - x)(\partial \ln f_A/\partial T)_p + x(\partial \ln f_B/\partial T)_p\},\end{aligned} \tag{52}$$

$$\begin{aligned}V_m^E &= \Delta_M V_m - \Delta_M V_m^{id} = \Delta_M V_m \\ &= RT\{(1 - x)(\partial \ln f_A/\partial p)_T + x(\partial \ln f_B/\partial p)_T\}.\end{aligned} \tag{53}$$

Activity coefficients and thence excess Gibbs functions can be calculated from the results of measurements at given temperature of the pressure and the compositions of the liquid (or solid) mixture and of the coexisting gas phase. Excess enthalpies can be measured calorimetrically and excess volumes dilatometrically. The experimental methods used to measure excess functions and the results obtained will be reviewed in a subsequent volume.

6 Standard Thermodynamic Functions of a Solute and of the Solvent in a Liquid (or Solid) Solution

In this Section we denote the solvent in a liquid (or rarely solid) solution by A and the solutes by B, C, ... The molality m_B of a solute B is defined as n_B/n_AM_A, where n_B denotes the amount of solute substance B, n_A the amount of solvent substance A, and M_A the molar mass of A. We shall deal here only with molality-based definitions and not with those based on concentration.[1, 2] The superscript $^\infty$ attached to the symbol for a property of a solution denotes the limiting value of the property at infinite dilution: $\Sigma_B m_B \to 0$.

Solutes.—The standard chemical potential $\mu_B^\ominus(\text{solute}, T)$ of a solute substance B at temperature T is defined by the relation:

$$\mu_B^\ominus(\text{solute}, T) = \{\mu_B - RT\ln(m_B/m^\ominus)\}^\infty + \int_p^{p^\ominus} V_B^\infty\,dp, \qquad (54)$$

where $\mu_B = \mu_B(\text{solute}, T, p, m_B, m_C, \ldots)$ and $V_B = V_B(\text{solute}, T, p, m_B, m_C, \ldots)$ are the chemical potential and partial molar volume respectively of the solute substance B in a solution containing molalities m_B, m_C, ..., of the solute substances B, C, ..., in a solvent substance A (which may itself be a mixture) at temperature T and pressure p, and where $p^\ominus$ is a standard value of pressure usually chosen to be 101.325 kPa and $m^\ominus$ is a standard value of molality usually chosen to be 1 mol kg^{-1}.

Formulae for all the other standard thermodynamic functions of a solute substance B follow from equation (54). In particular,

$$S_B^\ominus(\text{solute}, T) = \{S_B + R\ln(m_B/m^\ominus)\}^\infty - \int_p^{p^\ominus} (\partial V_B^\infty/\partial T)_p\,dp, \qquad (55)$$

$$H_B^\ominus(\text{solute}, T) = H_B^\infty + \int_p^{p^\ominus} \{V_B^\infty - T(\partial V_B^\infty/\partial T)_p\}\,dp, \qquad (56)$$

$$C_{p,B}^\ominus(\text{solute}, T) = C_{p,B}^\infty - \int_p^{p^\ominus} T(\partial^2 V_B^\infty/\partial T^2)_p\,dp. \qquad (57)$$

When $p^\ominus$ is chosen to be 101.325 kPa, and when p is close to atmospheric pressure, the integrals in equations (54)—(57) make only small contributions to the standard functions and are usually neglected.

According to Technical Note 270:[1] 'The standard state for a solute in aqueous solution is taken as the hypothetical ideal solution of unit molality.' Non-aqueous 'solutions' are treated as mixtures in Technical Note 270.

The activity coefficient γ_B of a solute substance B in a solution containing molalities m_B, m_C, ..., of the solute substances B, C, ..., is a number defined by the relation:

$$\mu_B(\text{solute}, T, p, m_B, m_C, \ldots) = \{\mu_B(\text{solute}, T, p, m_B, m_C, \ldots) - RT\ln(m_B/m^\ominus)\}^\infty + RT\ln(m_B\gamma_B/m^\ominus). \quad (58)$$

It follows from this definition that

$$\gamma_B^\infty = 1, \quad (\text{all B}). \quad (59)$$

Formulae for all other partial molar thermodynamic functions of B expressed in terms of the activity coefficient γ_B follow from equation (58). In particular,

$$S_B(\text{solute}, T, p, m_B, m_C, \ldots) = \{S_B(\text{solute}, T, p, m_B, m_C, \ldots) + R\ln(m_B/m^\ominus)\}^\infty - R\ln(m_B\gamma_B/m^\ominus) - RT(\partial \ln \gamma_B/\partial T)_p, \quad (60)$$

$$H_B(\text{solute}, T, p, m_B, m_C, \ldots) = H_B^\infty(\text{solute}, T, p) - RT^2(\partial \ln \gamma_B/\partial T)_p, \quad (61)$$

$$V_B(\text{solute}, T, p, m_B, m_C, \ldots) = V_B^\infty(\text{solute}, T, p) + RT(\partial \ln \gamma_B/\partial p)_T. \quad (62)$$

A solution of solute substances B, C, ..., in a solvent substance A for which $\gamma_B = \gamma_C = \ldots = 1$ is called an *ideal-dilute solution.*[2] Any real solution behaves more and more nearly like an ideal-dilute solution the smaller the sum $\Sigma_B m_B$ of the molalities of all the solutes. The implications of that statement are quite different for solutions of non-electrolytes and for solutions of electrolytes. For solutions of non-electrolytes γ_B will not usually differ from unity by more than 1 per cent unless $\Sigma_B m_B$ is greater than about 1 mol kg^{-1}; for solutions of electrolytes γ_B will usually differ from unity by more than 1 per cent even when $\Sigma_B m_B$ is as low as 10^{-4} mol kg^{-1}.

Eliminating the term $\{\ldots\}^\infty$ between equations (54) and (58) we obtain

$$\mu_B^\ominus(\text{solute}, T) = \mu_B - RT\ln(m_B\gamma_B/m^\ominus) + \int_p^{p^\ominus} V_B^\infty\, dp, \quad (63)$$

which is the most useful form for insertion into equation (6) to obtain an equilibrium constant.

Formulae for all the other standard thermodynamic functions follow from equation (63). In particular,

$$S_B^\ominus(\text{solute}, T) = S_B + R\ln(m_B\gamma_B/m^\ominus) + RT(\partial \ln \gamma_B/\partial T)_p - \int_p^{p^\ominus} (\partial V_B^\infty/\partial T)_p\, dp, \quad (64)$$

$$H_B^\ominus(\text{solute}, T) = H_B + RT^2(\partial \ln \gamma_B/\partial T)_p + \int_p^{p^\ominus} \{V_B^\infty - T(\partial V_B^\infty/\partial T)_p\}\, dp. \quad (65)$$

In principle, the activity coefficients γ_B of solute substances B in a solution can be directly determined from the results of measurements at given temperature of the pressure and the compositions of the liquid (or solid) solution and of the coexisting gas phase. In practice, this method fails unless the solutes have volatilities comparable with that of the solvent. The method therefore usually fails for electrolyte solutions, for which measurements of γ_B are, in practice, much more important than for non-electrolyte solutions. Three practical methods are available. If the osmotic coefficient of the solvent has been measured over a sufficient range of molalities, the activity coefficients γ_B can be calculated; the method is outlined below under the sub-heading Solvent. The ratio γ''_B/γ'_B of the activity coefficients of a solute B in two solutions, each saturated with respect to solid B in the same solvent but with different molalities of other solutes, is equal to the ratio m'_B/m''_B of the molalities (solubilities expressed as molalities) of B in the saturated solutions. If a justifiable extrapolation to $\Sigma_B m_B \rightarrow 0$ can be made, then the separate γ_B's can be found. The method is especially useful when B is a sparingly soluble salt and the solubility is measured in the presence of varying molalities of other more soluble salts. Finally, the activity coefficient of an electrolyte can sometimes be obtained from e.m.f. measurements on galvanic cells. The measurement of activity coefficients and analysis of the results both for solutions of a single electrolyte and for solutions of two or more electrolytes will be dealt with in a subsequent volume. Unfortunately, few activity coefficients have been measured in the usually multi-solute solutions relevant to chemical reactions in solution.

Solvent.—The standard chemical potential $\mu^{\ominus}_A$(solvent, T) of the solvent A at temperature T is defined by the relation:

$$\mu^{\ominus}_A(\text{solvent}, T) = \mu^*_A + \int_p^{p^{\ominus}} V^*_A \, \mathrm{d}p, \qquad (66)$$

where $\mu^*_A = \mu^*_A(\text{solvent}, T, p)$ and $V^*_A = V^*_A(\text{solvent}, T, p)$ are the chemical potential and molar volume respectively of the *pure* solvent substance A at the temperature T and pressure p, and where $p^{\ominus}$ is a standard value of pressure. We note that equation (66) for a solvent in a solution is of exactly the same form as equation (33) for any pure liquid or solid or for any component of a liquid or solid mixture.

Formulae for all the other standard thermodynamic functions of the solvent substance A follow from equation (66). In particular,

$$S^{\ominus}_A(\text{solvent}, T) = S^*_A - \int_p^{p^{\ominus}} (\partial V^*_A/\partial T)_p \, \mathrm{d}p, \qquad (67)$$

$$H^{\ominus}_A(\text{solvent}, T) = H^*_A + \int_p^{p^{\ominus}} \{V^*_A - T(\partial V^*_A/\partial T)_p\} \, \mathrm{d}p, \qquad (68)$$

$$C^{\ominus}_{p,A}(\text{solvent}, T) = C^*_{p,A} - \int_p^{p^{\ominus}} T(\partial^2 V^*_A/\partial T^2)_p \, \mathrm{d}p. \qquad (69)$$

When $p^\ominus$ is chosen to be 101.325 kPa, and when p is close to atmospheric pressure, the integrals in equations (66)—(69) make only small contributions to the standard functions and are usually neglected.

The osmotic coefficient ϕ of a solvent substance A in a solution containing molalities m_B, m_C, ..., of the solute substances B, C, ..., is a number defined by the relation:

$$\mu_A(\text{solvent}, T, p, m_B, m_C, \ldots) = \mu_A^*(\text{solvent}, T, p) - \phi RTM_A\Sigma_B m_B, \quad (70)$$

where M_A denotes the molar mass of the solvent substance A. It can be shown with the help of the Gibbs–Duhem equation that $\phi = 1$ for an ideal-dilute solution, defined as above as one in which $\gamma_B = \gamma_C = \ldots = 1$.

Formulae for all other partial molar thermodynamic functions of A expressed in terms of the osmotic coefficient ϕ follow from equation (70). In particular,

$$S_A(\text{solvent}, T, p, m_B, m_C, \ldots) = S_A^*(\text{solvent}, T, p) + \phi RM_A\Sigma_B m_B + (\partial\phi/\partial T)_p RTM_A\Sigma_B m_B, \quad (71)$$

$$H_A(\text{solvent}, T, p, m_B, m_C, \ldots) = H_A^*(\text{solvent}, T, p) + (\partial\phi/\partial T)_p RT^2M_A\Sigma_B m_B, \quad (72)$$

$$V_A(\text{solvent}, T, p, m_B, m_C, \ldots) = V_A^*(\text{solvent}, T, p) - (\partial\phi/\partial p)_T RTM_A\Sigma_B m_B. \quad (73)$$

Eliminating μ_A^* between equations (66) and (70), we obtain

$$\mu_A^\ominus(\text{solvent}, T) = \mu_A + \phi RTM_A\Sigma_B m_B + \int_p^{p^\ominus} V_A^*\,dp, \quad (74)$$

which is the most useful form for insertion into equation (6) to obtain an equilibrium constant.

Formulae for all the other standard thermodynamic functions follow from equation (74). In particular,

$$S_A^\ominus(\text{solvent}, T) = S_A - \phi RM_A\Sigma_B m_B - (\partial\phi/\partial T)_p RTM_A\Sigma_B m_B - \int_p^{p^\ominus} (\partial V_A^*/\partial T)_p\,dp, \quad (75)$$

$$H_A^\ominus(\text{solvent}, T) = H_A - (\partial\phi/\partial T)_p RT^2M_A\Sigma_B m_B + \int_p^{p^\ominus} \{V_A^* - T(\partial V_A^*/\partial T)_p\}\,dp. \quad (76)$$

The osmotic coefficient ϕ of the solvent substance A in a solution containing molality m_B of an involatile solute substance B can be determined from measurements of the vapour pressures p_σ of the solution and p_σ^* of the pure solvent according to the relation:

$$\phi(T, p_\sigma^*, m_B) = (RTM_A m_B)^{-1}\int_{p_\sigma}^{p_\sigma^*} \{V_A^*(\text{g}, T, p) - V_A(\text{solvent}, T, p, m_B)\}\,dp, \quad (77)$$

or, in approximate but more familiar form:

$$\phi \approx (M_A m_B)^{-1} \ln(p_\sigma^*/p_\sigma). \tag{78}$$

The vapour pressures p_σ and p_σ^* can be measured directly (see Dr Ambrose's review in Chapter 7 of experimental methods for measurements of vapour pressures of pure liquids) or, instead, p_σ^* and the difference $(p_\sigma^* - p_\sigma)$ can be measured directly, with some possible advantages. Alternatively, isopiestic measurements can be used to obtain ϕ if $\phi_R(m_R)$ is already known for a 'reference' solute R. At isopiestic equilibrium of the solvent between a solution having molality m_B of B and a solution having molality m_R of R we have

$$\mu_A(\text{solvent}, T, p, m_B) = \mu_A(\text{solvent}, T, p, m_R), \tag{79}$$

so that according to equation (70)

$$\phi(m_B) = \phi_R(m_R)\, m_R/m_B. \tag{80}$$

If $\phi_R(m_R)$ is known, and if the 'isopiestic ratio' m_R/m_B has been measured, $\phi(m_B)$ can thus be calculated.

If a membrane is available which is permeable to the solvent A but impermeable to the solute B, the osmotic coefficient is given in terms of the osmotic pressure Π of the solution by the relation:

$$\phi(T, p, m_B) = (M_A m_B)^{-1} \int_p^{p+\Pi} \{V_A(\text{solvent}, T, p, m_B)/RT\}\, \mathrm{d}p, \tag{81}$$

or, in approximate but more familiar form:

$$\phi \approx (M_A m_B)^{-1} \Pi V_A^*/RT. \tag{82}$$

Osmotic pressure measurements are especially useful for studying solutions of macromolecules.

The osmotic coefficient can also be calculated from the lowering $(T_f^* - T_f)$ of the freezing temperature from T_f^* for the pure solvent to T_f for the freezing of pure solvent from a solution, according to the relation:

$$\phi(T_f^*, p, m_B) = (M_A m_B)^{-1} \int_{T_f}^{T_f^*} \{\Delta_s^l H_A^*(T, p)/RT^2\}\, \mathrm{d}T$$
$$- (M_A m_B)^{-1} \int_{T_f}^{T_f^*} [\{H_A^*(\text{solvent}, T, p) - H_A(\text{solvent}, T, p, m_B)\}/RT^2]\, \mathrm{d}T, \tag{83}$$

where $\Delta_s^l H_A^*$ denotes the enthalpy of melting of pure A and

$$(H_A^* - H_A) = -(\partial \Delta_S H/\partial n_A)_{T,p,n_B},$$

where $\Delta_S H$ is the enthalpy of dissolution, which will be briefly discussed in Section 10. The second integral in equation (83) is usually omitted, in which case the left-hand side becomes $\phi(T_f, p, m_B)$; the osmotic coefficient is then obtained at a slightly different temperature for each molality. With ample accuracy for most purposes, the first integral in (83) can be expressed in the form:

$$\int_{T_f}^{T_f^*}\{\Delta_s^l H_A^*(T,p)/RT^2\}\,dT = \{\Delta_s^l H_A^*(T_f^*,p)/RT_f^*\}\{(T_f^* - T_f)/T_f^*\} + \{\Delta_s^l H_A^*(T_f^*,p)/RT_f^* - \Delta_s^l C_{p,A}^*(T_f^*,p)/2R\}\{(T_f^* - T_f)/T_f^*\}^2. \quad (84)$$

Equation (83) can be expressed in the approximate but more familiar form:

$$\phi \approx (M_A m_B)^{-1}(\Delta_s^l H_A^*/RT_f^*)(T_f^* - T_f)/T_f^*. \quad (85)$$

Measurements of osmotic coefficients by this route suffer from the disadvantage that they can be made only at the freezing temperature of the solvent, but are nevertheless of considerable importance, especially for electrolyte solutions.

Equations (77)—(83) can easily be generalized for solutions containing more than one solute. Equations (78), (82), and (85) are still often used with ϕ set equal to unity for the measurement of approximate molar masses of unknown solutes, especially of non-electrolytes. The measurement and analysis of osmotic coefficients will be reviewed in a subsequent volume.

Whenever all but one of ϕ and the γ_B's for a solution have been measured over a wide enough range of molalities, the remaining one can be calculated by integration of the relation:

$$d\{(\phi - 1)\Sigma_B m_B\} = \Sigma_B m_B\, d\ln\gamma_B, \quad (T, p \text{ constant}), \quad (86)$$

which for a single solute (or a single electrolyte) can be written for example in the form:

$$\ln\gamma_B = (\phi - 1) + \int_0^{m_B}(\phi - 1)\,d\ln m_B, \quad (T, p \text{ constant}). \quad (87)$$

Solubility.—The thermodynamic condition for a solution of a solute B in a solvent A saturated with respect to pure solid B at temperature T and pressure p is

$$\mu_B(\text{solute}, T, p, m_{B,\text{sat}}) = \mu_B^*(\text{s}, T, p), \quad (88)$$

where $m_{B,\text{sat}}$ denotes the molality of B in a saturated solution, that is to say the solubility expressed as a molality. Substituting from equations (33) and (63) into (88), we obtain

$$\mu_B^{\ominus}(\text{solute}, T) - \mu_B^{\ominus}(\text{s}, T) = -RT\ln(m_{B,\text{sat}}\gamma_{B,\text{sat}}/m^{\ominus}) + \int_p^{p^{\ominus}}\{V_B^{\infty}(\text{solute}, T, p) - V_B^*(\text{s}, T, p)\}\,dp, \quad (89)$$

and thence

$$S_B^{\ominus}(\text{solute}, T) - S_B^{\ominus}(\text{s}, T) = R\ln(m_{B,\text{sat}}\gamma_{B,\text{sat}}/m^{\ominus}) + RT\{\partial\ln(m_{B,\text{sat}}\gamma_{B,\text{sat}})/\partial T\}_p - \int_p^{p^{\ominus}}[\{\partial V_B^{\infty}(\text{solute}, T, p)/\partial T\}_p - \{\partial V_B^*(\text{s}, T, p)/\partial T\}_p]\,dp, \quad (90)$$

$$H^{\ominus}_{B}(\text{solute}, T) - H^{\ominus}_{B}(\text{s}, T) = RT^2\{\partial \ln(m_{B,\text{sat}}\,\gamma_{B,\text{sat}})/\partial T\}_p$$
$$+ \int_{p}^{p^{\ominus}} [V^{\infty}_{B}(\text{solute}, T, p) - T\{\partial V^{\infty}_{B}(\text{solute}, T, p)/\partial T\}_p]\,\mathrm{d}p$$
$$- \int_{p}^{p^{\ominus}} [V^{*}_{B}(\text{s}, T, p) - T\{\partial V^{*}_{B}(\text{s}, T, p)/\partial T\}_p]\,\mathrm{d}p, \qquad (91)$$

$$0 = - RT\{\partial \ln(m_{B,\text{sat}}\,\gamma_{B,\text{sat}})/\partial p\}_T - \{V^{\infty}_{B}(\text{solute}, T, p) - V^{*}_{B}(\text{s}, T, p)\}. \qquad (92)$$

Equation (90) can be used with measured values of the solubility, its temperature coefficient, the activity coefficient $\gamma_{B,\text{sat}}$ in the saturated solution, and its temperature coefficient [together with $V_B(T, p)$ for the solute at infinite dilution and for the pure solid if the integral is not negligible] to obtain the standard entropy of a dissolved solute (such as an electrolyte) from that of the pure solid substance at the same temperature and pressure. Alternatively, $S^{\ominus}_{B}(\text{solute}, T)$ can be obtained from the fewer quantities on the right-hand side of equation (89) together with a calorimetrically determined value of $\{H^{\ominus}_{B}(\text{solute}, T) - H^{\ominus}_{B}(\text{s}, T, p)\}$ (see Section 10).

7 Standard Equilibrium Constants of Reactions: Part II

The thermodynamic criterion of chemical equilibrium (^{e}) for the general reaction:

$$0 = \Sigma_B \nu_B \text{B}, \qquad (93)$$

is the relation:

$$\Sigma_B \nu_B \mu^{e}_{B} = 0. \qquad (94)$$

In view of equation (94), equation (6) may be rewritten in the form:

$$RT \ln K^{\ominus} = \Sigma_B \nu_B (\mu^{e}_{B} - \mu^{\ominus}_{B}). \qquad (95)$$

The standard equilibrium constant $K^{\ominus}$ of any chemical reaction can now be obtained by substituting into equation (95) the appropriate formula for $(\mu^{e}_{B} - \mu^{\ominus}_{B})$ for each reacting substance B. The formulae are: for a gaseous substance G according to equation (21):

$$\mu^{e}_{G} - \mu^{\ominus}_{G} = RT \ln(x^{e}_{G} p^{e}/p^{\ominus}) + \int_{0}^{p^{e}} (V^{e}_{G} - RT/p)\,\mathrm{d}p, \qquad (96)$$

for a substance M present in a liquid or solid mixture according to equation (42):

$$\mu^{e}_{M} - \mu^{\ominus}_{M} = RT \ln(x^{e}_{M} f^{e}_{M}) + \int_{p^{\ominus}}^{p^{e}} V^{*}_{M}\,\mathrm{d}p, \qquad (97)$$

for a pure liquid or solid substance P according to equations (42) and (38):

$$\mu^{e}_{P} - \mu^{\ominus}_{P} = \int_{p^{\ominus}}^{p^{e}} V^{*}_{P}\,\mathrm{d}p, \qquad (98)\dagger$$

† It is the often relatively negligible magnitude of this quantity which leads to the advice to 'leave out' pure solids and liquids in elementary treatments of equilibrium constants.

for a solute B in a solution according to equation (63):

$$\mu_B^e - \mu_B^\ominus = RT\ln(m_B^e\gamma_B^e/m^\ominus) + \int_{p^\ominus}^{p^e} V_B^\infty\,dp, \tag{99}$$

and for the solvent A in a solution according to equation (74):

$$\mu_A^e - \mu_A^\ominus = -\,\phi^e RTM_A\Sigma_B m_B^e + \int_{p^\ominus}^{p^e} V_A^*\,dp. \tag{100}$$

Reaction in a Gas Mixture.—For the reaction:

$$0 = \Sigma_G \nu_G G, \tag{101}$$

in a gas mixture we obtain

$$K^\ominus = \{\Pi_G(x_G^e)^{\nu_G}\}\,(p^e/p^\ominus)^\Sigma \exp\left\{\Sigma_G\nu_G\int_0^{p^e}(V_G^e/RT - 1/p)\,dp\right\}. \tag{102}$$

If the pressure p^e of the equilibrium mixture is low enough, then the factor exp{...} in equation (102) may be omitted. A value of $K^\ominus$ can then be calculated from measured values of the equilibrium mole fractions x_B^e and of the equilibrium pressure p^e. If the initial amounts n_B^i are known, then it is sufficient to determine p^e plus any one of the x_B^e's, chosen for the availability of a convenient method of chemical or physical (*e.g.* spectrophotometric) analysis. Provided that $\Sigma \neq 0$, the x_G^e's can be determined (but still only if the pressure is small enough) if the volume V^e is measured, in addition to the pressure, from the relation:

$$p^e V^e = \Sigma_G n_G^e RT = (\Sigma_G n_G^i + \xi^e\Sigma_B\nu_B)\,RT, \tag{103}$$

where ξ^e is the equilibrium extent of reaction defined by

$$n_G^e = n_G^i + \nu_G\xi^e. \tag{104}$$

Instead of the static methods implied above for determining the equilibrium mole fractions and pressure, flow methods can be used in which the reactant gases are made to flow at measured rates through a reaction chamber (for example, a long tube) in which they are allowed a sufficient residence time to reach equilibrium. The effluent gas is then analysed. Flow methods are especially useful for reactions which are slow in the absence of a solid catalyst; the reaction chamber is then loosely packed with solid catalyst.

Experimental methods for the study of equilibria in gas reactions will be reviewed in a subsequent volume.

If the pressure cannot be made low enough to justify the omission of the factor exp{...} from equation (102), then in addition to an analysis of the equilibrium mixture, measurements are needed of the molar volume of the gas mixture over a range of mole fractions of each of the substances round the equilibrium mole fractions, and all this over a range of pressures between 0 and the equilibrium pressure p^e. Such a formidable programme has not to the author's knowledge been carried right through for any mixture relevant to a chemical reaction.

If the pressure is low, but not quite low enough, then the molar volume V_m may be written in the form:

$$V_m = RT/p + \Sigma_i x_i^2 B_{ii} + \Sigma_i \Sigma_{j \neq i} x_i x_j B_{ij}, \qquad (105)$$

where B_{ii} is the second virial coefficient of the pure gas i and B_{ij} is the 'mixed' second virial coefficient arising from interactions of a molecule i with a molecule j. From equation (105) we obtain for the partial molar volume V_G of the gas G the relation:

$$V_G = RT/p + B_{GG} - (1 - x_G)^2 B_{GG} - \Sigma_{i \neq G} x_i^2 B_{ii} + 2(1 - x_G)\Sigma_{i \neq G} x_i B_{iG} - \Sigma_{i \neq G} x_i \Sigma_{j \neq i,G} x_j B_{ij}, \qquad (106)$$

which can be used, if all the B's are known, to carry out the integration in equation (96). The second virial coefficients B_{ii} of the pure substances have been measured for many substances and can be found in secondary thermodynamic tables.[3] The 'mixed' second virial coefficients B_{ij} have been measured for few pairs of substances. Their values may be estimated from B_{ii} and B_{jj} if B_{ii}, B_{jj}, and B_{ij} may reasonably be expected all to obey the same principle of corresponding states.[4] More usually, Lewis and Randall's empirical rule:

$$B_{ij} = (B_{ii} + B_{jj})/2, \qquad (107)$$

is used to reduce equation (106) to the strikingly simple form:

$$V_G = RT/p + B_{GG}. \qquad (108)$$

Unfortunately, such experimental evidence as there is suggests that Lewis and Randall's rule is seldom obeyed.

Conversely, if $K^\ominus$ is known, whether from measurements of the equilibrium yield at a pressure low enough to ensure that the factor $\exp\{\ldots\}$ in equation (102) may safely be omitted, or from calorimetric measurements of $\Delta H^\ominus$ (see Section 8) and of $\Delta S^\ominus$ (see Section 9) according to equation (15), or by use of a statistical-mechanical formula from spectroscopic measurements of molecular quantities usually supplemented by a calorimetric value of $\Delta H^\ominus$ at any one temperature, then the yield at some other pressure can be calculated only if the integral in equation (102) can be evaluated.

The statistical-mechanical formula for $K^\ominus$ for a gas reaction is

$$\ln K^\ominus(T) = \Sigma_B \nu_B \ln(q_B/V) + (\Sigma_B \nu_B)\ln(kT/p^\ominus) - \Delta\varepsilon_0/kT, \qquad (109)$$

where q_B is the partition function for a molecule B and $\Delta\varepsilon_0 = \Sigma_B \nu_B \varepsilon_{0,B}$ is the change of zero-temperature molecular energy for the reaction. When, as is usual, the value of $\Delta\varepsilon_0$ has not been obtained spectroscopically, equation (8) can be used to eliminate $\Delta\varepsilon_0$ from equation (109) in favour of a value of $\Delta H^\ominus(T_1)$ at any one temperature T_1 (usually 298.15 K). Equation

[3] J. H. Dymond and E. B. Smith, 'The Virial Coefficients of Gases. A Critical Compilation', Clarendon Press, Oxford, 1969.

[4] E. A. Guggenheim and M. L. McGlashan, *Proc. Roy. Soc.*, 1951, **A206**, 448.

(109) then becomes

$$\ln K^{\ominus}(T) = \Sigma_{B}\nu_{B}\ln(q_{B}/V) + (\Sigma_{B}\nu_{B})\ln(kT/p^{\ominus})$$
$$+(T_1^2/T)\Sigma_{B}\nu_{B}\{\mathrm{d}\ln(q_{B}/V)/\mathrm{d}T\}_{T=T_1} + (\Sigma_{B}\nu_{B})(T_1/T) - \Delta H^{\ominus}(T_1)/RT. \quad (110)$$

The evaluation of the partition function q_B of a molecule B from spectroscopic measurements is reviewed by Drs Frankiss and Green in Chapter 8.

Reactions involving Gases and Pure Solids or Liquids.—For the reaction:

$$0 = \Sigma_{G}\nu_{G}G(g) + \Sigma_{P}\nu_{P}P(s \text{ or } l), \quad (111)$$

involving a gas phase containing the substances G and solid or liquid phases of the pure substances P, we substitute equations (96) and (98) into (95) and obtain

$$K^{\ominus} = \{\Pi_{G}(x_{G}^{e})^{\nu_G}\}(p^{e}/p^{\ominus})^{\Sigma_G}\left[\exp\left\{\Sigma_{G}\nu_{G}\int_{0}^{p^e}(V_{G}^{e}/RT - 1/p)\,\mathrm{d}p\right\}\right]$$
$$\times\left[\exp\left\{\Sigma_{P}\nu_{P}\int_{p^{\ominus}}^{p^e}(V_{P}^{*}/RT)\,\mathrm{d}p\right\}\right]. \quad (112)$$

If it is sufficiently accurate to omit the last factor [...], equation (112) reduces to the equation obtained by 'leaving out the condensed phases' or, in American parlance, 'putting the activities of the condensed phases equal to unity'. The usual omission of that factor can cause serious errors when $p^{e} \gg p^{\ominus}$. For example, the omission of that factor for the reaction:

$$CO(g) + H_2(g) = C(s) + H_2O(g),$$

at say 50 MPa and 500 K leads to an error of about 6 per cent in $K^{\ominus}$ when $p^{\ominus}$ is chosen, as usual, to be 0.101 325 MPa.

Especially in metallurgical applications, some of the solid or liquid phases might be mixtures such as alloys of continuously variable composition; equation (97) must then be used in place of (98) to obtain $K^{\ominus}$. High-temperature and metallurgical equilibria involving gases, solids, and liquids are reviewed by Drs Kubaschewski, Spencer, and Dench in Chapter 9.

Whether or not the solid or liquid phases are pure substances or mixtures, the equilibrium constant $K^{\ominus}$ can also be obtained from calorimetrically determined values of $\Delta H^{\ominus}$ and $\Delta S^{\ominus}$ according to equation (15).

Reactions in Liquid or Solid Mixtures.—For the reaction:

$$0 = \Sigma_{M}\nu_{M}M, \quad (113)$$

in a liquid or solid mixture we substitute equation (97) into equation (95) and obtain

$$K^{\ominus} = \{\Pi_{M}(x_{M}^{e}f_{M}^{e})^{\nu_M}\}\exp\left\{\Sigma_{M}\nu_{M}\int_{p^{\ominus}}^{p^e}(V_{M}^{*}/RT)\,\mathrm{d}p\right\}. \quad (114)$$

The factor exp{...} is usually close to unity and is usually omitted.

In spite of many measurements of the activity coefficients f_M of substances in binary, and to a lesser extent in multi-component, mixtures (a field which will be dealt with in a subsequent volume) and in spite of considerable advances in the theory of mixtures of simple molecules (likewise), the equilibrium activity coefficients f_M^e have been measured for few if any reacting mixtures. Nor can theory yet help us much to predict their values. The best we can usually do is to measure the equilibrium mole fractions x_M^e and to use the approximation:

$$K^{\ominus} \approx \Pi_M(x_M^e)^{\nu_M}, \tag{115}$$

which results from the usually unreliable assumption that the reacting mixture is ideal (see Section 5).†

It is true that we can determine $K^{\ominus}$ for such a reaction with high accuracy from calorimetric values of $\Delta H^{\ominus}$ and $\Delta S^{\ominus}$ according to equation (15), but we are usually then little better off because we are unable to reverse the calculation by using equation (114) to calculate accurate values of yields. From a severely chemical point of view this is the main reason for the considerable effort currently being devoted to theory and measurements in the field of liquid mixtures.

Reactions in Solutions.—For the reaction:

$$0 = \Sigma_B \nu_B B + \nu_A A, \tag{116}$$

of solutes B and solvent A in solution, we substitute equations (99) and (100) into (95) and obtain

$$K^{\ominus} = \{\Pi_B(m_B^e \gamma_B^e/m^{\ominus})^{\nu_B}\}\exp(-\nu_A \phi^e M_A \Sigma_B m_B^e)$$
$$\times \exp\left\{\int_{p^{\ominus}}^{p^e}(\Sigma_B \nu_B V_B^{\infty}/RT + \nu_A V_A^*/RT)\,dp\right\}. \tag{117}$$

Equation (117) is general: when $\nu_A = 0$, as it is for example for the reaction:

$$Ag^+(aq) + 2NH_3(aq) = [Ag(NH_3)_2]^+(aq),$$

one simply puts $\nu_A = 0$; for a reaction like

$$CH_3CO_2H(aq) + H_2O(l) = CH_3CO_2^-(aq) + H_3O^+(aq),$$

one puts $\nu_A = -1$. When, as is often nearly enough so, the equilibrium pressure p^e is close to the standard pressure $p^{\ominus}$, the integral in equation (117) is negligible and the second exp{...} may be, and usually is, omitted. Less accurate, but still good enough for many purposes, is the usual omission of the factor $\exp(-\nu_A \phi^e M_A \Sigma_B m_B^e)$ even when $\nu_A \neq 0$. As an example of the error involved, when $\nu_A = 1$, $\phi^e = 1$, $M_A = 0.018\ \text{kg mol}^{-1}$, and $m_B^e = 1\ \text{mol kg}^{-1}$, that factor differs from unity by about 2 per cent.

† The results of the famous study by Berthelot and St. Gilles of the equilibrium of the reaction:

$$CH_3CO_2C_2H_5 + H_2O = CH_3CO_2H + C_2H_5OH$$

(in mixtures with all four substances present in comparable amounts and treated on a par, not in dilute solutions with water as solvent) happened to conform roughly with equation (115). In view of the key role of this study in the growth of physical chemistry, it is interesting to speculate on how the subject might have grown had mixtures of those four substances not turned out to be, unexpectedly as it now seems to us, roughly ideal.

For reactions in solutions of non-electrolytes, for example:

$$N_2O_4(\text{in } CCl_4) = 2NO_2(\text{in } CCl_4),$$

the activity coefficients γ_B^{e} can confidently be set equal to unity with useful accuracy at mole fractions up to say 0.1 of solute (molality about 0.7 mol kg^{-1} in CCl_4 or 5.5 mol kg^{-1} in H_2O). Such reactions are usually studied spectrophotometrically or, if they are slow enough or can be quenched, by analysis.

For reactions involving ions, but for which the ionic strength at equilibrium is small enough (say 0.001 mol kg^{-1} or less), for example:

$$CH_3CO_2H(aq) + H_2O(l) = CH_3CO_2^-(aq) + H_3O^+(aq),$$

for which the ionic strength is about 0.003 mol kg^{-1} when $\Sigma_B m_B = 0.1$ mol kg^{-1}, the activity coefficients may still be put equal to unity with useful accuracy. Alternatively, the Debye–Hückel limiting law or one of its extensions can be used to calculate $K^{\ominus}$ from the m_B^{e}'s or to calculate the m_B^{e}'s from $K^{\ominus}$. The equilibrium constant of such a reaction can be obtained from measurements of the e.m.f.'s of galvanic cells, of electric conductivities, of the distribution of a non-electrolyte solute between the solution in question and a virtually immiscible solvent in which the ionic solutes are insoluble, or relatively by the use of indicators.

For reactions in solutions having ionic strengths greater than about 0.001 mol kg^{-1} the activity coefficients in equation (117) may not be ignored, and the Debye–Hückel theory is far from accurate. Many activity coefficients of single electrolytes have been measured by the methods mentioned in Section 6 and can be found in secondary thermodynamic tables.[5] Unfortunately, however, activity coefficients are seldom available for the multi-electrolyte solutions relevant to chemical reactions in which the ionic strength at equilibrium is appreciably greater than about 0.001 mol kg^{-1}. From a severely chemical point of view this is the main reason for the considerable effort currently being devoted to theoretical and experimental studies of multi-electrolyte solutions. For many such reactions $K^{\ominus}$ can be calculated from the results of e.m.f. measurements on galvanic cells. For reactions for which $K^{\ominus}$ is neither much greater nor much less than 1, $K^{\ominus}$ can be calculated, except for the activity coefficients, from the measured extent of the reaction at equilibrium.

The experimental methods used to obtain values of $K^{\ominus}$ for reactions in solutions, especially in electrolyte solutions, will be reviewed in a subsequent volume.

As before, $K^{\ominus}$ can be obtained with high accuracy for reactions in solution from calorimetric values of $\Delta H^{\ominus}$ and $\Delta S^{\ominus}$ according to equation (15). Unless activity and osmotic coefficients are available, calculable with confidence, or negligibly different from unity, however, we cannot use equation (117) to calculate accurate values of yields even when a value of $K^{\ominus}$ is known.

[5] R. A. Robinson and R. H. Stokes, 'Electrolyte Solutions', 2nd edn. (revised), Butterworths, London, 1970.

8 Standard Enthalpy Changes

The calculation of a standard enthalpy change $\Delta H^{\ominus}$ from the results of a calorimetric experiment involves many small corrections.

For simplicity, we consider as an example a reaction:

$$0 = \Sigma_{\mathrm{G}}\nu_{\mathrm{G}}\mathrm{G}, \tag{118}$$

confined to a single gaseous phase. We suppose that amounts of substance $n_{\mathrm{G}}^{\mathrm{i}}$ are caused to react in a thermally insulated bomb calorimeter at temperature T_1 and pressure p_1 so as to give final amounts $n_{\mathrm{G}}^{\mathrm{f}}$ at temperature T_2 and pressure p_2, the $n_{\mathrm{G}}^{\mathrm{f}}$'s and $n_{\mathrm{G}}^{\mathrm{i}}$'s being related by

$$n_{\mathrm{G}}^{\mathrm{f}} = n_{\mathrm{G}}^{\mathrm{i}} + \nu_{\mathrm{G}}\xi, \tag{119}$$

where ξ is the extent of reaction. We then have

$$U(T_2, p_2, \xi) - U(T_1, p_1, 0) = 0, \tag{120}$$

where U denotes thermodynamic energy $(H - pV)$. Next we suppose that the temperature of the calorimeter is returned to T_1 and that the pressure is then found (or calculated) to be p_3, and next that the quantity of (electrical) work w is measured which causes the temperature to change exactly to T_2, the pressure then being found to be p_4. We then have

$$U(T_2, p_4, \xi) - U(T_1, p_3, \xi) = w. \tag{121}$$

Subtracting equation (121) from (120) we obtain

$$\{U(T_1, p_3, \xi) - U(T_1, p_1, 0)\} = -w + \{U(T_2, p_4, \xi) - U(T_2, p_2, \xi)\}, \tag{122}$$

which we rewrite in the form:

$$\{H(T_1, p_3, \xi) - H(T_1, p_1, 0)\} = -w + \{H(T_2, p_4, \xi) - H(T_2, p_2, \xi)\}$$
$$+ \{p_3V(T_1, p_3, \xi) - p_1V(T_1, p_1, 0)\} - \{p_4V(T_2, p_4, \xi) - p_2V(T_2, p_2, \xi)\}. \tag{123}$$

In deriving equations (120)—(123) we have assumed either that the thermal insulation of the calorimeter is perfect or, more realistically, that corrections have accurately been made for heat leaks.

We next assume either that the energy of the fabric of the calorimeter remains unchanged during each part of the experiment, or less drastically, that any change is exactly the same in each part of the experiment and so has cancelled out in equations (122) and (123). We may then write

$$H(T, p, \xi) = \Sigma_{\mathrm{G}}(n_{\mathrm{G}}^{\mathrm{i}} + \nu_{\mathrm{G}}\xi)\, H_{\mathrm{G}}(T, p, \xi), \tag{124}$$

and

$$V(T, p, \xi) = \Sigma_{\mathrm{G}}(n_{\mathrm{G}}^{\mathrm{i}} + \nu_{\mathrm{G}}\xi)\, V_{\mathrm{G}}(T, p, \xi), \tag{125}$$

where H_{G} and V_{G} are the partial molar enthalpy and partial molar volume respectively of the substance G. By use of equation (23) we rewrite equation (124) in the form:

$$H(T, p, \xi) = \Sigma_{\mathrm{G}}(n_{\mathrm{G}}^{\mathrm{i}} + \nu_{\mathrm{G}}\xi)\, H_{\mathrm{G}}^{\ominus}(T)$$
$$+ \Sigma_{\mathrm{G}}(n_{\mathrm{G}}^{\mathrm{i}} + \nu_{\mathrm{G}}\xi)\int_0^p [V_{\mathrm{G}}(T, p, \xi) - T\{\partial V_{\mathrm{G}}(T, p, \xi)/\partial T\}_p]\, \mathrm{d}p. \tag{126}$$

Substituting equation (126) into (123) and rearranging we obtain finally

$$\xi\Sigma_G\nu_G H_G^{\ominus}(T_1) = \xi\Delta H^{\ominus}(T_1) = -w$$
$$-\Sigma_G(n_G^i + \nu_G\xi)\int_0^{p_3}[V_G(T_1, p, \xi) - T_1\{\partial V_G(T_1, p, \xi)/\partial T\}_p]\,dp$$
$$+\Sigma_G n_G^i\int_0^{p_1}[V_G(T_1, p, 0) - T_1\{\partial V_G(T_1, p, 0)/\partial T\}_p]\,dp$$
$$+\Sigma_G(n_G^i + \nu_G\xi)\int_{p_2}^{p_4}[V_G(T_2, p, \xi) - T_2\{\partial V_G(T_2, p, \xi)/\partial T\}_p]\,dp$$
$$+\Sigma_G(n_G^i + \nu_G\xi)\{p_3V_G(T_1, p_3, \xi) - p_4V_G(T_2, p_4, \xi) + p_2V_G(T_2, p_2, \xi)\}$$
$$-\Sigma_G n_G^i p_1 V_G(T_1, p_1, 0). \quad (127)$$

The five terms after the first in equation (127) can all be evaluated if enough is known about $V_G(T, p, \xi)$ for each substance G and the given n_G^i's. If the pressures were all low enough to allow the gas mixtures to be regarded as perfect, equation (127) would reduce to

$$\Delta H^{\ominus}(T_1) = -w/\xi + (\Sigma_G\nu_G)RT_1. \quad (128)$$

Had we taken as our example the complete combustion of a sample of say $C_6H_5CO_2H$(s) in a bomb calorimeter at room temperature to form a gas mixture O_2(g) + CO_2(g) + H_2O(g) and a liquid solution of O_2(solute) + CO_2(solute) in H_2O(solvent), the analysis would have been a good deal more complicated. In particular, we should then have had to know or measure also the thermal and volumetric properties of solutions of O_2 + CO_2 in H_2O(l) and we should have had to use the appropriate recipe for $H_B^{\ominus}$ from Sections 4—6 according as the substance B was a component of a gas mixture, a solid, or the solvent in a solution. Nor would the analysis have been less complicated (except for extremely dilute solutions) had we taken as our example a quantitative reaction between two aqueous electrolyte solutions brought about by causing the two solutions to mix in a thermally insulated constant-pressure ('open') calorimeter, even if we had taken precautions to prevent the large errors which can result from the evaporation or condensation of small amounts of solvent into or from the air space in the calorimeter when the composition and hence the equilibrium vapour pressure change as a result of the reaction.

Combustion and reaction calorimetry are reviewed by Dr Head in Chapter 3. The present Section will perhaps have helped the reader to appreciate the fantastic achievements exemplified by Dr Head's reference on page 101 to the values of the standard enthalpy of combustion of $C_6H_5CO_2H$(s) at 298.15 K measured in three different laboratories.

The quantity usually tabulated in primary thermodynamic tables is $\Delta_f H_B^{\ominus}(T)$, the standard enthalpy of formation of the substance B at temperature T (usually 298.15 K or 0). The standard enthalpy of formation at temperature T is defined as the standard enthalpy change of the reaction in which the substance B is formed from its elements each in the state most stable at that temperature. Exceptionally, the state chosen for phosphorus is the white form, which has been better characterized than the more stable red form. For electrolytes in aqueous solution the standard enthalpies of formation are made up from strictly additive but strictly inseparable contributions from the several ions. Economy is effected by tabulating a value only for each ion rather than for each electrolyte, the value being based upon the arbitrary convention that the standard enthalpy of formation of $H_3O^+(aq)$ is taken as zero. In Technical Note 270 [1] the same states of the elements have been chosen to define $\Delta_f H^\ominus(T \to 0)$ as were chosen to define $\Delta_f H^\ominus(298.15\ \mathrm{K})$ except for bromine and mercury, for which the solids are chosen to define $\Delta_f H^\ominus(T \to 0)$ and the liquids to define $\Delta_f H^\ominus(298.15\ \mathrm{K})$. Such conventions, varying from one set of tables to another, form a trap for the unwary; it is well to follow Dr Herington's advice on page 57 to look for the zeros as a help to interpreting the symbols at the heads of columns in any table.

9 Standard Entropies

The standard entropies found in primary thermodynamic tables are values of

$$S_B^\ominus(\text{state}, T) - S_B(\text{s}, T \to 0), \tag{129}$$

corrected for a few substances, such as CO, N_2O, NO, H_2O, CH_3D, and $Na_2SO_4 \cdot 10H_2O$, to the values which it is believed would have been obtained if all orientational metastability had been resolved as the temperature was lowered during the experiments, so that the solid at the lowest temperature studied was in each case a perfectly ordered one. Values of $\Delta S^\ominus(T)$ including such corrections lead, with corresponding values of $\Delta H^\ominus(T)$ and on the assumption that $\Delta S(\text{s}, T \to 0) = 0$, to standard equilibrium constants $K^\ominus$ which agree with those obtained directly.

For a gaseous substance B the standard entropy $S_B^\ominus(\text{g}, T)$ is given by the relation:

$$\begin{aligned} S_B^\ominus(\text{g}, T) = {} & S_B(\text{s}, T \to 0) + \int_0^T C_{\sigma,B}(\text{s}\alpha, T)\,\mathrm{d}\ln T + \int_{T_1}^{T_{\alpha\beta}} C_{\sigma,B}(\text{s}\alpha, T)\,\mathrm{d}\ln T \\ & + \Delta_{\text{s}\alpha}^{\text{s}\beta} H_B(T_{\alpha\beta}) + \int_{T_{\alpha\beta}}^{T_f} C_{\sigma,B}(\text{s}\beta, T)\,\mathrm{d}\ln T + \Delta_{\text{s}\beta}^{\text{l}} H_B(T_f) \\ & + \int_{T_f}^{T_b} C_{\sigma,B}(\text{l}, T)\,\mathrm{d}\ln T + \Delta_{\text{l}}^{\text{g}} H_B(T_b) + R\ln\{p_\sigma(T_b)/p^\ominus\} \\ & + \int_0^{p_\sigma(T_b)} [\{\partial V_B(\text{g}, T_b, p)/\partial T\}_p - R/p]\,\mathrm{d}p + \int_{T_b}^{T} C_{p,B}^\ominus(\text{g}, T)\,\mathrm{d}\ln T, \end{aligned} \tag{130}$$

where T_1 denotes the lowest temperature at which a measurement of heat capacity was made, C_σ the heat capacity at saturation of solid or liquid, sα and sβ two crystalline forms of B having (sα + sβ + g) triple-point temperature $T_{\alpha\beta}$ and enthalpy of transition $\Delta_\alpha^\beta H_B$, T_f the (sβ + l + g) triple-point temperature and $\Delta_{\text{s}\beta}^{\text{l}} H_B$ the enthalpy of melting, T_b a convenient temperature (often 298.15 K) at which the enthalpy of evaporation is

$\Delta_{l}^{g}H_{B}(T_{b})$ and the vapour pressure is $p_{\sigma}(T_{b})$, $p^{\ominus}$ a standard value of pressure (usually chosen to be 101.325 kPa), V_{B} molar volume, and $C_{p}^{\ominus}$ standard heat capacity at constant pressure. If more than two crystalline forms are found, or if there is only one, obvious additions or omissions must be made to or from equation (130). If $S_{B}^{\ominus}(s, T)$ or $S_{B}^{\ominus}(l, T)$ is required instead of $S_{B}^{\ominus}(g, T)$, the upper limit of the integral in the fourth or sixth term on the right-hand side of equation (130) should be replaced by T, the remaining terms should be omitted, and the term [see equation (34)]:

$$-\int_{p_{\sigma}(T)}^{p^{\ominus}} \{\partial V_{B}(\text{s or l}, T, p)/\partial T\}_{p}\, dp, \tag{131}$$

should be added, where $p_{\sigma}(T)$ is the equilibrium vapour pressure of the solid or liquid at temperature T. If the temperature T_{1} is low enough (typically 2 or 3 K), the first term in equation (130) can be evaluated either by smooth extrapolation to $T \rightarrow 0$ or by using the values of C_{σ} at a few of the lowest temperatures to find a value of the coefficient a in Debye's formula $C_{\sigma} = aT^{3}$ and then evaluating the integral as $aT_{1}^{3}/3$. The penultimate term in equation (130) can usually be approximated, with sufficient accuracy, to $p_{\sigma}(dB/dT)_{T=T_{b}}$. The last term, zero if $T = T_{b}$ (as is often so), can be evaluated either by statistical-mechanical calculation if the molecule is simple enough (see Chapter 8) or from flow-calorimetric measurements (see Chapter 6) of $C_{p,B}(g, T, p)$ with the help of equation (24).

The heat capacity at saturation C_{σ}(s or l) is defined as the limit as $\delta T \rightarrow 0$ of $w_{\sigma}/\delta T$ where w_{σ} is the work which must be done (usually electrically) on the solid or liquid so as to raise its temperature by δT in the presence of a vanishingly small amount of vapour, that is to say with the pressure varying with temperature along the (s + g) or (l + g) saturation line. The quantity C_{σ} is usually more closely related[6] to the quantity actually measured in modern heat-capacity calorimeters than the heat capacity at constant pressure C_{p}, to which it is related through the formula:

$$\begin{aligned} &C_{\sigma}(\text{s or l}, T) - C_{p}(\text{s or l}, T, p_{\sigma}) \\ &\quad = -\{\partial V_{m}(\text{s or l}, T, p_{\sigma})/\partial T\}_{p=p_{\sigma}} \Delta_{\text{s or l}}^{g} H(T, p_{\sigma})/\Delta_{\text{s or l}}^{g} V(T, p_{\sigma}), \end{aligned} \tag{132}$$

where p_{σ} is as before the vapour pressure of solid or liquid at temperature T. The difference (132) vanishes for the solid as $T \rightarrow 0$; even for a typical liquid (benzene, say) at its normal boiling temperature, the difference amounts only to about 0.1 J K^{-1} mol^{-1} compared with a C_{σ} of about 130 J K^{-1} mol^{-1}.

Heat capacity calorimetry is reviewed by Dr Martin in Chapter 4.

For a solute substance B, $S_{B}^{\ominus}(\text{solute}, T)$ is obtained from $S_{B}^{\ominus}(s, T)$ by use of equation (90). For aqueous electrolytes the standard entropy is made up from strictly additive but strictly inseparable contributions from the

[6] 'Experimental Thermodynamics, Volume I, Calorimetry of Non-reacting Systems', ed. J. P. McCullough and D. W. Scott, Butterworths, London, 1968, especially Chapter 5.

several ions. Economy is effected by tabulating a value only for each ion rather than for each electrolyte, the value being based upon the arbitrary convention that the standard entropy of $H_3O^+(aq)$ is taken as zero.

10 Dissolution

The enthalpy of dissolution (or of solution) $\Delta_S H$ of an amount n_B of the solid solute substance B in an amount n_A of the solvent substance A can be determined calorimetrically and is given by

$$\Delta_S H(T,p,n_B,n_A) = n_A\{H_A(\text{solvent},T,p,n_B/n_A) - H_A^*(\text{solvent},T,p)\} + n_B\{H_B(\text{solute},T,p,n_B/n_A) - H_B^*(\text{s},T,p)\}. \tag{133}$$

From equation (133) we deduce

$$(\partial\Delta_S H/\partial n_A)_{T,p,n_B} = H_A(\text{solvent},T,p,n_B/n_A) - H_A^*(\text{solvent},T,p) = (\partial\phi/\partial T)_p RT^2 n_B/n_A, \tag{134}$$

$$(\partial\Delta_S H/\partial n_A)^{\infty}_{T,p,n_B} = 0, \tag{135}$$

$$(\partial\Delta_S H/\partial n_B)_{T,p,n_A} = H_B(\text{solute},T,p,n_B/n_A) - H_B^*(\text{s},T,p) = \{H_B^{\ominus}(\text{solute},T) - H_B^{\ominus}(\text{s},T)\} - RT^2(\partial\ln\gamma_B/\partial T)_p - \int_p^{p^{\ominus}}\{V_B^{\infty} - T(\partial V_B^{\infty}/\partial T)_p\}\,\mathrm{d}p + \int_p^{p^{\ominus}}\{V_B^* - T(\partial V_B^*/\partial T)_p\}\,\mathrm{d}p, \tag{136}$$

$$(\partial\Delta_S H/\partial n_B)^{\infty}_{T,p,n_A} = \{H_B^{\ominus}(\text{solute},T) - H_B^{\ominus}(\text{s},T)\} - \int_p^{p^{\ominus}}\{V_B^{\infty} - T(\partial V_B^{\infty}/\partial T)_p\}\,\mathrm{d}p + \int_p^{p^{\ominus}}\{V_B^* - T(\partial V_B^*/\partial T)_p\}\,\mathrm{d}p. \tag{137}$$

Equation (137) is especially useful (the integrals are usually negligible) to find $\{H_B^{\ominus}(\text{solute},T) - H_B^{\ominus}(\text{s},T)\}$ from the limiting slope as $n_B \to 0$ of a plot of the enthalpy of solution $\Delta_S H$ against n_B at given n_A.

Enthalpies of dissolution will be reviewed in a subsequent volume.

The volume of dissolution $\Delta_S V$, defined by substitution of V for H in equation (133), leads to some of the quantities which have appeared, albeit in small terms, in previous equations. In particular [see equation (92)],

$$(\partial\Delta_S V/\partial n_B)^{\infty}_{T,p,n_A} = V_B^{\infty}(\text{solute},T,p) - V_B^*(\text{s},T,p). \tag{138}$$

11 Other Aspects of Chemical Thermodynamics

Until now this Chapter has been concerned with a brief introductory review of the formulae and experimental methods of chemical thermodynamics only from the severely chemical point of view that the sole object of chemical thermodynamics is the quantitative study of the yields of chemical reactions. Even from this limited point of view, we have seen the need for studies of the thermodynamic properties of pure substances and of non-reacting mixtures and solutions.

As well as the subjects touched upon in Sections 1—10, chemical thermodynamicists have many other interests. (It would be easy to justify each of these even from the severely chemical point of view outlined above, but this is left as an exercise for the reader.) Among those other interests are: gas imperfections and the study of the interaction energies of molecules through statistical-mechanical formulae; phase equilibria, as expressed in two-dimensional sections or projections (or three-dimensional models) of $(p, T, V_m, x_B, x_C, \ldots)$ diagrams; the critical state both for pure substances and for mixtures; statistical-mechanical theories of fluid mixtures, of solutions of macromolecules, and of electrolyte solutions; many properties of pure substances which do not happen to have been mentioned explicitly in Sections 1—10 such as expansivities and compressibilities of pure substances and of mixtures; the extent of additivity of bond energies and their use in estimating unmeasured enthalpies of reaction; the principle of corresponding states and deviations from the principle and their use for estimating unmeasured quantities; the use of other correlations to extrapolate into the unknown; lattice energies of crystals and the interaction energies of atoms or ions or molecules in crystals; first-order (phase) transitions and higher-order transitions in solids; the effects of changes of electric, magnetic, and gravitational fields on the thermodynamic properties of pure substances and of mixtures; fluid-to-fluid surfaces and the dependence of surface tension on temperature, pressure, and composition; experimental and statistical-mechanical studies of the adsorption of gases on solids; Onsager's reciprocal relations and the extension of thermodynamics to steady states.

All these fields as well as those previously mentioned and those reviewed in the present volume, will be reviewed from time to time in subsequent volumes of these Specialist Periodical Reports on Chemical Thermodynamics.

2
Thermodynamic Quantities, Thermodynamic Data, and their Uses

BY E. F. G. HERINGTON

Theory and practice are so closely interwoven in the study and applications of chemical thermodynamics that it is not easy to divide the subject into suitable sections for discussion. The scope of the present Chapter is indicated by the title, and the headings of the various sections are chosen so as to indicate the subdivisions of the topic which are used for an orderly presentation.

1 Introduction

In order that the reader may appreciate the nature of thermodynamic quantities, thermodynamic data, and their uses, it is necessary for him to know the present state of thermodynamic theory and be aware of sources of information on the fundamental background of classical and statistical thermodynamics. This Chapter does not aim to give a short course in thermodynamics nor does it review in detail the history of the foundation of the subject with which such names as Joseph Black (1728—99), Count Rumford (1753—1814), Sadi Carnot (1796—1832), James Joule (1818—89), and Lord Kelvin (1824—1907) are associated.

Thermodynamics has its origins in the study of engines and in the relation between heat and energy, but chemical thermodynamics did not develop until the nineteenth century and was much advanced by J. Willard Gibbs, who published a monograph entitled 'The Equilibrium of Heterogeneous Substances' in 1876. A history of J. Willard Gibbs has recently been reprinted.[1] Since that time, thermodynamics has provided a firm foundation for so many branches of chemistry that it will not be possible to review all aspects of chemical thermodynamics in this article. Indeed, no attempt will be made to describe the thermodynamics of electrolytes or of irreversible processes. Fortunately, the thermodynamics of electrolytes is well summarized in books (see, for example, refs. 2, 3, and 4). Irreversible thermo-

[1] L. P. Wheeler, 'Josiah Willard Gibbs: The History of a Great Mind', Archon Books, Hamden, Connecticut, 1970.
[2] H. S. Harned and B. B. Owen, 'The Physical Chemistry of Electrolytic Solutions', 3rd edn., Reinhold, New York, 1958.
[3] H. S. Harned and R. A. Robinson, 'Equilibrium Properties of Electrolyte Solutions: Vol. 2, Multicomponent Electrolyte Solutions', Pergamon Press, Oxford, 1968.
[4] R. A. Robinson and R. H. Stokes, 'Electrolyte Solutions', 2nd edn., Butterworth Scientific Publications, London, 1959.

dynamics, which has not yet fulfilled its first promise to be a useful practical discipline, is somewhat set apart from the main stream of classical and statistical thermodynamic theory and therefore will not be considered in the present context.

Numerous text-books on thermodynamics have been written which use very different approaches to the subject because there is no general agreement on how thermodynamics should be taught. Probably there never will be any general agreement on the best method for the teaching of thermodynamics because some aspects of the subject attract abstract thinkers while other aspects attract the attention of the most practical of workers.

This lack of unification can be illustrated by the following example. Margenau and Murphy[5] state that the most satisfactory formulation of the laws of thermodynamics is probably that of Carathéodory, based on the properties of Pfaffian differential equations, yet the Carathéodory formulation is dealt with in such a cursory manner (p. 98) in the revision of Lewis and Randall's 'Thermodynamics', by Pitzer and Brewer,[6] that it is not listed in either the name or the subject index. Nevertheless, many practical workers in this country and in America will undoubtedly study and use this modern version of Lewis and Randall's book.

Statistical thermodynamics, by means of which the values of thermodynamic functions for the gas state can be calculated from the properties of individual molecules, is a more recent development than classical thermodynamics. In particular, statistical thermodynamics illuminates the physical meaning of entropy and therefore there are teachers who believe that chemical thermodynamics should be taught by combining traditional thermodynamics with statistical thermodynamics. However, those who favour the historical approach in teaching and those who consider that traditional thermodynamics is a pure subject regard with abhorrence the contamination, by the introduction of statistical thermodynamics, of what they regard as the pure stream of classical thermodynamics.

Lack of agreement on teaching methods is only one of the reasons why so many thermodynamic text-books are written. Thermodynamics can be appreciated and used at various levels of understanding. Several successful practising thermodynamicists have confessed to the writer that when they were college students they were able to produce excellent answers to thermodynamic questions but it was not until years later that they felt they appreciated the true relation between the parts and the whole of thermodynamics. At this point some people feel they must produce a new thermodynamic text-book, thus adding yet another volume to an already extensive list. Text-books have also been written so that the writer could have a document which uses the symbols he favours. Unfortunately, in the past,

[5] H. Margenau and G. M. Murphy, 'The Mathematics of Physics and Chemistry', D. Van Nostrand, New York, 1943; 2nd edn., 1956.

[6] K. S. Pitzer and L. Brewer (revision of Lewis and Randall), 'Thermodynamics', McGraw-Hill, New York, 1961.

different symbols have sometimes been used for the same quantity and the same symbol has been used for different quantities, which has naturally led to confusion. Some authors (for example, see Fowler and Guggenheim [7]) have listed some of the symbols used by others but it is to be hoped (and one trusts not in vain) that in the future the recommendations [8] of the International Union of Pure and Applied Chemistry will be followed universally. Nevertheless an experienced reader, on picking up a thermodynamic script, will always study the definition of the symbols employed. Rather surprisingly, he will thus find that many American writers still use F for Gibbs energy instead of G, in spite of the fact that Gibbs was an American!

There are several distinct methods employed for the discussion of thermodynamics in text-books which are of interest to chemists, although in teaching there are hybrids between these methods. The oldest approach used cycles, thus revealing the origin of the subject in the theory of the efficiency of heat engines. Perhaps this approach reached its peak with Butler's book,[9] where the weakness of the whole treatment is revealed. The subject is first developed in terms of cycles and then is redeveloped in terms of defined thermodynamic functions. There is at present a consensus of opinion that the use of cycles is not the best method. Another approach can be illustrated by the work of Guggenheim,[10] where the subject is elegantly and concisely developed by the use of thermodynamic functions. This treatment, which closely follows the original work of Willard Gibbs, is attractive to a reader of a somewhat mathematical turn of mind but tends to repel the more practical student, who is happier with worked numerical examples. The first edition of the book by Lewis and Randall [11] has had an immense impact because it deals with practical calculations, and indeed many people feel that it is because of this book that America is now in the forefront in its use and in its application of thermodynamics in industry. The modern version of this book, revised by K. S. Pitzer and L. Brewer,[6] illustrates the recent trend to use a combination of classical and statistical thermodynamics for instruction.

2 Bibliography of Sources of Theory

Anybody would be foolhardy to claim that it is possible to list all the currently available books on chemical thermodynamics that are likely to be used by English-speaking chemists. Nevertheless, if the full significance of the nature of thermodynamic quantities and data is to be appreciated,

[7] R. H. Fowler and E. A. Guggenheim, 'Statistical Thermodynamics', Cambridge University Press, Cambridge, 1939.
[8] *Pure Appl. Chem.*, 1970, **21**, 13.
[9] J. A. V. Butler, 'Chemical Thermodynamics', 5th edn., MacMillan, London, 1962.
[10] E. A. Guggenheim, 'Thermodynamics: An Advanced Treatment for Chemists and Physicists', North-Holland Publishing Co., Amsterdam, 1949; 5th edn., 1967.
[11] G. N. Lewis and M. Randall, 'Thermodynamics and the Free Energy of Chemical Substances', McGraw-Hill, New York, 1923.

suitable books must be consulted. This section lists many of the textbooks on chemical and statistical thermodynamics that are likely to be consulted by chemists. Partington,[12] in 1949, supplied a good, but not complete, bibliography of works on chemical thermodynamics up to that date, and will be used as the starting point for the following résumé. Partington's book is a monumental work 'in the pseudo-Teutonic tradition' to use the phraseology of the author. The second section (118 pp.) gives an account of the principles of thermodynamics and starts with the three full pages of bibliography in small print referred to above. Later sections are concerned with statistical mechanics and the properties of gases.

Guggenheim's book,[10] although entitled an advanced treatment, in fact discusses basic principles very thoroughly. The book is addressed equally to physicists and to chemists, but not to engineers. Statistical thermodynamics is introduced early in the development of the subject and the text makes considerable use of a function called absolute activity. The study of systems in electric and magnetic fields is more detailed than usual. The whole treatment is elegant but there are fewer numerical examples than some people find useful. Other books first published in 1949 include those by Sears [13] and Smith.[14]

Of books published [15–28] in 1950, it should be remarked that the one by Hildebrand and Scott [19] on 'Solubility of Non-electrolytes' has provoked

[12] J. R. Partington, 'An Advanced Treatise on Physical Chemistry (Fundamental Principles, and the Properties of Gases)', Longmans, Green, and Co., London, vol. 1, 1949.

[13] F. W. Sears, 'An Introduction to Thermodynamics', Addison-Wesley, Cambridge, Mass., 1949.

[14] J. M. Smith, 'Introduction to Chemical Engineering Thermodynamics', McGraw-Hill, New York, 1949.

[15] I. Prigogine and R. Defay, 'Thermodynamique Chimique', 2nd edn., Edition Desoer, Liège, 1950.

[16] G. R. Fitterer, 'Metallurgical Thermodynamics', Univ. Pittsburgh, Met. Eng. Dept., Oakmont, Pa., 1950.

[17] G. Gourdet and A. Proust, 'Les Diagrammes Thermodynamiques', H. Dunod, Paris, 1950.

[18] E. O. Hercus, 'Elements of Thermodynamics and Statistical Mechanics', University Press, Melbourne, 1950.

[19] J. H. Hildebrand and R. L. Scott, 'Solubility of Non-electrolytes', 3rd edn., D. Van Nostrand, Princeton, New Jersey, 1950.

[20] L. Holleck, 'Physikalische Chemie und ihre rechnerische Anwendung: Thermodynamik', Springer-Verlag, Berlin, 1950.

[21] F. E. W. Wetmore and D. J. LeRoy, 'Principles of Phase Equilibria', McGraw-Hill, New York, 1950.

[22] I. M. Klotz, 'Chemical Thermodynamics', Prentice-Hall, New York, 1950.

[23] K. Nesselmann, 'Die Grundlagen der angewandten Thermodynamik', Springer-Verlag, Berlin, 1950.

[24] W. J. Peck and A. J. Richmond, 'Applied Thermodynamic Problems for Engineers', Longmans, Green, and Co., New York, 1950.

[25] F. D. Rossini, 'Chemical Thermodynamics', J. Wiley and Sons, New York, 1950.

[26] E. Schmidt, 'Thermodynamics. Principles and Applications', 3rd edn., Oxford University Press, New York, 1950.

[27] J. H. van der Waals, 'Thermodynamic Properties of Mixtures of Alkanes Differing in Chain Length', D. B. Centen, Amsterdam, 1950.

[28] J. R. Partington, 'Thermodynamics', 4th edn., Constable and Co., London, 1950.

much work on the measurement and interpretation of solubilities of gases, liquids, and solids. Rossini's book[25] places much emphasis on practical applications of thermodynamics, particularly in the petroleum industry. It contains information on data and in addition lists journals extant in 1950 that contain thermodynamic information.

Of the books published[29–39] in 1951, those in refs. 31, 33, 34, and 38 have stood the test of time particularly well. Findlay's[31] 'The Phase Rule' has had an enormous impact on the understanding of heterogeneous equilibria ever since the appearance of the first edition in 1904. Roberts' 'Heat',[34] like Findlay's 'Phase Rule', has been revised and brought up to date by later workers. Zemansky's 'Heat and Thermodynamics'[38] has proved to be very popular and has run to several editions, as has Porter's book.[39]

Guggenheim's book[40] is chiefly concerned with the theory of non-polar solutions, based on various approximations to the quasi-crystalline model of the liquid phase. The book is rather narrowly conceived in that it deals mostly with successive approximations to the combinatory problem. Allis and Herlin's book[41] deals with both thermodynamic and statistical mechanics, in accord with the title.

Lumsden's 'Thermodynamics of Alloys'[42] develops the thermodynamics of metals from the fundamental laws up to the applications of statistical mechanics. The concept of entropy is approached *via* randomness rather than *via* Carnot cycles. Applications of the theory to the correlation of the properties of pure metals are shown and the theories of solution are developed. Solid solutions, and the statistical mechanics of liquids, liquid solutions, and imperfect crystals, are also considered.

Ubbelohde's book[43] provides a concise introduction to modern thermodynamics and in particular treats some subjects not often found in elemen-

29 C. L. Brown, 'Basic Thermodynamics', McGraw-Hill, New York, 1951.

30 C. Fabry, 'The Elements of Thermodynamics', F. Muller, London, 1951.

31 A. Findlay, 'The Phase Rule', 9th edn., revised by A. M. Campbell and N. O. Smith, Dover Publications, New York, 1951.

32 G. A. Hawkins, 'Thermodynamics', 2nd edn., J. Wiley and Sons, New York, 1951.

33 O. Kubaschewski and E. Ll. Evans, 'Metallurgical Thermochemistry', Butterworth–Springer, London, 1951.

34 'Roberts' Heat and Thermodynamics', 4th edn., ed. A. R. Miller, Blackie and Sons, London, 1951.

35 M. A. Paul, 'Principles of Chemical Thermodynamics', McGraw-Hill, New York, 1951.

36 G. A. Hawkins, 'Thermodynamics', 2nd edn., J. Wiley and Sons, New York, 1951.

37 J. E. Ricci, 'The Phase Rule and Heterogeneous Equilibrium', D. Van Nostrand, New York, 1951.

38 M. W. Zemansky, 'Heat and Thermodynamics', 3rd edn., McGraw-Hill, New York, 1951.

39 A. W. Porter, 'Thermodynamics', 4th edn., J. Wiley and Sons, New York, 1951.

40 E. A. Guggenheim 'Mixtures: The Theory of Equilibrium Properties of some Simple Classes of Mixtures, Solutions and Alloys', Oxford University Press, Oxford, 1952.

41 W. P. Allis and M. A. Herlin, 'Thermodynamics and Statistical Mechanics', McGraw-Hill, New York, 1952.

42 J. Lumsden, 'Thermodynamics of Alloys', The Institute of Metals, London, 1952.

43 A. R. Ubbelohde, 'An Introduction to Modern Thermodynamical Principles', 2nd edn., Oxford University Press, Oxford, 1952.

tary text-books such as, for example, tests of the Third Law, the statistical theory of solids and gases, some calculations on thermodynamic functions, and the statistical consequences of molecular flexibility.

In Wagner's book [44] the term 'alloy' is used to include solutions of non-metals in liquid iron. The book opens with a discussion of activity relations and free energy diagrams for heterogeneous solutions. The theories of disordered structures, ordered structures, thermodynamic functions, and phase diagrams are next considered. Data are reviewed. Finally, chemical equilibria involving alloys with molten salts and silicates are discussed.

Of those books [45–47] which appeared in 1954, comments will be made on the large work by Hirschfelder *et al.*[46] and on Everett's translation [47] and revision of a Belgian work. The large work [46] presents a statistical mechanical study of the properties of fluids and discusses equilibrium and non-equilibrium properties, together with intermolecular forces. Gases at low pressure are treated in terms of virial coefficients, the calculation of which is dealt with in detail for a variety of intermolecular fields. Equations of state for dense gases and liquids are given and an appendix contains a large collection of numerical tables. The text-book by Everett [47] is a translation and revision of the 1950 edition of the text in the French language. The book treats chemical equilibria, and in particular pays more attention than usual to phase equilibria. Thus the stability of phases, azeotropy, and indifferent states are considered at some length. The methods employed are based on those of Gibbs and de Donder, so that the 'degree of advancement' or 'extent of reaction' plays an important role in the discussions.

Denbigh's 'Principles of Chemical Equilibrium' [48] is divided into three parts, in which the principles of thermodynamics, reaction and phase equilibria, and statistical mechanics and its relation with thermodynamics are developed. The book particularly takes note of the practical side of the subject and gives numerous illustrations of applications to experiment.

Guggenheim's 'Boltzmann's Distribution Law' [49] treats the subject of the title in its widest aspects. The factorization of the partition function is discussed and equations are given for the partition function for some simple degrees of freedom. The results are used to outline the properties of ideal monatomic gases, and in later sections free energy, total energy,

[44] C. Wagner, 'Thermodynamics of Alloys', translated by S. Mellgren and J. H. Westbrook, Addison-Wesley, Cambridge, Mass., 1952.

[45] M. Dole, 'Introduction to Statistical Thermodynamics', Prentice-Hall, New York, 1954.

[46] J. O. Hirschfelder, C. F. Curtiss, and R. B. Bird, 'Molecular Theory of Gases and Liquids', J. Wiley and Sons, New York, 1954.

[47] I. Prigogine and R. Defay, translated and revised by D. H. Everett, 'Chemical Thermodynamics', Longmans, Green, and Co., London, 1954.

[48] K. G. Denbigh, 'Principles of Chemical Equilibrium', Cambridge University Press, Cambridge, 1955.

[49] E. A. Guggenheim, 'Boltzmann's Distribution Law', North-Holland Publishing Co., Amsterdam, 1955.

and conditions of equilibrium are discussed. The properties of crystals, ideal diatomic gases, and phase and chemical equilibria are also studied, and the book ends with accounts of Fermi–Dirac and Bose–Einstein statistics.

'Thermodynamik der Mischphasen', by Haase,[50] covers the whole of chemical thermodynamics, in spite of the title. The contents are revealed by the names of the chapters, *viz.* principles of thermodynamics, equilibrium conditions, differential equations for co-existing phases, gases, condensed phases, non-electrolyte solutions, electrolyte solutions, and mixed crystals. Münster's 'Statistische Thermodynamik'[51] is a basic text and reference book for physicists and physical chemists on statistical mechanics and treats the general methods of thermodynamics with application to gases, crystals, melts, and solutions. There is a mathematical appendix. Pippard's text[52] is for use by advanced students of physics. Bray and White's volume,[53] which is aimed at the honours biochemistry student and the postgraduate entering biochemistry, is designed for a reader whose acquaintance with mathematics does not extend beyond the ideas of simple calculus. A chapter entitled 'Free Energy and Metabolism' considers oxidation and reduction reactions and speaks of 'high energy compounds'.

'The Molecular Theory of Solutions', by Prigogine,[54] includes thermodynamic considerations, and Zemansky's book[55] is a later edition of the book mentioned earlier.[38] Wilson's volume[56] is of a fairly advanced level and requires a fairly high standard of mathematical ability. It is mainly designed for the use of physicists. Chapters include accounts of partial differentiation, and the book approaches entropy through Carnot cycles but also describes Carathéodory's axiomatic approach. Superconductivity and solutions are considered thermodynamically. Caldin's introduction[57] is designed for chemistry undergraduates. A student who has mastered the text should be well prepared to go on to more advanced work. Chisholm and de Borde's book[58] develops the equations for Bose–Einstein, Fermi–Dirac, and classical statistics by unusual routes and then applies the

[50] R. Haase, 'Thermodynamik der Mischphasen', Springer-Verlag, Berlin, 1956.

[51] A. Münster, 'Statistische Thermodynamik', Springer-Verlag, Berlin, Göttingen, and Heidelberg, 1956.

[52] A. B. Pippard, 'Elements of Classical Thermodynamics', Cambridge University Press, Cambridge, 1957.

[53] H. G. Bray and K. White, 'Kinetics and Thermodynamics in Biochemistry', J. and A. Churchill, London, 1957.

[54] I. Prigogine, 'The Molecular Theory of Solutions', North-Holland Publishing Co., Amsterdam, 1957.

[55] M. W. Zemansky, 'Heat and Thermodynamics', McGraw-Hill, New York, 1957.

[56] A. A. Wilson, 'Thermodynamics and Statistical Mechanics', Cambridge University Press, Cambridge, 1957.

[57] E. F. Caldin, 'An Introduction to Chemical Thermodynamics', Oxford University Press, Oxford, 1958.

[58] J. S. Chisholm and A. H. de Borde, 'An Introduction to Statistical Mechanics', Pergamon Press, Oxford, 1958.

equations to such subjects as the heat capacities of ortho- and para-hydrogen, heat capacities of solids, chemical and vapour pressure constants, the law of mass action, and the electronic heat capacity of metals.

Landau and Lifshitz [59] derive the principal thermodynamic quantities and relations associated with the macroscopic state. Perfect and real gases, condensed systems, solutions, and chemical reactions are then examined.

In spite of the title [60] of volume 12 of the 'Handbuch der Physik – Encyclopedia of Physics', the book contains only two articles concerning thermodynamics out of the total of six articles. One article, by Rowlinson, is on the properties of real gases (72 pp.) and another, by Mayer, is on the theory of real gases (133 pp.). Rowlinson's article gives an account of the equilibrium properties of pure and mixed gases and shows the advantage of choosing density rather than pressure as the independent variable in discussions of p, V, T data. Mayer's article is concerned with statistical mechanics and treats theories of imperfect gases and of condensation.

Wall [61] seeks to establish means for the calculation of the free energies of reactions, but assumes that the reader already knows something of the basic laws of thermodynamics. However, the principles of thermodynamics and statistical mechanics are presented.

Woodbridge-Constant's work [62] is intended for American first-year graduate students in physics but might also be of interest to British chemistry students. Subjects treated include the kinetic theory of gases and classical and quantum statistical mechanics. A large number of examples (with answers) are given. Aston and Fritz's book [63] is of general interest to chemists. Everett [64] assumes only the minimum of mathematical knowledge, and for the most part nothing more than simple algebra and geometry is employed. The first seven chapters form an elementary survey of some of the simplest applications of thermodynamics. At first each problem is set out in three ways, in relation to a mechanical analogy, in a graphical form, and in an algebraic form. The equivalence of these three treatments is stressed so that, when in more complicated problems the first two methods become inapplicable, the student has sufficient confidence to use the third alone. Chapter 11, which deals with the laws of thermodynamics, is intended to introduce the study of more advanced textbooks, and derives nearly all the equations used earlier.

[59] L. D. Landau and E. M. Lifshitz, 'Statistical Physics', translated from the Russian by E. Peierls and R. F. Peierls, Pergamon Press, London, 1958.

[60] 'Handbuch der Physik – Encyclopedia of Physics, Vol. 12. Thermodynamics of Gases', ed. S. Flügge, Springer-Verlag, Berlin, 1958.

[61] F. T. Wall, 'Chemical Thermodynamics, a Course of Study', W. H. Freeman and Co., San Francisco, 1958.

[62] F. Woodbridge-Constant, 'Theoretical Physics – Thermodynamics, Electromagnetism, Waves and Particles', Addison-Wesley, Reading, Mass., 1958.

[63] J. G. Aston and J. J. Fritz, 'Thermodynamics and Statistical Thermodynamics', J. Wiley and Sons, New York, 1959.

[64] D. H. Everett, 'An Introduction to the Study of Chemical Thermodynamics', Longmans, Green, and Co., London, 1959; 2nd edn., 1971.

Concerning refs. 65—68, it may be noted that ref. 67 applies to a later edition of a book mentioned earlier,[34] while Callen's volume[68] gives mnemonic diagrams to represent thermodynamic functions.

Schrödinger's treatment[69] is very elegant and of an advanced nature. Bransom's text[70] is based on a course of eighty lectures delivered in the Department of Chemical Engineering of the University of Birmingham, presented to develop physical concepts fundamental to physical chemistry and heat engineering. The titles of some of the chapters, *e.g.* 'Some Applications to Gas-handling', and 'Distillation', illustrate the applied approach.

Kirkwood and Oppenheim's book[71] is based on lectures by Kirkwood from notes taken by Oppenheim, Karplus, and Rich. The purpose of the book is to present a rigorous and logical discussion of fundamentals, but it does not treat statistical thermodynamics. The treatment of the Second Law discusses Carathéodory's principle.

The approach of Landsberg's book[72] to thermodynamics can most succinctly be summarized by a quotation from the author's preface, 'The abstract nature of the subject is always allowed to become fully evident in this book'. The volume should therefore be of the greatest value to somebody already familiar with thermodynamics and interested in the conceptual foundations of the subject. 'Thermodynamics with Quantum Mechanical Illustrations' is by the same author.[73] Bikerman's book is concerned with surfaces.[74]

The 'Scientific Papers of J. Willard Gibbs' are available in a Dover edition.[75] The Royal Institute of Chemistry Monographs for Teachers, No. 5, is a 45 page booklet[76] which discusses equilibrium in ideal and real systems and the determination of equilibrium constants by experiment and by calculation from Gibbs energy data. Of books listed in refs. 77—79, that

65 R. Vogel, 'Die heterogenen Gleichgewichte', Academische Verlagsgesellschaft, Leipzig, 1959.

66 E. F. Obert, 'Concepts of Thermodynamics', McGraw-Hill, New York, 1960.

67 J. K. Roberts, 'Heat and Thermodynamics', revised by A. R. Miller, 5th edn., Blackie and Sons, London, 1960.

68 H. B. Callen, 'Thermodynamics', J. Wiley and Sons, New York, 1960.

69 E. Schrödinger, 'Statistical Thermodynamics', 2nd edn., Cambridge University Press, Cambridge, 1960.

70 H. Bransom, 'Applied Thermodynamics', D. Van Nostrand, New York, 1961.

71 J. G. Kirkwood and I. Oppenheim, 'Chemical Thermodynamics', McGraw-Hill, New York, 1961.

72 P. T. Landsberg, 'Thermodynamics (Monographs in Statistical Physics and Thermodynamics)', vol. 2, ed. I. Prigogine, Interscience, New York, 1961.

73 P. T. Landsberg, 'Thermodynamics with Quantum Mechanical Illustrations', J. Wiley and Sons, New York, 1961.

74 J. J. Bikerman, 'Contributions to the Thermodynamics of Surfaces', published by the author, Cambridge, Mass., 1961.

75 'The Scientific Papers of J. Willard Gibbs', Dover Publications, New York, 1961.

76 P. G. Ashmore, 'Principles of Chemical Equilibrium', Monographs for Teachers, No. 5, Royal Institute of Chemistry, London, 1962.

77 N. Davidson, 'Statistical Mechanics', McGraw-Hill, New York, 1962.

78 V. M. Faires, 'Thermodynamics', 4th edn., MacMillan, New York, 1962.

79 J. H. Hildebrand and R. L. Scott, 'Regular Solutions', Prentice-Hall, Englewood Cliffs, New Jersey, 1962.

by Faires[78] is designed for the engineer, with the contents revealed by chapter headings such as 'Compressors and Expanders', 'Internal Combustion Engines', 'Gas Turbines and Jet Engines', and 'Power from Steam'. Reference 79 develops themes presented in a book discussed earlier.[19]

The Second Symposium on Thermophysical Properties[80] records, in the thermodynamic section, 10 review papers, 9 on the subject of experimental investigation, and 4 papers on computational methods.

Swalin's volume[81] gives an introduction to the thermodynamics and statistical mechanics of solids for students of metallurgy, but the section which introduces thermodynamics is to be regarded as a refresher course for somebody who has already studied the subject. Later, the thermodynamics of phase diagrams are discussed. Andrews' book[82] 'Equilibrium Statistical Mechanics' presents the basic theory and discusses ideal gases. Appropriate thermodynamic equations for real gases are given in Gonikberg's book.[83]

Hargreaves' book,[84] which is primarily intended for Higher National Certificate and Bachelor of Science students, presents thermodynamic functions in a pictorial way. Mahan's elementary book[85] is clearly written and gives a classical account of thermodynamic laws, with entropy introduced as a macroscopic quantity. Jancel's book,[86] on the other hand, is highly mathematical and will probably be of interest only to those concerned with the foundations of statistical mechanics.

Topic 10 of volume 5 of the 'International Encyclopedia of Physical Chemistry and Chemical Physics'[87] gives a clear account of the perfect gas. Following an introductory chapter on thermodynamics, there is a chapter on the measurement of heat capacities which surveys the principles and limitations of various experimental methods. A large part of the work deals with the calculation of thermodynamic properties, including consideration of the ortho- and para-states, residual entropy, hindered internal

80 'Progress in International Research on Thermodynamic and Transport Properties', ed. J. F. Masi and D. H. Tsai (papers presented at the Second Symposium on Thermophysical Properties, 24—26 January, 1962, at Princeton University). American Society of Mechanical Engineers and Academic Press, New York, 1962.

81 R. A. Swalin, 'Thermodynamics of Solids', J. Wiley and Sons, New York, 1962.

82 F. C. Andrews, 'Equilibrium Statistical Mechanics', J. Wiley and Sons, New York, 1963.

83 M. G. Gonikberg, 'Chemical Equilibria and Reaction Rates at High Pressures', Academy of Sciences of the U.S.S.R., Moscow, 1960; translated by Israel Program for Scientific Translations, Jerusalem, 1963.

84 G. Hargreaves, 'Elementary Chemical Thermodynamics', 2nd edn., Butterworth Scientific Publications, London, 1963.

85 B. H. Mahan, 'Elementary Chemical Thermodynamics', W. A. Benjamin, New York, 1963.

86 R. Jancel, 'Les Fondements de la Mécanique Statistique Classique et Quantique', Gauthier-Villars, Paris, 1963.

87 J. S. Rowlinson, 'The Perfect Gas (Topic 10)', vol. 5 of the International Encyclopedia of Physical Chemistry and Chemical Physics, Pergamon Press, Oxford, 1963.

rotation, and electronic contributions to thermodynamic properties. Mixed gases and the calculation of equilibrium constants are also treated.

Buckingham [88] deals with the laws and their applications in a classical manner, with emphasis on the properties of ideal and real gases and solutions. Excess functions for mixtures are examined very thoroughly, and equilibria are treated by the use of chemical potentials. The final chapter considers the influence of external fields on thermodynamic properties.

Of the next eleven books listed,[89–99] comments will be passed on only five. The text by Eyring *et al.*[91] includes consideration of the thermodynamic properties of crystals, black-body radiation, dielectric, diamagnetic, and paramagnetic properties of matter, real gases, equilibrium properties of liquids, liquid mixtures, and surface chemistry.

The first part (237 pp.) of Klotz's book,[94] which is issued separately under the title 'Introduction to Chemical Thermodynamics', gives an account of the laws and basic equations of classical thermodynamics as applied to systems of constant composition. The second part [94] is concerned with systems of variable composition. The book has a good treatment of the Third Law and its chemical applications but reaction equilibrium is covered only to a limited extent.

The treatment of chemical thermodynamics by Krestovnikov and Vigdorovich,[95] although primarily intended for students of metallurgy, treats thermodynamics as a whole but is much concerned with phase equilibria.

The monograph by van Ness [97] is written from the viewpoint of an engineer and represents the author's own efforts to develop the known

[88] A. D. Buckingham, 'The Laws and Applications of Thermodynamics', Pergamon Press, Oxford, 1964.

[89] H. S. Green and C. A. Hurst, 'Order–Disorder Phenomena', J. Wiley and Sons, New York, 1964.

[90] J. Coull and E. B. Stuart, 'Equilibrium Thermodynamics', J. Wiley and Sons, New York, 1964.

[91] H. Eyring, D. Henderson, B. J. Stover, and E. M. Eyring, 'Statistical Mechanics and Dynamics', J. Wiley and Sons, New York, 1964.

[92] H. L. Frisch, 'The Equation of State of the Classical Hard Sphere Fluid', Advances in Chemical Physics, vol. VI, ed. I. Prigogine. J. Wiley and Sons, New York, 1964.

[93] J. O. Hirschfelder, C. F. Curtiss, and R. B. Bird, 'Molecular Theory of Gases and Liquids', J. Wiley and Sons, New York, 1964.

[94] I. M. Klotz, 'Chemical Thermodynamics', W. A. Benjamin, New York, 1964.

[95] A. N. Krestovnikov and V. N. Vigdorovich, 'Chemical Thermodynamics', State Publishing House for Literature on Ferrous and Non-ferrous Metallurgy, Moscow, 1962; translated by Israel Program for Scientific Translations, Jerusalem, 1964.

[96] M. Nelkon, 'Heat', 2nd edn., Blackie and Sons, London, 1964.

[97] H. C. van Ness, 'Classical Thermodynamics of Non-electrolyte Solutions', Pergamon Press, Oxford, 1964.

[98] A. I. Veinik, 'Thermodynamics: A Generalized Approach', Publishing House of the Higher, Secondary, Specialized, and Vocational Education of the Belorussian S.S.R., Minsk, 1961; translated by Israel Program for Scientific Translations, Jerusalem, 1964.

[99] Plenary Lectures, Symposium on Thermodynamics and Thermochemistry, IUPAC, Lund, Sweden, 1963, Butterworths Scientific Publications, London, 1964.

theories concerned with the properties of vapour and liquid solutions of non-electrolytes.

The Plenary Lectures, Symposium on Thermodynamics and Thermochemistry,[99] is a stiff-backed edition of the report in *Pure and Applied Chemistry*, 1964, **8**, No. 2, and includes the following subjects: excursions in chemical thermodynamics from the past into the future (Rossini); key heat of formation data (Skinner); calorimetry of combustion and related reactions – inorganic reactions (Holley); calorimetry of combustion and related reactions – organic compounds (Cox); calorimetry of non-reacting systems with particular emphasis on solution and mixing processes (McGlashan); thermochemistry of step-wise complex formation (Leden); biochemical calorimetry (Wadsö); progress in the calorimetry and thermodynamics of phase and ordering transitions (Westrum).

Allen's book[100] discusses equilibrium constants, entropy in terms of disorder, and formulae for the derivation of values of entropy from heat capacity measurements. An important part of the book discusses energy changes in relation to the Periodic classification and to chemical bonding. Gibbs energies of formation are plotted for some series of compounds (*e.g.* oxides and chlorides) against atomic number. There is a thermodynamic study of the problem of selection of a catalyst for the Deacon process.

Bent's introduction[101] to classical and statistical thermodynamics uses microscopic disorder to discuss entropy. The book is divided into five parts, namely, classical thermodynamics, free energy and phase stability, statistical thermodynamics, applications, and the role of mathematics in thermodynamics.

The three volumes edited by De Boer and Uhlenbeck, published[102] during the period 1962—65, contain a series of essays and review articles. Volume 1 includes essays on the problems of a dynamical theory in statistical physics and on the theory of linear graphs, with applications to the theory of the virial development of the properties of gases, while Volume II includes essays on the imperfect gas and on a group theory of the liquid state. Volume III presents the doctoral thesis of Kahn on the theory of the equation of state. The content of Reiss' book[103] is indicated by the title.

Brout's book[104] assumes a substantial knowledge of statistical and quantum mechanics and there is little description of experimentally observed phase-transitions. The work is frankly biased and presents the

[100] J. A. Allen, 'Energy Changes in Chemistry', Blackie and Sons, London, 1965.
[101] H. A. Bent, 'The Second Law. An Introduction to Classical and Statistical Thermodynamics', Oxford University Press, Oxford, 1965.
[102] E. J. De Boer and G. E. Uhlenbeck, 'Studies in Statistical Mechanics', North-Holland Publishing Co., Amsterdam, vol. I, 1962; vol. II, 1964; and vol. III, 1965.
[103] H. Reiss, 'Scaled Particle Methods in the Statistical Thermodynamics of Fluids', Advances in Chemical Physics, vol. IX, ed. I. Prigogine, Interscience, 1965.
[104] R. H. Brout, 'Phase Transitions', W. A. Benjamin, New York, 1965.

author's own views rather than a systematic treatment. The contents are indicated by the chapter headings: Ising Model; Condensation; Freezing; Ferromagnetism; Superconductivity; Bose–Einstein Condensation; and Theoretical Refinement.

Switz's translation [105] of Fisher's Russian text (published 1961) deals with general theory, distribution functions, superposition approximations, surface phenomena, freezing and evaporation, and machine calculations using a rigid-sphere model. The supplement deals with developments since 1961, series expansion of thermodynamic quantities, and distribution functions obtained by the Mayers and their collaborators.

The relevance of thermodynamics to biochemistry is the theme of Lehminger's work,[106] which discusses such topics as the 'high-energy bond' but recognizes that the living cell is an open system that exchanges matter with its surroundings.

The second edition of Wall's book,[107] in which he has now adopted the thermodynamic symbols recommended by IUPAC, provides a good introduction for students. Buchdahl's work [108] places emphasis on mathematical and physical theory but includes consideration of the perfect gas and critical phenomena, and chemical reactions which may be of interest to chemists, although the text is a subsidiary one rather than a primary work.

The aim of Denbigh's volume [109] is to develop the general theory of equilibrium, including its statistical treatment, with problems to illustrate numerous practical applications. It is assumed that the student is already very familiar with the concepts of temperature and heat and that he already has some knowledge of thermodynamics. It is recognized that thermodynamics needs to be studied several times over at advancing levels. Part I presents a traditional development of the subject, Part II presents reaction and phase equilibria, and Part III develops statistical thermodynamics.

Dugdale [110] provides an introduction to entropy which considers both thermodynamics and statistics. Raja Gopal [111] introduces elementary concepts of heat capacities such as lattice heat capacity, and electronic heat capacity.

A recent text, 'The Phase Rule',[112] prepared essentially for the use of students for Grad. R.I.C., H.N.C., and pass-degree courses, considers

[105] I. Z. Fisher, 'Statistical Theory of Liquids', translated by T. M. Switz, University of Chicago Press, Chicago, 1965.

[106] A. L. Lehminger, 'Bioenergetics: The Molecular Basis of Biological Energy Transformations', W. A. Benjamin, New York, 1965.

[107] F. T. Wall, 'Chemical Thermodynamics: A Course of Study', 2nd edn., W. H. Freeman and Co., San Francisco and London, 1965.

[108] H. A. Buchdahl, 'The Concepts of Classical Thermodynamics', Cambridge University Press, Cambridge, 1966.

[109] K. G. Denbigh, 'The Principles of Chemical Equilibrium', Cambridge University Press, Cambridge, 1966.

[110] J. S. Dugdale, 'Entropy and Low Temperature Physics', Hutchinson, London, 1966.

[111] E. S. Raja Gopal, 'Specific Heats at Low Temperatures', Butterworths, London, 1966.

[112] F. D. Ferguson and T. K. Jones, 'The Phase Rule', Butterworths, London, 1966.

one-, two-, and three-component systems and experimental methods. Examination questions are provided. Guggenheim has provided a text [113] for teachers and has published,[114] in expanded form, eleven lectures he delivered in 1963 while he was a George Fisher Baker Lecturer at Cornell University. The book has a short chapter on the grand partition function and provides examples of the applications of statistical mechanics to subjects of interest to chemists, such as the condensed phases of the inert gases, the Born cycle for sodium and potassium chloride, solutions of electrolytes, and adsorption of gases on solids.

Although this Chapter does not consider electrolytes in detail, reference will be made to Lumsden's book [115] because he treats molten salts as examples of high-melting substances.

The first half of Linford's book [116] on energetics in biology introduces thermodynamics, and the second part deals with electrical, chemical, and radiant energy. The work can be read in conjunction with Lehminger's book, described earlier.[106]

Pryde has supplied an excellent treatment [117] of modern theories of the liquid state, but, although the text includes sections on the fundamentals of classical and statistical thermodynamics, it does not provide an introduction to thermodynamics.

Tisza [118] has attempted a reformulation of macroscopic and statistical thermodynamics, with particular emphasis on the study of critical phenomena. There are a hundred pages of mainly historical introduction and much of the rest of the book is a reprint of papers such as those which appeared in *Annalen der Physik* in 1961 and 1963.

Several of the books [119–135] published in 1967 are of an introductory type. Thus Wyatt's book [132] is aimed at the sixth form and undergraduates.

113 E. A. Guggenheim, 'Elements of Chemical Thermodynamics', Monographs for Teachers, No. 21, Royal Institute of Chemistry, London, 1966.

114 E. A. Guggenheim, 'Applications of Statistical Mechanics', Oxford University Press, Oxford, 1966.

115 J. Lumsden, 'Thermodynamics of Molten Salt Mixtures', Academic Press, London, 1966.

116 J. H. Linford, 'An Introduction to Energetics with Applications in Biology', Butterworths, London, 1966.

117 J. A. Pryde, 'The Liquid State', Hutchinson University Library, London, 1966.

118 L. Tisza, 'Generalized Thermodynamics', M.I.T. Press, Cambridge, Mass., 1966.

119 A. Bellemens, V. Mathot, and M. Simon, 'Statistical Mechanics of Mixtures – The Average Potential Model', Advances in Chemical Physics, vol. XI, ed. I. Prigogine, Interscience, 1967.

120 O. Kubaschewski, E. Ll. Evans, and C. B. Alcock, 'Metallurgical Thermochemistry', 4th ed., Pergamon Press, Oxford, 1967; Japanese edition, Sangyo Toshy Co., Tokio, 1968.

121 W. F. Luder, 'A Different Approach to Thermodynamics', Reinhold, London, 1967.

122 'Intermolecular Forces', Advances in Chemical Physics, vol. XII, ed. J. O. Hirschfelder, Interscience, New York, 1967.

123 H. J. G. Hayman, 'Statistical Thermodynamics: An Introduction to its Foundations', Elsevier, London, 1967.

124 E. A. Guggenheim, 'Thermodynamics: An Advanced Treatment for Chemists and Physicists', 5th edn., J. Wiley and Sons, New York, 1967.

The work may be regarded as a short cut to some of the larger works. Thompson's volume[133] is suitable for 'A'-level standard and provides a good introduction; Hughes[134] employs the multiple-choice programmed approach to present the fundamentals of thermodynamics and includes treatment of enthalpies of formation, free energy, entropy, and probability. Williamson's book[135] is at the undergraduate level but is not self-contained in its basic thermodynamic arguments, because it relies on external sources. Subjects that are dealt with include definitions, observed behaviour, thermodynamic correlations (Hildebrand's solubility parameter, Guggenheim's quasi-chemical and corresponding state treatments, and the cell model), and experimental methods.

Of the books[136–141] published in 1968, that by Battino and Wood[138] provides an introductory course in classical thermodynamics for students of biology, chemistry, and engineering, but the applications and problem exercises are mainly chemical. The Third Law is discussed in the last three pages. Nash's introduction[139] to statistical thermodynamics has two main sections, in the first of which the underlying ideas are carefully examined and in the second of which the applications of partition functions (mainly to gases) are studied. Hill's treatment[140] is designed to combine physical chemistry with biochemistry and molecular biology, and requires the reader to have prior knowledge of thermodynamics and statistical mechanics. Rice[141] views some of the topics from several angles, and the

[125] R. H. Fowler, 'Statistical Mechanics. The Theory of the Properties of Matter in Equilibrium', 2nd edn., Cambridge University Press, Cambridge, 1967.

[126] R. P. Baumann, 'An Introduction to Equilibrium Thermodynamics', Prentice-Hall, London, 1967.

[127] H. A. Bent, 'The Second Law. An Introduction to Classical and Statistical Thermodynamics', Oxford University Press, New York, 1967.

[128] A. R. Allnatt, 'Equilibrium Statistical Mechanics. Statistical Mechanics of Point-defect Interaction in Solids', Advances in Chemical Physics, vol. XI, ed. I. Prigogine, J. Wiley and Sons, New York, 1967.

[129] 'Advances in Chemical Physics', vol. XI', ed. I. Prigogine, Interscience, London, 1967.

[130] J. L. Margrave, 'The Characterization of High Temperature Vapours', J. Wiley and Sons, New York, 1967.

[131] I. M. Klotz, 'Energy Changes in Biochemical Reactions', Academic Press, New York, 1967.

[132] P. A. H. Wyatt, 'Energy and Entropy in Chemistry', MacMillan, London, 1967.

[133] J. J. Thompson, 'Concepts in Chemistry: An Introduction to Chemical Energetics', Longmans, Green, and Co., London, 1967.

[134] D. E. P. Hughes, 'Chemical Energetics', Chatto and Windus, London, 1967.

[135] A. G. Williamson, 'An Introduction to Non-electrolyte Solutions', Oliver and Boyd, Edinburgh and London, 1967.

[136] C. J. Adkins, 'Equilibrium Thermodynamics', McGraw-Hill, Toronto, 1968.

[137] E. A. Mason and T. H. Spurling, 'The Virial Equation of State', Pergamon Press, Oxford, 1968.

[138] R. Battino and S. E. Wood, 'Thermodynamics – An Introduction', Academic Press, London, 1968.

[139] L. K. Nash, 'Elements of Statistical Thermodynamics', Addison-Wesley, Cambridge, Mass., 1968.

[140] T. L. Hill, 'Thermodynamics for Chemists and Biologists', Addison-Wesley, Cambridge, Mass., 1968.

[141] O. K. Rice, 'Statistical Mechanics, Thermodynamics and Kinetics', W. H. Freeman and Co., San Francisco and London, 1968.

physical significance of the formulae is discussed, with applications of theory to the needs of chemistry students.

Of books published[142–151] in 1969, that by Mayer[142] gives a formal and rigorous approach to statistical mechanics, with a few examples of application and with little comparison between theory and experiment. Morton and Beckett's 'Basic Thermodynamics'[144] deals with sixth-form, O.N.C., and undergraduate courses, but only about one-sixth of the book is directly related to chemical aspects. However, applications to chemistry include the use of the concepts of free energy and activity and the determination of equilibrium constants. Appendices treat units, bond dissociation energies, standard electrode potentials, questions, and sources of useful equipment, and provide further information.

Rock states that he has written his book[145] because 'he enjoys thinking, teaching, and writing about thermodynamics'. Preceding the first chapter is a mnemonic diagram giving the interrelation of some thermodynamic equations. The book is designed for undergraduates, and the first five chapters introduce the usual thermodynamic functions through the First and Second Laws. Chapters 6 and 7 are concerned with phase changes. Later chapters deal with the Third Law and with statistical mechanics. Chapters are followed by suggestions for further reading and by problems.

Warn gives an account[146] of the gas laws, with the First Law treated from a chemical standpoint and with entropy treated from the point of view of spontaneous change and randomization.

Haase and Schönert's contribution[149] gives an elegant presentation of the fundamental theory of solid–liquid equilibria in one-, two-, and three-component systems, but the treatment presupposes considerable previous knowledge of the subject. Münster's book[150] is a substantially revised and enlarged English version of the German edition that was first published in 1956.

Rowlinson's aim[151] is to integrate theory and practice in the study of liquid mixtures with the theory for pure liquids.

142 J. E. Mayer, 'Equilibrium Statistical Mechanics', International Encyclopedia of Physical Chemistry and Chemical Physics, Topic 8, vol. 1, Pergamon Press, Oxford, 1969.

143 R. E. Dickinson, 'Molecular Thermodynamics', W. A. Benjamin, New York, 1969.

144 A. S. Morton and P. J. Beckett, 'Basic Thermodynamics', Butterworths, London, 1969.

145 P. A. Rock, 'Chemical Thermodynamics – Principles and Applications', Collier-Macmillan, London, 1969.

146 J. R. W. Warn, 'Concise Thermodynamics in SI Units', Van Nostrand–Reinhold, London, 1969.

147 G. C. Pimentel and R. D. Spratley, 'Understanding Chemical Thermodynamics', Holden-Day, San Francisco, 1969.

148 C. Truesdell, 'Rational Thermodynamics', McGraw-Hill, New York, 1969,

149 R. Haase and H. Schönert, 'Solid–Liquid Equilibrium', International Encyclopedia of Physical Chemistry and Chemical Physics, Pergamon Press, Oxford, 1969.

150 A. Münster, 'Statistical Thermodynamics', vol. 1, Springer-Verlag, Berlin, 1969.

151 J. S. Rowlinson, 'Liquids and Liquid Mixtures', 2nd edn., Butterworths, London, 1969.

The year 1970 saw the publication of the second edition [152] of Guggenheim's monograph and of an Annual Review [153] with a section on the thermodynamics of solid surfaces. Halberstadt [154] has made a translation of Münster's 'Classical Thermodynamics', a book which is not for the beginner but is suitable for the advanced student. The development of the laws is presented classically, after Joule, Clausius, and Carnot, and in an axiomatic manner after Carathéodory. The treatment is concentrated and formal.

Coopersmith [155] has presented an article largely concerned with the mathematical description of thermodynamic functions near a critical point. Reisman [156] has included classical treatments and modern theories applicable to the study of condensed phase–vapour equilibria.

Nash's 'Elements of Classical Thermodynamics' was first published in 1962 and his 'Elements of Statistical Thermodynamics' appeared in 1968. His new book [157a] contains revised versions of these earlier books in the same volume, but as separate texts. Chemical equilibria are treated in detail but physical equilibria are not strongly treated. The book is intended for the American first-year student.

A book by Smith and Harris [157b] is the successor to 'Worked Examples in Engineering Thermodynamics' and as such will appeal chiefly to the engineer but contains much information useful to chemical engineers and physical chemists.

Books and oral teaching are not the only methods for conveying information that are now available, because films and film strips are prepared for teaching purposes. For example, a film by Porter [158] is part of a series 'The Laws of Disorder' and is concerned with entropy, which is introduced as a statistical concept. The tendency of matter to mix is discussed and the nature of spontaneous and chemical changes are considered. The laws of probability and the concept of heat as disordered motion are depicted. The Second Law of thermodynamics is introduced. The Petroleum Films Bureau has produced two films [159] on energy in chemistry, illustrated by

[152] E. A. Guggenheim, 'Elements of Chemical Thermodynamics', 2n edn., Monographs for Teachers, No. 12, Royal Institute of Chemistry, London, 1970.

[153] G. D. Holsey and C. M. Greenlief, 'Equilibrium at Solid Surfaces', *Ann. Rev. Phys. Chem.*, 1970, **21**, 129.

[154] A. Münster, 'Classical Thermodynamics', translated by E. S. Halberstadt, Wiley–Interscience, London and New York, 1970.

[155] M. H. Coopersmith, 'The Thermodynamic Description of Phase Transitions', Advances in Chemical Physics, vol. XVII, ed. Prigogine and Rice, Interscience, New York, 1970.

[156] A. Reisman, 'Phase Equilibria, Basic Principles, Applications and Experimental Techniques', Academic Press, New York, 1970.

[157a] L. K. Nash, 'Elements of Classical and Statistical Thermodynamics', Addison-Wesley, Reading, Mass., 1970.

[157b] H. Smith and J. Harris, 'Thermodynamic Problems in SI Units', Macdonald and Co., London, 1972.

[158] 'Entropy', A film by G. Porter, being one of the series 'The Laws of Disorder', I.C.I. Film Library, Millbank, London S.W.1.

[159] Energy in Chemistry (1) and (2), Esso, 1969, colour 16 mm., Petroleum Films Bureau, 4 Brook Street, London W.1.

experiments which use polystyrene cups as calorimeters for the demonstration of enthalpies of reaction. The second film relates energy to the work which can be obtained from a chemical reaction and includes the demonstration of a fuel cell which works on methanol in an alkaline solution of hydrogen peroxide.

Longman's have produced three 8 mm film loops [160] on the subject of energetics. The treatment is based on the textbook 'Introduction to Chemical Energetics' by Thompson.[133]

Reference will finally be made in this section to the International Conference on Thermodynamics, 1970. The proceedings must be consulted [161] for details, but the scope of the meeting can be indicated by the titles of the sections: Foundations; Phase, Surfaces, and Thermodynamic Limit; Thermo-mechanics; Irreversibility and Quantum Mechanics; Statistical Thermodynamics in Astrophysics and Relativity; Pedogogical Papers. As can be seen from the titles, the meeting was not very closely connected with chemical thermodynamics. However, the summary by Zemansky (p. 549) of the discussion shows once again that there is yet no agreement on the methods to be used for the teaching of thermodynamics in any discipline.

3 Units, Symbols, and Measuring Scales

The International System of Units (called SI) has now been universally adopted for scientific publications. The document defining the system is published by the International Bureau of Weights and Measures,[162] but a version in English, prepared jointly by the National Physical Laboratory, U.K., and the National Bureau of Standards, U.S.A., is now available.[163]

In SI, the base units are as follows: the unit of length is the metre (m); the unit of mass is the kilogram (kg); the unit of time is the second (s); the unit of electric current is the ampere (A); the unit of thermodynamic temperature is the kelvin (K); the unit of luminous intensity is the candela (cd); and the unit of amount of substance is the mole (mol). The units chosen are a good choice except that the name kilogram is unfortunate for base unit of mass, because this name erroneously suggests that the gram is the base unit, with kilogram a multiple (*i.e.* 10^3 times) of it. However, the rule for the formation of decimal fractions and multiples of mass is that

160 Longman Loops: 'Chemistry. 16. Energetics I. Heat Energy Changes'; '17. Energetics II. Entropy Changes'; '18. Energetics III. Spontaneous Changes', Longman, London, 1970.

161 Proceedings of the International Conference on Thermodynamics, Cardiff, 1—4th April, 1970, *Pure Appl. Chem.*, 1970, **22**, No. 3—4.

162 J. Terrien and J. de Boer, 'Le Système International d'Unités', Bureau International des Poids et Mésures, Paris, 1970.

163 'The International System of Units SI', translation approved by the International Bureau of Weights and Measures of its publication 'Le Système International d'Unités', prepared jointly and published independently by the National Physical Laboratory, U.K. (Her Majesty's Stationery Office, London, 1970) and the National Bureau of Standards, U.S.A. (N.B.S. Special Publication 330, 1970, U.S. Government Printing Office, Washington, D.C.).

these should be constructed by adding the approved prefixes (see below) to gram and not to kilogram. Even the spelling of kilogram in the English language, with kilogramme as an alternative, is unresolved and is currently under review. The fractions, with symbols, are as follows: 10^{-1}, d; 10^{-2}, c; 10^{-3}, m; 10^{-6}, μ; 10^{-9}, n; 10^{-12}, p; 10^{-15}, f; 10^{-18}, a. The multiples, with symbols, are as follows: 10^{1}, da; 10^{2}, h; 10^{3}, k; 10^{6}, M; 10^{9}, G; 10^{12}, T. In SI there are derived units and supplementary units as well as base units, and some comments will be passed on those which are of particular interest in the study of thermodynamics.

Of the derived units, it may be noted that the unit for force is the newton (N), which, expressed in terms of SI base units, is $\mathrm{m\,kg\,s^{-2}}$; the unit for density or mass density is kilogram per cubic metre ($\mathrm{kg\,m^{-3}}$); the unit for concentration (of amount of substance) is mole per cubic metre ($\mathrm{mol\,m^{-3}}$); the unit for specific volume is cubic metre per kilogram ($\mathrm{m^3\,kg^{-1}}$). Some other derived units have special names; for example, the name for the unit of energy, work, or quantity of heat is the joule (J), equal to the newton metre (N m). The unit of power is the watt (W), equal to $\mathrm{J\,s^{-1}}$. The unit for heat capacity and entropy is the joule per kelvin ($\mathrm{J\,K^{-1}}$); the unit for molar energy is joule per mole ($\mathrm{J\,mol^{-1}}$); and for molar entropy and molar heat capacity is joule per kelvin mole ($\mathrm{J\,K^{-1}\,mol^{-1}}$).

Some comment [164, 165] must be made on the quantities molecular weight, relative molecular mass, and molar mass. Most chemists are well acquainted with the concept of 'molecular weight', which they usually compute from the chemical formula of a substance using 'atomic weights'. Atomic weights (more strictly relative atomic masses) are nowadays based on the value 12 ascribed as the atomic weight of the isotope 12 of carbon. Molecular weights which are derived from relative atomic masses are thus dimensionless. The General Conference of Weights and Measures (CGPM) gave in 1967, and confirmed in 1969, the following definition of the mole, 'The mole is the amount of substance of a system which contains as many elementary entities as there are atoms in 0.012 kilogram of carbon 12', with a Note 'when the mole is used, the elementary entities must be specified and may be atoms, molecules, ions, electrons, other particles, or specified groups of such particles'. It follows that the SI unit of molar mass is the kilogram per mole ($\mathrm{kg\,mol^{-1}}$). The practical chemist should note that this means that to obtain a value for the molar mass divided by the SI unit he must take the molecular weight (*i.e.* the sum of the atomic weights) and divide by one thousand. This is the quantity to be used in calculations (*e.g.* in calculating the rate of effusion) where the other units are SI. The unit of molar mass with which the chemist is better acquainted, *viz.* gram per mole, will continue to be used as an accepted decimal fraction of the SI unit.

[164] J. T. Edsall, *Nature*, 1970, **228**, 888.
[165] G. Hyde, *Nature*, 1970, **229**, 142.

The document cited above [162] is the definitive statement of SI, but many other publications have appeared suggesting the style of usage of SI in different fields, and a few examples may be listed.[166–174]

The adoption of SI does, however, present thermodynamicists with certain problems which chiefly arise because of the large number of data generated before the adoption of SI. The SI unit of energy is the joule. This presents little difficulty in the study of chemical thermodynamics because the thermochemical calory (cal_{th}), defined as $1\ cal_{th} = 4.184$ J exactly, has been in use for many years. As conversion from thermochemical calories to joules requires merely multiplication by the correct factor, conversion is easy. However, chemists who in very rough calculations are in the habit of putting the number 2 in equations when they see R (R being approximately 2 cal K^{-1} mol^{-1}) will have now to remember to use 8 (R being approximately 8 J K^{-1} mol^{-1})!

The situation with respect to pressure in SI units and thermodynamics is, however, not so happy. The SI pressure unit is the newton per square metre (N m^{-2}) or pascal (Pa) and the quantity previously called the standard atmosphere is equal to 101.325 kN m^{-2}. The difficulty arises because for many purposes (for example in the definition of enthalpies of formation) the standard pressure has been chosen as the standard atmosphere. Thus not only is the number 101.325 involved between the old and new systems, but no universal factor can be applied to change the value of a thermodynamic quantity for a real vapour from a pressure of 101.325 kN m^{-2} to say 100 kN m^{-2}. Therefore it appears likely that for many years in the future thermodynamicists will have to head their tables, 'Standard pressure 101.325 kPa (formerly known as 1 standard atmosphere)'.

The above comments are made to introduce the subject of units (and symbols) but the definitive documents on symbols for chemists are the Royal Society's [175] 'Quantities, Units, and Symbols' (1971) and the IUPAC 'Manual of Symbols and Terminology for Physicochemical Quantities and Units'.[176]

[166] 'The Use of SI Units', PD 5686, British Standards Institution, 1972; The 'International System of Units (SI)', British Standard 3763, 1970.
[167] 'Letter Symbols, Signs, and Abbreviations', British Standard 1991, Part I, 1967.
[168] 'Units and Standards of Measurement employed at the National Physical Laboratory, I. Mechanics (4th Edition)', Her Majesty's Stationery Office, London, 1967.
[169] The Royal Society Conference of Editors, 'Metrication in Scientific Journals', The Royal Society, London, 1968.
[170] M. L. McGlashan, 'Physicochemical Quantities and Units', Monographs for Teachers, No. 15, Royal Institute of Chemistry, London, 2nd edn., 1971.
[171] P. Anderton and P. H. Bigg, 'Changing to the Metric System – Conversion Factors, Symbols and Definitions', 3rd edn., Her Majesty's Stationery Office, London, 1969.
[172] 'The Use of SI Units', British Standards Institution, PD5686, 1969.
[173] G. H. Rayner and A. E. Dooke, 'SI Units in Electricity and Magnetism', Her Majesty's Stationery Office, London, 1970.
[174] P. Vigoureux, 'Electric Units and Standards', Her Majesty's Stationery Office, London, 1970.
[175] 'Quantities, Units, and Symbols', A report by the Symbols Committee of the Royal Society, London, 1971.
[176] M. L. McGlashan, 'Manual of Symbols and Terminology for Physicochemical Quantities and Units', Butterworths, London, 1970.

However, the practical worker needs more than a set of definitions of units and symbols, because for practical work scales for measurement have to be realized. Consider, for example, the position with regard to temperature. The absolute thermodynamic scale has a single fixed point and for this purpose the triple point of water has been chosen and has been assigned the temperature 273.16 K, exactly. The present working scale [177] is called the International Practical Temperature Scale (IPTS) of 1968, and this replaced the International Practical Temperature Scale of 1948. Both the 1948 and 1968 scales (IPTS-48 and IPTS-68) were based on the assigned values of the temperatures of a number of reproducible equilibrium states (defining fixed points) and on standard instruments calibrated at those temperatures. Interpolation between the fixed-point temperatures is provided by formulae used to establish the relation between indications of the standard instruments and values of the International Practical Temperature Scale. The intention in the use of IPTS-48, IPTS-68, and the earlier International Temperature Scale of 1927 was to provide practical scales which gave as nearly as possible thermodynamic temperatures. As the art of temperature measurement has been refined, so have new scales been necessary. The IPTS-68 differs from the IPTS-48 in the following ways. The lower bound of the scale is now 13.81 K instead of 90.18 K. The values assigned to the fixed points which define it are modified where necessary to conform as nearly as possible to thermodynamic temperatures as determined by continuing metrology; for details see ref. 177. IPTS-68 distinguishes between the International Practical Kelvin Temperature, with the symbol T_{68}, and the International Practical Celsius Temperature, with the symbol t_{68}; the relation between these quantities is $t_{68} = T_{68} - 273.15$ K.

The differences between the 1948 and 1968 scales are shown (ref. 177) in the Table.

As can be seen, it is essential in precise work to declare which temperature scale has been employed. Rossini [178] discusses IPTS-68 and considers the problem of conversion of temperature and calorimetric data obtained under the previous scale.

In exact thermochemical work the calibration for energy can nearly always be traced back to the use of electrical units. Whatever the definitions of ohm and volt may be, practical standards are used, and these are adjusted at rare intervals by international agreement. For example, as a result of international comparisons, adjustments [174] were made on 1 January 1969, so that the 'NPL ohm' was increased by 3.7 parts in 10^6 and the 'NPL volt' was decreased by 13 parts in 10^6. These changes are of such a magnitude that they have a small but significant effect on precise measurements of the enthalpy of combustion of very pure benzoic acid samples.

[177] 'The International Practical Temperature Scale of 1968 (National Physical Laboratory)', Her Majesty's Stationery Office, London, 1969.

[178] F. D. Rossini, *J. Chem. Thermodynamics*, 1970, **2**, 447; see also *Pure Appl. Chem.*, 1970, **22**, 557.

Table* *Approximate differences* $(t_{68} - t_{48})/\mathrm{K}$, *between the values of Celsius temperature given by the IPTS of* 1968 *and the IPTS of* 1948

$t_{68}/°C$	0	−10	−20	−30	−40	−50	−60	−70	−80	−90	−100
−100	0.022	0.013	0.003	−0.006	−0.013	−0.013	−0.005	0.007	0.012		
−0	0.000	0.006	0.012	0.018	0.024	0.029	0.032	0.034	0.033	0.029	0.022

$t_{68}/°C$	0	10	20	30	40	50	60	70	80	90	100
0	0.000	−0.004	−0.007	−0.009	−0.010	−0.010	−0.010	−0.008	−0.006	−0.003	0.000
100	0.000	0.004	0.007	0.012	0.016	0.020	0.025	0.029	0.034	0.038	0.043
200	0.043	0.047	0.051	0.054	0.058	0.061	0.064	0.067	0.069	0.071	0.073
300	0.073	0.074	0.075	0.076	0.077	0.077	0.077	0.077	0.077	0.076	0.076
400	0.076	0.075	0.075	0.075	0.074	0.074	0.074	0.075	0.076	0.077	0.079
500	0.079	0.082	0.085	0.089	0.094	0.100	0.108	0.116	0.126	0.137	0.150
600	0.150	0.165	0.182	0.200	0.23	0.25	0.28	0.31	0.34	0.36	0.39
700	0.39	0.42	0.45	0.47	0.50	0.53	0.56	0.58	0.61	0.64	0.67
800	0.67	0.70	0.72	0.75	0.78	0.81	0.84	0.87	0.89	0.92	0.95
900	0.95	0.98	1.01	1.04	1.07	1.10	1.12	1.15	1.18	1.21	1.24
1000	1.24	1.27	1.30	1.33	1.36	1.39	1.42	1.44			

$t_{68}/°C$	0	100	200	300	400	500	600	700	800	900	1000
1000		1.5	1.7	1.8	2.0	2.2	2.4	2.6	2.8	3.0	3.2
2000	3.2	3.5	3.7	4.0	4.2	4.5	4.8	5.0	5.3	5.6	5.9
3000	5.9	6.2	6.5	6.9	7.2	7.5	7.9	8.2	8.6	9.0	9.3

* With acknowledgements to 'International Practical Temperature Scale of 1968', Her Majesty's Stationery Office, London, 1969.

In precise thermodynamic work it must also be noted that relative atomic weights (usually called 'atomic weights') are revised every two years (in odd numbered years) by the International Commission on Atomic Weights, so that when values for molar quantities are listed it is essential to record the values of atomic weights used.

Some reference must now be made to estimates of error and uncertainty to be attached to the value of any thermodynamic quantity that has been experimentally determined. A measured value without an attendant indication of uncertainty is of little or no use. However, the treatment of errors is a complicated subject and in thermodynamics is made even more difficult because of the great expense and time necessary to carry out replicate experiments. It is impossible in this article to review the vast literature on statistics and errors, so only a few comments will now be made. It is strongly recommended that the meaning of $\pm$ values should be precisely defined in all instances. For example, it should be stated whether the $\pm$ value is a standard deviation and if so whether it refers to the mean or a single result. The number of degrees of freedom should also be shown. If the quantity quoted has been calculated by combining the results of a number of observations on different quantities, the method used for the propagation of errors should be stated. Similarly, if a weighted mean is presented, the weighting scheme should be recorded. Some of these topics, as they affect thermochemists, are discussed in papers by Rossini [179, 180] and by Rossini and Deming.[181]

4 Measurement of Thermodynamic Quantities

Thermodynamics is an unusual subject in that there are precise and mathematically exact relations between many apparently unconnected measured properties. For example, anybody unaware of thermodynamics might be surprised to find that there is a relation between the change of boiling temperature of a pure substance with change in pressure and the enthalpy of vaporization (Clausius–Clapeyron relation). Or again, a young student may be amazed to find that there is a relation (and a very direct relation) between the elevation in boiling temperature and the depression in freezing temperature of a volatile solvent when an involatile material is dissolved in the solvent.

Thermodynamics is what Einstein called a 'theory of principle', that is thermodynamics [182] starts from empirically observed general properties and deduces from them results 'that apply to every case which presents itself' without making any assumptions regarding the constituents. The universality of thermodynamics so impressed Einstein that he stated, 'It

[179] F. D. Rossini, *Chem. Rev.*, 1936, **18**, 233.

[180] 'Experimental Thermochemistry: Measurement of Heats of Reaction', ed. F. D. Rossini, Chapter 14, Interscience, New York, 1956.

[181] F. D. Rossini and W. E. Deming, *J. Washington Acad. Sci.*, 1939, **29**, 416.

[182] M. J. Klein, *Science*, 1967, **157**, 510.

is the only physical theory of universal content concerning which I am convinced that, within the framework of applicability of its basic concepts, it will never be overthrown'. It follows therefore that in principle one can measure certain thermodynamic quantities and can derive other quantities by application of equations based on thermodynamic theory. The set for measurement could be arbitrarily chosen, but in practice certain quantities are more easily measured than others and so these are usually studied experimentally.

If the importance of carrying out measurements on well-characterized materials had not been so often overlooked in the past, it might be considered unnecessary to emphasize that precise measurements are meaningful only if made on specimens of controlled and known composition. If a property of a single substance is being measured, not only is it necessary that the specimen should be of high purity (*viz.* contain a high mole fraction of the substance under investigation) but the nature and mole fractions of contaminants must be known. The criteria to be applied are that the contaminants should be of such a nature and present at such low mole fractions as to have a negligible effect on the property to be measured.

To illustrate this philosophy, consider the measurement of energies of combustion. In bomb calorimetry the presence of a small amount of water in the sample of an organic compound undergoing combustion in oxygen can produce serious errors in the derived specific energy of combustion if this is calculated on the mass of sample taken. On the other hand, a small amount of an isomeric organic compound may have very little effect on the observed energy of combustion because the specific energies of combustion of the main component and the contaminant may be very similar. Thus for example, the presence of a small amount of *p*-xylene in *o*-xylene will introduce only a small error into the measured energy of combustion of *o*-xylene. An impurity may have a greater effect on some methods of measurement than on others. For example, traces of air may seriously upset the measurement of vapour pressure when a static method is used but may have little effect on the result obtained if a dynamic method is employed. In general, thermodynamic measurements are now made on samples of 99.9 moles per cent purity or better.

Other articles in this book discuss methods for the measurement of a range of thermodynamic properties, and so details of measurement procedures will not be given here. However, a few sources of general information on thermodynamic measurement will be cited.

Weissberger's series of volumes [183] present an extensive review of methods used in physical chemistry and include accounts of methods used in the study of many thermodynamic properties.

[183] 'Technique of Organic Chemistry', ed. A. Weissberger, vol. I. 'Physical Methods', Interscience, New York, 1960.

The International Union of Pure and Applied Chemistry publication [184] 'Experimental Thermochemistry: Measurement of Heats of Reaction' is written by experts for experts, and deals chiefly with organic compounds. The volume opens with a chapter entitled 'General Principles of Modern Thermochemistry', by Rossini, taken mostly from a book [25] by that author (1950) entitled 'Chemical Thermodynamics'. Other chapters discuss the calibration of calorimeters for flame and bomb reactions, and the combustion of oxygen, nitrogen, sulphur, chlorine, bromine, and iodine compounds. Fifty pages are taken in the discussion of microcalorimetry of slow reactions.

The second volume in this series,[185] edited by Skinner, treats many kinds of combustion calorimetry, fluorine reaction calorimetry, reaction calorimetry, high-temperature calorimetry, solution calorimetry, heats of mixing, microcalorimetry, and biochemical reactions.

Calvet and Prat's book (translated by Skinner) [186] is concerned with microcalorimetry.

The book on thermochemistry [187] by Cox and Pilcher gives a detailed discussion of the principles of the accurate determination of enthalpies of formation and enthalpies of vaporization (or sublimation) of organic and organometallic compounds. The effects of impurities on enthalpies of reaction are also considered and criteria for the purity of substances under investigation are discussed.

McCullough and Scott [188a] have edited a volume on behalf of the IUPAC Commission on Thermodynamics and Thermochemistry, devoted to the study of the calorimetry of non-reacting systems. The 24 contributors have supplied chapters covering the temperature range from below 20 K to very high temperatures. Temperature scales are discussed, but it is unfortunate that, although the book appeared in 1968, the article on temperature scales and temperature measurement was prepared too early to include reference to the IPTS-68.

Considerable interest is being shown in differential thermal analysis and differential scanning calorimetry,[188b, 188c] but more work is required to examine the accuracy of results obtained with these non-equilibrium thermal techniques.

[184] 'Experimental Thermochemistry: Measurement of Heats of Reaction', ed. F. D. Rossini, Interscience, New York, 1956.

[185] 'Experimental Thermochemistry. vol. II', ed. H. A. Skinner, Interscience, New York, 1962.

[186] E. Calvet and H. Prat, 'Recent Progress in Microcalorimetry', translated by H. A. Skinner, Pergamon Press, Oxford, 1963.

[187] J. D. Cox and G. Pilcher, 'Thermochemistry of Organic and Organometallic Compounds', Academic Press, London, 1970.

[188a] 'Experimental Thermodynamics, vol. 1: Calorimetry of Non-reacting Systems', ed. J. P. McCullough and D. W. Scott, Butterworths, London, 1968.

[188b] 'Thermal Analysis: vol. 1, Instrumentation, Organic Materials and Polymers; vol. 2, Inorganic Materials and Physical Chemistry', ed. R. F. Schwenker and P. D. Garn, Academic Press, New York and London, 1969.

[188c] 'Differential Thermal Analysis: vol. 1', ed. R. C. Mackenzie, Academic Press, New York and London, 1970.

As mentioned earlier in this section, it is often possible to derive values for some thermodynamic properties by combining measurements made on other properties. However, care may be necessary in making the choice of the most reliable route to obtain data. As examples, consider the methods available for obtaining the standard entropy of a material. Measurements of heat capacity from the lowest temperature up to the temperature of interest (see Chapter 4) can provide values of the standard entropy provided the material satisfies the conditions necessary for the Third Law to be applicable. However, there may sometimes be doubt whether the Third Law is obeyed, and then a different route must be sought. If a suitable chemical equilibrium can be studied over a range of temperatures, then the equilibrium constant of the reaction may be measured. The standard enthalpy of reaction can then be found by means of the Second Law, from the equation:

$$\mathrm{d}\ln K_p/\mathrm{d}T = \Delta H_\mathrm{r}^\circ/RT^2,$$

where K_p is the equilibrium constant and $\Delta H_\mathrm{r}^\circ$ is the standard enthalpy of the reaction. By combining the values of $\Delta G_\mathrm{r}^\circ$ (the standard Gibbs energy change for the reaction) with $\Delta H_\mathrm{r}^\circ$, the standard entropy change of the reaction can be found, and hence the standard entropy of the compound of interest may be calculated if the standard entropies of all the other reactants and products are known.

Alternatively, the entropy change of a reaction can be found by combining a value of the Gibbs energy change for the reaction, calculated from a measurement of the equilibrium constant at one temperature, with the enthalpy change for the reaction measured separately, or calculated from calorimetric measurements on the reactants and products. The entropy of the compound of interest can then be calculated from the entropy of the reaction by using values for the entropy of all the other reactants and products.

The entropy of a substance in the gas phase can often be calculated by the use of statistical thermodynamics. For this purpose, information on the structure of the molecule is required. The fundamental frequencies of vibration must be established, and values for the barrier heights of any restricted rotations must be obtained (see Chapter 8).

The nature of the compound and the availability of data often determine which of the thermodynamic routes is chosen for calculation, and the accuracy of the result will depend on the accuracy of the data employed. The Second Law route described above requires a differentiation, and therefore the values of the equilibrium constant must be very accurately known if accurate values of $\Delta H_\mathrm{r}^\circ$ are to be obtained by this method.

A prime object of thermodynamic measurement is to obtain a consistent and accurate set of data on elements, compounds, mixtures, and alloys so that the data can be combined to produce information on equilibria and energy changes for systems which have not yet been studied, and for systems partially studied but on which more information is required. The next section lists some sources of collected data.

5 Tabulations of Thermodynamic Data

For practical applications it is necessary to have values of thermodynamic functions available, and in particular it is necessary to have values of enthalpy, entropy, heat capacity, and Gibbs energy for a wide range of temperatures and pressures. The critical evaluation of data is a difficult and important art, often made more difficult by the failure of authors of original papers to supply necessary information. Some comments on this subject can be found in a recent article.[189]

The results of recent measurements are described in other Chapters of this volume, but here collections of data will be reviewed.

Reactions are often carried out at pressures in the neighbourhood of 101.325 kPa, and in this pressure region the thermodynamic properties of condensed phases (*i.e.* liquids and solids) do not change very rapidly with small changes in pressure. As a result, it is usual to adopt the following standard states when recording thermodynamic quantities:

(*a*) the standard state for a gas is the hypothetical ideal gas at a pressure of 101.325 kPa (formerly called 1 standard atmosphere);

(*b*) the standard state for a liquid is that of the pure substance under a pressure of 101.325 kPa (formerly called 1 standard atmosphere);

(*c*) the standard state for a solid is that of the pure crystalline substance, under a pressure of 101.325 kPa (formerly called 1 standard atmosphere).

Investigators must take care to read the foreword of the particular table they use so that they know which standard state has been employed, because most thermodynamic properties are calculated with respect to convenient scales. For example, the standard enthalpy of formation of a compound, ΔH_f°, is almost always quoted for a temperature of 298.15 K, and the enthalpy of formation of an element in its standard state must by definition be zero. It is therefore practically useful to look at a table and find, for an element, where a zero entry occurs. For example, the following values might appear: C(graphite), $\Delta H_f^\circ = 0.000\ \text{kcal}_{\text{th}}\ \text{mol}^{-1}$; C(diamond), $\Delta H_f^\circ = 0.4532\ \text{kcal}_{\text{th}}\ \text{mol}^{-1}$. It is clear that C(graphite) is the standard state adopted for carbon in the table under consideration. Entropy, on the other hand, is usually defined by taking as zero the entropy, at $T = 0$, of the crystalline form in which all the molecules are orientated regularly. Because many of the extant tables have used thermochemical calories, care will also have to be taken in the future to see that values taken from different tables are corrected to the same units.

The simplest thermodynamic tables list only values for 298.15 K, but more detailed tabulations present data for a wide temperature range.

[189] 'Critical Data in the Physical Sciences', National Standard Reference Data System, September, 1970.

There have been fashions in the method chosen for presentation of data, and the methods employed have rather naturally reflected the type of computation facilities available. Earlier workers favoured power series in T (or sometimes in Celsius temperature), but these tabulations were tedious to use and the equations were often poorly conditioned, so that the final result depended heavily upon partial cancellations of large terms. With the advent of statistical mechanical calculations, tables of values for small intervals of temperature became common and are still produced. Such tables are easy to use and a value can often be interpolated readily. This presentation method preserves the high accuracy obtainable by the statistical thermodynamic treatment and is also convenient for the tabulation of data obtained by direct experiment. However, the current widespread availability of electronic digital computers has again encouraged the use of temperature series, but it is now possible to use a larger number of terms and there is a tendency to use well-conditioned orthogonal functions (*e.g.* Chebyshev series) to represent the data. Some difficulties still arise, however, because refinement in measurement has often revealed small but real 'bumps' in the data (as in the curve of heat capacity of solid against temperature for example) and these 'bumps' are sometimes difficult to represent by a convenient power series. A common procedure in these circumstances is to smooth the data (eliminating the 'bump'), fit a series to the smooth data, and then instruct the computer to add back suitable values in the correct position to take care of the 'bump'.

In the following account of some data sources, no attempt has been made to cover numerical values concerning mixtures, although a few sources of such information will be mentioned. Most text-books on thermodynamics present numerical data, and here only text-books which contain much numerical information will be mentioned. The quality of data is continuously being improved, so that later compilations often replace early ones; however, the older literature is still useful because in some instances it contains data not available elsewhere. Data sources will be discussed in roughly chronological order and, although many of the main sources are mentioned, no claim for completeness in coverage is made.

The first edition of Lewis and Randall's book [11] presents numerous values for thermodynamic functions in the text and a table of standard free energies of formation at 25 °C. It should be remarked that (Gibbs) free energy is now called Gibbs energy by IUPAC.

Kharasch presented a survey [190] of values for the enthalpies of combustion of organic compounds published up to 1929.

Parks and Huffman's 'Free Energies of Some Organic Compounds' [191] makes much use of power series for the representation of data, and at the date of publication (1932) was a pioneer effort.

[190] M. S. Kharasch, *J. Res. Nat. Bur. Stand.*, 1929, **2**, 359.

[191] G. S. Parks and H. M. Huffman, 'Free Energies of Some Organic Compounds', American Chemical Society Monograph No. 60, The Chemical Catalog Co. New York, 1932.

The International Critical Tables,[192] often referred to as ICT, were prepared by numerous contributors over a number of years. The index lists substances by name and under the name tabulates the references to the properties. Thermodynamic properties that are dealt with include heat capacity, enthalpy, entropy, enthalpies of combustion, solution, and formation, free energy, melting and boiling temperatures, vapour pressure, critical constants, depression of freezing temperature, and elevation of boiling temperature.

Kelley (sometimes with co-authors) has issued a large number of reviews of data on the thermodynamic properties of inorganic compounds in U.S. Bureau of Mines' Bulletins; for example, Bulletin 371 reviews[193] data extant in 1934 on high-temperature thermal properties. Bulletin 383 includes[194] values for gaseous and solid heat capacity and enthalpy of vaporization, and presents extensive tables giving the temperatures at which the vapour pressures equal 0.0001, 0.001, 0.01, 0.1, 0.25, 0.5, and 1.0 atm. Bulletin 384 presents[195] existing thermodynamic data for carbonates. As far as possible, decomposition pressure determinations are correlated with calorimetrically determined enthalpies of formation and entropy and with solubility and standard electrode potential data. The results of these studies are summarized in tables. Bulletin 394 revises[196] and supplements Bulletin 388 (published in 1935) and discusses entropies at 298.1 K, which are summarized in tables giving heat capacities at 10, 25, 50, 100, 150, 200, and 298.1 K, entropies calculated from the Third Law, entropies calculated by other methods, and recommended values. Bulletin 393 supplies[197] enthalpy of fusion values for inorganic substances, the values being obtained from available freezing temperatures of binary systems. Summarizing tables are provided which list enthalpies of fusion obtained from binary data, from vapour pressure measurements, and from direct measurements.

Kelley's collection of data on metal carbides and nitrides[198] includes thermodynamic data for methane and ammonia because some of the

[192] 'International Critical Tables of Numerical Data: Physics, Chemistry and Technology', ed. E. W. Washburn, in eight parts, McGraw-Hill, New York, 1926—33.

[193] K. K. Kelley, 'Contributions to the Data on Theoretical Metallurgy – II – High-temperature Specific-heat Equations for Inorganic Substances', U.S. Bureau of Mines, Bulletin 371, 1934.

[194] K. K. Kelley, 'Contributions to the Data on Theoretical Metallurgy – III – The Free Energies of Vaporization and Vapour Pressures of Inorganic Substances', U.S. Bureau of Mines, Bulletin 383, 1935.

[195] K. K. Kelley and C. T. Anderson, 'Contributions to the Data on Theoretical Metallurgy – IV – Metal Carbonates – Correlations and Applications of Thermodynamic Properties', U.S. Bureau of Mines, Bulletin 384, 1935.

[196] K. K. Kelley, 'Contributions to the Data on Theoretical Metallurgy – VI – A Revision of the Entropies of Inorganic Substances', U.S. Bureau of Mines, Bulletin 394, 1935.

[197] K. K. Kelley, 'Contributions to the Data on Theoretical Metallurgy – V – Heats of Fusion of Inorganic Substances', U.S. Bureau of Mines, Bulletin 393, 1936.

[198] K. K. Kelley, 'Contributions to the Data on Theoretical Metallurgy – VIII – The Thermodynamic Properties of Metal Carbides and Nitrides', U.S. Bureau of Mines, Bulletin 407, 1937.

reactions considered require values for these substances. Various values for the thermodynamic properties are discussed and data are then presented in the form of free-energy equations. Bulletin 406 includes data [199] on metallurgically important substances such as sulphur, sulphur oxides, hydrogen sulphide, metal sulphides, and metal sulphates, and the information is summarized in the form of free-energy relations.

Bichowsky and Rossini [200] present a collection of values for the thermodynamic properties of inorganic compounds and of C_1 and C_2 organic compounds. Various volumes of the Landolt–Börnstein tables [201] give much information on values of thermodynamic properties.

Several of the volumes of a series of tables called 'Annual Tables of Constants and Numerical Data' provide information on thermodynamic quantities. For example, volume 4 lists [202] specific heat capacities for the following classes of substances: elements, inorganic compounds, natural minerals, organic compounds (hydrocarbons, alcohols, phenols, acids, aldehydes, ketones, ethers, esters, halogen derivatives, nitrogen compounds, heterocyclic compounds, organometallic compounds), mixtures and solutions (gaseous solutions, alloys, technical materials), C_p/C_V for organic compounds, and miscellaneous information. Volume 3 gives tables [203] for the quantities named in the title for inorganic compounds, arranged under elements, and for organic compounds, partly arranged by formula and partly by class of compound. Volume 6 lists boiling temperatures [204] of elements, inorganic compounds, and organic compounds, with equations for vapour pressure. For mixtures the pressure, temperature, composition data are given for aqueous solutions of inorganic and organic compounds, for organic–inorganic compounds, and organic–organic compounds. Some data are given on ternary mixtures and on azeotropy. The following quantities are tabulated for gases: isotherms and virial coefficients, critical-point data, rectilinear diameters, Joule–Thomson effect, equations of state, and van der Waals coefficients. Volume 9 gives data [205] on concentration cells and standard potentials. Volume 10 provides [206] the following infor-

[199] K. K. Kelley, 'Contributions to the Data on Theoretical Metallurgy – VII – The Thermodynamic Properties of Sulphur and its Inorganic Compounds', U.S. Bureau of Mines, Bulletin 406, 1937.

[200] F. R. Bichowsky and F. D. Rossini, 'Thermochemistry of the Chemical Substances', Reinhold, New York, 1936.

[201] Landolt–Börnstein, 'Physikalisch-Chemische Tabellen', Julius Springer, Berlin, 5th edn., Hauptwerk, 1923; I. Erg. Band, 1927; II. Erg. Band, 1931; III. Erg. Band, 1936.

[202] 'Annual Tables of Constants and Numerical Data, 4, Heat Conductance (M. L. Brouty) and Specific Heat (F. Wolfers)', Hermann et Cie., Paris, 1937.

[203] 'Annual Tables of Constants and Numerical Data, 3, Free Energy, Heat Content, Entropy and Activity (J. Gueron and J. P. Mathieu)', Hermann et Cie., Paris, 1937.

[204] 'Annual Tables of Constants and Numerical Data, 6, Vapour Pressure, Boiling Temperatures (W. P. Jorissen and P. C. van Keekem); Gas Laws (W. H. Keesom, J. J. M. Van Santen, and J. Haantjes)', Hermann et Cie., Paris, 1937.

[205] 'Annual Tables of Constants and Numerical Data, 9, Electromotive Forces (H. S. Harned) and Oxidation–Reduction Potentials (G. A. Åkerlöf)', Hermann et Cie., Paris, 1937.

[206] 'Annual Tables of Constants and Numerical Data, 10, Thermochemistry (W. A. Roth)', Hermann et Cie., Paris, 1937.

mation on the properties of elements, inorganic, and organic compounds: enthalpies of fusion, transition, sublimation, vaporization, dissociation, and ionization; enthalpies of formation and reaction of inorganic compounds, oxides, and nitrides, and intermetallic, addition, and organic compounds; enthalpies of combustion, hydration, solution, and dilution. Volume 20 lists [207] the melting temperatures of elements, inorganic, organometallic, and organic compounds, with a table showing the effect of pressure. Data, including phase diagrams, are provided for mixtures of elements, element–inorganic compounds, inorganic–inorganic compounds, inorganic–organic compounds, and organic–organic compounds. Volume 27 includes [208] values of melting, freezing, and boiling temperatures, specific volumes, and densities. Volume 28 presents [209] numerous phase diagrams for metals and alloys. Volume 34 has thirteen pages devoted to cryometry [210] (*e.g.* molar depression of freezing temperature) and seven pages devoted to ebulliometry.

Stull has provided [211] extensive tables of the vapour pressures of pure substances. Mention should also be made of Lang's thermodynamic properties of metal oxides [212] and of charts by Hottel *et al.*[213]

Lecat has covered the literature on azeotropes [214] for the period 1932—1941 and includes data on vapour pressure. The second edition [215] contains information on 2450 systems.

The Bureau of Mines' Bulletin 476 revises [216] and elaborates data given in Bulletin 371 (published 1934) and contains information concerning elements and compounds, arranged alphabetically. Numerical data presented in both tabular and algebraic form are based on experimental measurements and on statistical mechanical calculations. The claim is made that the compilation covers all available high-temperature enthalpy and heat capacity data for inorganic substances up to the time of publication.

[207] 'Annual Tables of Constants and Numerical Data, 20, Melting Temperatures (Pure Substances and Binary Systems) (F. Meyer)', Masson et Cie., Paris, 1937.

[208] 'Annual Tables of Constants and Numerical Data, 27, Characteristic Constants of Organic Compounds (G. Kravtzoff)', Masson et Cie., Paris, 1939.

[209] 'Annual Tables of Constants and Numerical Data, 28, Metals and Alloys (collected and edited by T. F. Russell and revised by C. H. Desch)', Masson et Cie., Paris, 1939.

[210] 'Annual Tables of Constants and Numerical Data, 34, Cryometry, Ebulliometry (W. Swiętosławski, M. Lazniewski, and J. Pomorski)', Hermann et Cie., Paris, 1941.

[211] D. R. Stull, 'Vapour Pressures of Pure Substances', *Ind. and Eng. Chem.*, 1947, **39**, 517, 540.

[212] W. Lang, 'Die thermodynamischen Eigenschaften der Metalloxyde', Springer-Verlag, Berlin, 1949.

[213] H. C. Hottel, G. C. Williams, and C. N. Satterfield, 'Thermodynamic Charts for Combustion Processes', J. Wiley and Sons, New York, 1949.

[214] M. Lecat, 'L'Azéotropie, la Tension de Vapour des Mélanges de Liquids', Bibliographie Maurice Lamertin, Brussels, 1942.

[215] M. Lecat, 'Tables Azéotropiques, Tome Premier, Azéotropes binaires orthobares', 2nd edn., published by the author, Brussels, 1942.

[216] K. K. Kelley, 'Contributions to the Data on Theoretical Metallurgy – X – High-temperature Heat-content, Heat Capacity and Entropy Data for Inorganic Compounds', U.S. Bureau of Mines, Bulletin 476, 1949.

Bulletin 477 states [217] that the first Bulletin in this series, which appeared in 1932, contained a compilation of all the entropy values then available for the elements and inorganic substances and resulted in a list of some 150 substances. Revision followed in 1936 (Bulletin 394) and 1941 (Bulletin 280), when over 500 entropies were tabulated. Bulletin 477 constitutes further revision, so that 800 entropy values are available. Values for 25 °C are presented, and tables of low-temperature enthalpies (at 10, 25, 50, 100, 150, and 298.16 K) are given, with values of entropies at 25 °C calculated by the Third Law method, spectroscopically, and obtained by other means.

The articles by Brewer *et al.*,[218] and many books [19, 25, 31, 33–35, 42] already mentioned, contain many data.

The volume edited by Quill [219] contains ten papers by various authors, dealing with liquid–solid equilibria, temperature–composition diagrams of metal–metal halide systems, and properties of the elements, carbides, sulphides, silicides, phosphides, and halides.

Sage has for many years studied the properties of gases and liquids, and in his book with Lacey lists [220] the thermodynamic properties of the lighter paraffin hydrocarbons and nitrogen.

Ribaud [221] has presented tabulations of Gibbs energy for saturated hydrocarbons and some other organic compounds, and Brinkley and Lewis [222] have considered the thermodynamics of combustion gases. Latimer's book [223] contains an extensive collection of redox potentials.

The National Bureau of Standards' Circular 500 on the selected values of chemical thermodynamic properties [224] consists of tables of data divided into two series, with the first giving enthalpies and Gibbs energies of formation, entropies, and heat capacities at 25 °C, and the second series giving data on phase transitions, including the temperature, enthalpies, entropies, and heat capacities related to the change. The substances for which data are listed include inorganic materials and organic substances

[217] K. K. Kelley, 'Contributions to the Data on Theoretical Metallurgy – XI – Entropies of Inorganic Substances – Revision (1948) of Data and Methods of Calculation', U.S. Bureau of Mines, Bulletin 477, 1950.

[218] L. Brewer, L. A. Bramley, P. W. Gilles, and N. L. Lafgren, in 'The Chemistry and Metallurgy of Miscellaneous Materials – Thermodynamics', ed. L. L. Quill, McGraw-Hill, New York, 1950.

[219] 'The Chemistry and Metallurgy of Miscellaneous Materials – Thermodynamics', ed. L. L. Quill, McGraw-Hill, New York, 1950.

[220] B. H. Sage and W. N. Lacey, 'Thermodynamic Properties of the Lighter Paraffin Hydrocarbons and Nitrogen', American Petroleum Institute, New York, 1950.

[221] G. Ribaud, Publication No. 266 Scientifiques et Techniques du Ministère de l'air France, 1952.

[222] S. R. Brinkley and B. Lewis, 'The Thermodynamics of Combustion: General Considerations', U.S. Bureau of Mines, Bulletin 4806, 1952.

[223] W. M. Latimer, 'The Oxidation States of the Elements and their Potentials in Aqueous Solutions', 2nd edn., Prentice-Hall, New York, 1952.

[224] F. D. Rossini, D. D. Wagman, W. H. Evans, S. Levine, and I. Jaffe, 'Selected Values of Chemical Thermodynamic Properties' (Circular of the National Bureau of Standards, No. 500), United States Government Printing Office, Washington, D.C., 1952.

containing one or two carbon atoms. Thermochemical calories are used throughout.

Dreisbach[225] presents tables on the pressure, volume, temperature relations of organic compounds, using Cox charts for families of compounds. In the last chapter of Gaydon's book,[226] dissociation energies for 275 molecules are listed.

The Coal Tar Data Book[227] includes values of boiling and freezing temperatures, the variation of boiling temperature with pressure, density, heat capacity, critical constants, enthalpies of fusion, of vaporization, and of combustion, and azeotropic constants for common substances that occur in the products of coal carbonization. 'Selected Values of Physical and Thermodynamic Properties of Hydrocarbons and Related compounds', by Rossini *et al.*,[228] consists of 1050 pages, of which nearly 800 are tables of data. Properties listed include boiling temperature, the variation of boiling temperature with pressure, density, freezing temperature, molar volume, critical properties, vapour pressure, enthalpies of vaporization and fusion, entropy of vaporization, enthalpies of combustion and formation, free energy of formation, entropy, 'enthalpy function' and 'free energy function', and logarithm of the equilibrium constant of formation. The compounds listed include many classes of hydrocarbons and certain sulphur compounds.

Coughlin[229] has presented data on approximately 170 inorganic oxides, arranged in three different ways: (i) tables of enthalpy and free-energy-of-formation values at 25 °C, at phase-change temperatures, and at even intervals of 100 K from 400 to 2000 K (and occasionally to higher temperatures); (ii) equations giving free energies of formation as functions of temperature; (iii) approximate equations giving free energies as linear functions of temperature. Jordan's tables[230] of vapour pressure present data in the form of tables, equations, and graphs for the following classes of compounds: hydrocarbons, halogen compounds, alcohols, oxygen compounds, acids, acid chlorides and anhydrides, esters, nitrogen compounds, and phenols, and organic compounds containing arsenic, selenium, sulphur, silicon, antimony, boron, gallium, lead, phosphorus, tin, thallium, beryllium, and zinc.

225 R. R. Dreisbach, 'Pressure–Volume–Temperature Relationships of Organic Compounds', 3rd edn., Handbook Publishers, Sandusky, Ohio, 1952.

226 A. G. Gaydon, 'Dissociation Energies and Spectra of Diatomic Molecules', 2nd edn., revised, Chapman and Hall, London, 1953.

227 'Coal Tar Data Book', Coal Tar Research Association, Gomersal, 1953.

228 F. D. Rossini, K. S. Pitzer, R. L. Arnett, R. M. Brown, and G. C. Pimentel, 'Selected Values of Physical and Thermodynamic Properties of Hydrocarbons and Related Compounds, Tables of the American Petroleum Institute Research, Project 44', Carnegie Press, Pittsburgh, Penn., 1953.

229 J. P. Coughlin, 'Contributions to the Data on Theoretical Metallurgy – XII – Heats and Free Energies of Formation of Inorganic Oxides', U.S. Bureau of Mines, Bulletin 542, 1954.

230 T. E. Jordan, 'Vapour Pressure of Organic Compounds', Interscience, New York, 1954.

The monograph by Aubert and Sivolobov[231] presents data on 69 hydrocarbons, 10 alcohols, 2 ethers, graphite, and thiophen for the solid, liquid, and vapour states in the temperature range 90 to 1400 K. Methods for the calculation of vapour heat capacities are given and the principal types of heat capacity anomaly in the condensed phases are presented. Tables giving melting temperature, enthalpies of transition, and entropies are provided.

'Tables of Thermal Properties of Gases', by Hilsenrath *et al.*,[232] include the thermodynamic properties of air, argon, carbon dioxide, carbon monoxide, hydrogen, nitrogen, oxygen, and steam.

Din[233] was the editor of a series of books designed to provide reliable thermodynamic data for industrially important gases. Temperature–entropy diagrams were chosen as the most generally useful graphical presentations and these are supplemented by tables of entropy, enthalpy, volume, heat capacity at constant pressure and at constant volume, and Joule–Thomson coefficients. Unfortunately, there is no consistency in the choice of units, although the thermochemical calorie is employed. The report on each substance (*i.e.* ammonia, carbon dioxide, carbon monoxide, air, argon, acetylene, ethylene, and propane) consists of a brief introduction, a survey of experimental data, a description of methods used for the thermodynamic calculations, and a set of tables.

The 'Selected Values of Properties of Chemical Compounds', issued since 1955 in loose-leaf form,[234] includes values for density, critical constants, vapour pressure, enthalpy, entropy, enthalpies of transition, usion, and vaporization, enthalpy of formation, 'Gibbs energy function', heat capacity, and logarithm of equilibrium constant of formation.

In 'Thermodynamic Properties of the Elements', by Stull and Sinke,[235] the authors have chosen $T = 298.15$ K instead of $T = 0$ as the standard reference temperature for the enthalpy tabulations. However, values of $\{H^\circ(298.15\text{ K}) - H^\circ(0)\}$ are given to allow calculation of $\{G^\circ - H^\circ(0)\}/T$ from the values of $\{G^\circ - H^\circ(298.15\text{ K})\}/T$ listed. Values of S° and C_p° are tabulated for values of T above 298.15 K. The melting and boiling

231 M. Aubert and N. Sivolobov, 'Chaleurs Spécifiques et Entropies de Hydrocarbures et de Quelques Combustibles Liquides', Publication No. 297 Scientifiques et Techniques du Ministère de l'Air France, 1955.

232 J. Hilsenrath, C. W. Beckett, W. S. Benedict, L. Fano, H. J. Hoge, J. F. Masi, R. L. Nutall, Y. S. Touloukian, and H. W. Woolley, 'Tables of Thermal Properties of Gases', National Bureau of Standards Circular 564, United States Government Printing Office, Washington, D.C., 1955.

233 'Thermodynamic Functions of Gases, Vol. 1 (Ammonia, Carbon Dioxide, Carbon Monoxide); Vol. 2 (Air, Argon, Acetylene, Ethylene, Propane)', ed. F. Din, Butterworths Scientific Publications, London, 1956.

234 'Selected Values of Properties of Chemical Compounds', Manufacturing Chemists' Association Research Project, Semi-annual looseleaf, 1955—1966; 'Selected Values of Properties of Chemical Compounds', Thermodynamic Research Center Data Project, Texas A & M University, Texas, 1966—, continuing.

235 D. R. Stull and G. C. Sinke, 'Thermodynamic Properties of the Elements'. Advances in Chemistry Series, No. 18, American Chemical Society, Washington, D.C., 1956.

temperatures and enthalpies of fusion and vaporization are given for the solids and liquids. Separate tables are presented for the molecular species as ideal gases when appropriate. Woolley's data [236] include the thermodynamic functions for the real and ideal gas and the vapour pressure of liquid and solid nitrogen.

'Thermodynamic Data on Alloys', by Kubaschewski and Catterall,[237] critically surveys data on some 400 binary and ternary metallic systems, with the term 'alloy' interpreted to include transition-metal oxides, sulphides, carbides, nitrides, and phosphides. When available, values are given for the integral enthalpy and entropy of formation of the alloy, the relative partial molar enthalpy and entropy of one component, and the volume change on mixing.

A publication from the Argonne National Laboratory, by Glassner,[238] lists the thermochemical properties of oxides, fluorides, and chlorides up to 2500 K.

Field and Franklin [239] present critical compilations of data on gaseous ions up to 1956. In the first edition of Cottrell's book (1951), the values of bond energies and bond dissociation energies for carbon and nitrogen compounds were based on the 'low' values for the enthalpy of vaporization of carbon and the enthalpy of dissociation of nitrogen, but the second edition [240] uses the 'high' values. This edition also discusses the data on which the tabulations are based. Skinner's lecture,[241] which summarizes the results of measurements, discusses values for enthalpies of decomposition, combustion, hydrogenation, halogenation, hydrohalogenation, hydrolysis, and polymerization. Wall's book [61] gives a table of thermodynamic data in an appendix. As the object of Janz's book [242] is the provision of a collection of practical methods for the computation of thermodynamic properties, it will be discussed later. However, it contains extant data as a starting point for the computations.

The extensive tables by Charlot *et al.*[243] are classified under the chemical symbol for the element, and they list temperatures and concentrations at which the potentials were measured.

[236] H. W. Woolley, 'Thermodynamic Properties of Gaseous Nitrogen', National Advisory Committee for Aeronautics, Technical Note 3271, 1956.

[237] O. Kubaschewski and J. A. Catterall, 'Thermodynamic Data of Alloys', Pergamon Press, London, 1956.

[238] A. Glassner, 'The Thermochemical Properties of the Oxides, Fluorides, and Chlorides to 2500 °K', Argonne National Laboratory, Lemont, Ill., ANL-5750, 1957.

[239] F. H. Field and J. L. Franklin, 'Electron Impact Phenomena and the Properties of Gaseous Ions', Academic Press, New York, 1957.

[240] T. L. Cottrell, 'The Strengths of Chemical Bonds', 2nd edn., Butterworths Scientific Publications, London, 1958.

[241] H. A. Skinner, 'Modern Aspects of Thermochemistry', Royal Institute of Chemistry Lecture, 1958.

[242] G. J. Janz, 'Estimation of Thermodynamic Properties of Organic Compounds', Academic Press, New York, 1958.

[243] G. Charlot, D. Bézier, and J. Coulot, 'Tables de Constantes et Données Numériques. Potentials d'Oxydo-Réduction', Pergamon Press, London, 1958.

Brewer and Chandrasekharaiah's tables [244] of 'free energy functions' for gaseous monoxides have been calculated from spectroscopic results for 500 K intervals from room temperature to 3000 K.

Sinke's tables [245] of the thermodynamic properties of combustion products were prepared for the calculation of specific impulse and are not considered definitive or exhaustive. The tables list values of heat capacity, entropy, 'free energy function', enthalpy, enthalpy of formation, free energy of formation, and the logarithm of the equilibrium constant of formation from 298 to 6000 K at 100 K intervals.

Kobe and co-workers [246] have produced extensive lists of values of thermodynamic properties of hydrocarbons and derivatives, the data being distributed between 25 articles.

In addition to the thermodynamic properties of ions, Latimer [247] gives extensive tables for the properties of elements in the crystalline and gaseous states and for inorganic compounds in the form of crystals and in aqueous solution.

Holley *et al.*[248] present data for gases at pressures up to several hundred atmospheres and at temperatures to several hundred kelvins, at close intervals so as to allow for easy interpolation.

Kelley,[249] in Bulletin 584, has revised and extended Bulletin 476 (1949) and included data available to September 1958. Enthalpies are given in tabular and algebraic form.

The second volume of the book [250] by Hougen *et al.*, 'Chemical Process Principles', contains some tables of thermodynamic data. Tester,[251] on behalf of the British Petroleum Co. Ltd., has prepared 'Thermodynamic Properties of Methane'. Elliot and Gleiser [252] treat the thermochemistry of steelmaking.

244 L. Brewer and M. S. Chandrasekharaiah, 'Free Energy Functions for Gaseous Monoxides', Lawrence Radiation Laboratory, Berkeley, California, UCRL-8Y13, 1959.

245 G. C. Sinke, 'Thermodynamic Properties of Combustion Products', Dow Chemical Co., Midland, Michigan, Report No. AR-15-59, 1959.

246 K. A. Kobe and co-workers, *Petrol. Refiner*, 1949, **28** (1); **28** (2); **28** (3); **28** (5); **28** (7); **28** (10); **28** (11); 1950, **29** (1); **29** (2); **29** (3); **29** (5); **29** (7); **29** (9); **29** (12); 1951, **30** (4); **30** (6); **30** (8); **30** (11); **30** (12); 1954, **33** (8); **33** (11); 1957, **36** (10); **36** (12); 1958, **37** (7); 1959, **38** (12).

247 W. M. Latimer, 'Oxidation Potentials', 2nd edn., Prentice-Hall, New York, 1959.

248 C. E. Holley, jun., W. J. Warlton, and R. F. Ziegler, 'Compressibility Factors and Fugacity Coefficients calculated from the Beattie–Bridgeman Equation of State for Hydrogen, Nitrogen, Oxygen, Carbon Dioxide, Ammonia, Methane and Helium', Los Alamos Scientific Laboratory, University of California, LA-2271, 1959.

249 K. K. Kelley, 'Contributions to the Data on Theoretical Metallurgy – XIII – High-temperature Heat Content, Heat Capacity, and Entropy Data for the Elements and Inorganic Compounds', U.S. Bureau of Mines, Bulletin 584, 1959.

250 O. A. Hougen, K. M. Watson, and R. A. Ragatz, 'Chemical Process Principles, Part II, Thermodynamics', 2nd edn., J. Wiley and Sons, New York, 1959.

251 H. E. Tester, 'Thermodynamic Properties of Methane', issued by British Petroleum Co., 1959.

252 J. F. Elliot and M. Gleiser, 'Thermochemistry for Steelmaking, Vol. 1', Addison-Wesley, Reading, Mass., 1960.

In 1960, Springer-Verlag produced another volume [253] in the well-known Landolt–Börnstein series. This new volume includes the following data: vapour pressures of pure materials (173 pp.); properties of co-existent phases of pure materials (28 pp.); melting and allotropic conversion under pressure (47 pp.); liquid crystal transformation temperatures (48 pp.); vapour pressures of binary and ternary mixtures (430 pp.); and heterogeneous equilibrium reactions (63 pp.). There are also tabulations of freezing-temperature lowering (58 pp.), boiling-temperature elevation (12 pp.), and osmotic pressure (36 pp.).

Pitzer and Brewer's version [6] of Lewis and Randall's book contains many data within the text and in the appendices. For example, Appendix 7 gives tables of thermodynamic properties which include values of $-\{G^\circ - H^\circ(298.15\ \text{K})\}/T$ for the temperatures 298.15, 500, 1000, 1500, and 2000 K and values of $\{H^\circ(298.15\ \text{K}) - H^\circ(0)\}$ for solid, liquid, and gaseous elements, solid and liquid halides, solid oxides, sulphides, and related compounds, solid carbides and nitrides, and gaseous halides. There are also tables of free energies, based on $H^\circ(0)$, for the elements, gaseous carbon compounds, gaseous halides, gaseous oxides, hydrides, and related compounds in extensive appendices.

Rand and Kubaschewski [254] have provided a critical assessment of all the available thermochemical data for binary compounds of uranium and a consistent set of values, including in some instances estimates, for enthalpies of formation and standard entropies, enthalpies of transformation, heat capacities, vapour pressures, and free energies of formation.

Corruccini and Gniewek [255] give tables of the heat capacity and enthalpy of 28 metals, 3 alloys, 8 other inorganic substances, and 8 organic substances in the temperature range 1 to 300 K. Constants of the Debye–Sommerfeld equations are also given.

The National Bureau of Standards Report 6928 lists the properties [256] of compounds of lithium, beryllium, magnesium, and aluminium with hydrogen, oxygen, fluorine, chlorine, nitrogen, and carbon. Thermodynamic functions for the ideal gases, solids, and liquids are tabulated (*e.g.* values of 'free energy functions' and 'enthalpy functions', entropy, and heat capacity).

Kelley and King,[257] in Bulletin 592, revise data in Bulletins 350, 394,

[253] Landolt–Börnstein: 'Zahlenwerte und Funktionen aus Physik, Chemie, Astronomie, Geophysik, und Technik', Sechste Auflage, II. Band, 'Eigenschaften der Materie in ihre Aggregatzustand; 2. Teil, Gleichgewichte ausser Schmelzgleichgewichten', Springer-Verlag, Berlin, Göttingen, and Heidelberg, 1960.

[254] M. H. Rand and O. Kubaschewski, 'Thermochemical Properties of Uranium Compounds', AERE Report No. 3487, 1960.

[255] R. J. Corruccini and J. J. Gniewek, 'Specific Heats and Enthalpies of Technical Solids at Low Temperatures', National Bureau of Standards Monograph 21, 1960.

[256] Preliminary Report on the Thermodynamic Properties of Selected Light-element Compounds, National Bureau of Standards Report 6928, 1960.

[257] K. K. Kelley and E. G. King, 'Contributions to the Data on Theoretical Metallurgy – XIV – Entropies of the Elements and Inorganic Compounds', U.S. Bureau of Mines, Bulletin 592, 1961.

434, and 477 to give a list of 1300 entropy values; this new tabulation is considered by the authors to supersede those mentioned.

Scott and McCullough's review [258] of the thermodynamics of hydrocarbons and sulphur compounds includes lists of the following properties: 'free energy function', 'enthalpy function', enthalpy, entropy, heat capacity, enthalpy of formation, free energy of formation, and logarithm of equilibrium constant of formation at temperatures 0, 273.16, 298.16, 300, 400, 500, 600, 700, 800, 900, and 1000 K.

Haar *et al.*[259] include values for the ideal thermodynamic functions, heat capacity, enthalpy, free energy, and entropy of a very wide range of hydrides, deuterides, and tritides. Much information is also given on exchange reactions.

Johnson's compendium,[260] which tabulates 'best' values, includes density, expansivity, heat capacity, transition enthalpies, and phase equilibria. The temperature range is from near 0 to 110 K and the substances covered in the first part are helium, neon, argon, hydrogen, nitrogen, oxygen, air, carbon monoxide, fluorine, and methanol. The second part of the compendium deals with a number of substances used in low-temperature work.

In the first volume of Dreisbach's book,[261] the physical properties of 511 compounds are tabulated, with each compound occupying a single page. Among the thermodynamic properties listed are freezing and boiling temperatures, critical constants, molar volumes, heat capacities, and enthalpies of fusion and vaporization. The second volume [261] provides similar data for 476 organic straight-chain compounds and the third volume [261] provides data for 434 aliphatic compounds and 22 miscellaneous compounds.

Bernecker and Long [262] have surveyed the enthalpies of formation of gaseous ions and Kondratiev [263] treats the energies of dissociation of chemical bonds.

Din,[264] in the third volume edited by him on the thermodynamic functions of gases, has treated methane, ethane, and nitrogen, but whereas

258 D. C. Scott and J. P. McCullough, 'The Chemical Thermodynamic Properties of Hydrocarbons and Related Substances. Properties of 100 Linear Alkane Thiols, Sulphides, and Symmetrical Disulphides in the Ideal Gas State from 0 to 1000 K', U.S. Bureau of Mines, Bulletin 595, 1961.

259 L. Haar, A. S. Friedman, and C. W. Beckett, 'Ideal Gas Thermodynamic Functions and Isotope Exchange Functions for the Diatomic Hydrides, Deuterides and Tritides', National Bureau of Standards Monograph 20, 1961.

260 'Properties of Materials at Low Temperature (Phase 1). A Compendium', ed. V. J. Johnson, Pergamon Press, Oxford, 1961.

261 R. R. Dreisbach, 'Physical Properties of Chemical Compounds', Advances in Chemistry Series, No. 15, American Chemical Society, Washington, D.C., 1955; vol. II, Advances in Chemistry Series, No. 22, 1959; vol. III, Advances in Chemistry Series, No. 29, 1961.

262 R. R. Bernecker and F. A. Long, *J. Phys. Chem.*, 1961, **65**, 1565.

263 V. N. Kondratiev, 'Energies of Dissociation of Chemical Bonds', Publishing House of the Academy of Sciences, U.S.S.R., Moscow, 1962.

264 'Thermodynamic Functions of Gases, Vol. 3 (Methane, Nitrogen, Ethane)', ed. F. Din, Butterworths, London, 1961.

some statistical calculations are included, the data on nitrogen are not entirely consistent with the statistical mechanical treatment.

A new volume[265] of Landolt–Börnstein appeared in 1961. This deals with calorimetric quantities and is concerned with elements, alloys, and compounds, and with reaction enthalpies. Subjects covered include the experimental and theoretical basis of thermochemistry, standard values of molar enthalpies, entropies, enthalpies of formation, free energies of formation, and enthalpies of phase change. Planck, Einstein, and Debye functions, anharmonicity, and internal rotation are considered. The final section presents thermodynamic data for mixtures and solutions.

The 'Handbook of Thermophysical Properties of Solid Materials' covers results[266] published in the period 1940—1957, and among other properties reports on melting temperatures, densities, enthalpies of transition, heat capacities, and vapour pressures of the elements. The information is presented on data sheets and also graphically. The work is intended for use by engineers and thus only materials melting above 1000 °F are listed.

Chermin,[267] in a series of articles, discusses the thermodynamics of some classes of hydrocarbon and hydrocarbon derivatives. Edmister,[268] in 'Applied Hydrocarbon Thermodynamics', includes information on equations of state, Mollier charts for pure hydrocarbons, compression and expansion charts for gases, and details of petroleum distillation calculations. Green[269] has reviewed the thermodynamic properties of organic oxygen compounds and the thermodynamic properties[270] of the normal alcohols C_1 to C_{12}. Justice[271] has treated the thermodynamic properties and electronic energy levels of some rare-earth sesquioxides.

Horsley's[272] 'Azeotropic Data', with Part 1 published in 1952 and Part 2 published in 1962, provides a valuable catalogue of the subject of the title.

Lick and Emmons' publication[273] on the thermodynamic properties of helium at temperatures up to 50 000 K does not include data for the liquid

[265] Landolt–Börnstein, 'Zahlenwerte und Funktionen aus Physik, Chemie, Astronomie, Geophysik, und Technik', Sechste Auflage, II Band, 'Eigenschaften der Materie in ihre Aggregatzustand. 4. Teil, Kalorische Zustandgrössen', Springer-Verlag, Berlin, Göttingen, and Heidelberg, 1961.

[266] A. Goldsmith, T. E. Waterman, and H. J. Hirschhorn, 'Handbook of Thermophysical Properties of Solid Materials. Vol. 1. Elements', Pergamon Press, Oxford, 1961.

[267] H. A. G. Chermin, *Petrol. Refiner*, 1961, **40** (2), **40** (3); *Hydrocarbon Process, Petrol. Refiner*, 1960, **40** (6); 1961, **40** (9), **40** (10).

[268] W. C. Edmister, 'Applied Hydrocarbon Thermodynamics', Gulf Publishing Co., Houston, Texas, 1961.

[269] J. H. S. Green, 'Thermodynamic Properties of Organic Oxygen Compounds', *Quart. Rev.*, 1961, **15**, 125.

[270] J. H. S. Green, 'Thermodynamic Properties of the Normal Alcohols C_1—C_{12}', *J. Appl. Chem.*, 1961, 397.

[271] B. H. Justice, 'Thermodynamic Properties and Electronic Energy Levels of Eight Rare-earth Sesquioxides', University of Michigan, Ph.D. thesis, 1961, University Microfilms Inc., Ann Arbor, Michigan. See also B. H. Justice and E. F. Westrum, jun., *J. Phys. Chem.*, 1963, **67**, 339, 345, 659.

[272] L. H. Horsley, 'Azeotropic Data', American Chemical Society, Washington, D.C., Part I, 1952; Part 2, 1962.

[273] W. J. Lick and H. W. Emmons, 'Thermodynamic Properties of Helium to 50,000 °K' Harvard University Press, Cambridge, Mass., 1962.

near $T = 0$. Helium below 800 K is considered to be a perfect monatomic gas, but these tables allow for single and double ionization that arises at stellar temperatures and for the effect of pressure on the gas at these temperatures. Twelve pages of description are followed by 100 pages of tables giving density, energy, enthalpy, entropy, and heat capacity in dimensionless forms. Other tables give the fractions of the various species in the equilibrium mixtures. There are two enthalpy–entropy diagrams.

A Russian publication, now available in English,[274] lists the thermodynamic properties of many substances.

Karr [275] has provided values for the boiling temperature at atmospheric and reduced pressure, melting temperature, and density for 1024 compounds.

Mortimer's book [276] examines strain energies in saturated and unsaturated organic compounds; stabilization energies in non-aromatic compounds; strain and resonance energies in aromatic compounds; polymerization energies; molecular addition compounds; bond dissociation energies and enthalpies of formation of free radicals; metal–carbon and metal–halogen bonds; ionization energies in aqueous solution; and bond strengths in silicon, phosphorus, and sulphur compounds.

In 1962, the Bureau of Mines issued a reprint of Bulletins 383, 384, 393, and 406 in one volume,[277] as Bulletin 601.

The report [278] on the Symposium on Nuclear Materials, 1962, provides a useful source of information in the form of tables and graphs on important nuclear materials, with particular emphasis on activities and on vaporization processes.

Nesmeyanov [279] shows graphically all the results he has collected on the vapour pressures of the elements, together with his recommended values. The book also includes tables showing recommended values for the enthalpies of vaporization, the temperatures at which the elements exert pressures between 10^{-10} Torr and 1 atm, four-term equations, and vapour pressures at a number of rounded temperatures. Rand and Kubaschewski's 'Thermochemical Properties of Uranium Compounds' contains many data,[280] including heat capacity, solid transformations, melting temperature

[274] V. P. Glushko, L. V. Gurvich, G. A. Khachkuruzov, I. V. Veits, and V. A. Medvedev, 'Thermodynamic Properties of Individual Substances', vol. I, vol. II, Publishing House of the Academy of Sciences of the U.S.S.R., Moscow, 1962.

[275] C. Karr, 'Physical Properties of Aromatic Ethers – A Literature Survey of 1024 Compounds', National Bureau of Standards, IC 8079, 1962.

[276] C. T. Mortimer, 'Reaction Heats and Bond Strengths', Pergamon Press, London, 1962.

[277] K. K. Kelley, 'Contributions to the Data on Theoretical Metallurgy – XV – A Reprint of Bulletins 383, 384, 393, and 406', U.S. Bureau of Mines, Bulletin 601, 1962.

[278] Thermodynamics of Nuclear Materials – Proceedings of Symposium, IAEA, 1962.

[279] A. N. Nesmeyanov, 'Vapour Pressure of the Chemical Elements', Elsevier, Amsterdam, 1963.

[280] R. H. Rand and O. Kubaschewski, 'Thermochemical Properties of Uranium Compounds', Oliver and Boyd, Edinburgh, 1963.

and enthalpy of fusion, and vapour pressure of uranium, enthalpies of formation, and free energies and entropies of ions in solution. Data are included on halides, oxides, sulphides, nitrides, phosphides, carbides, hydrides, and on uranium systems containing bismuth, silicon, germanium, tin, lead, aluminium, gallium, indium, thallium, beryllium, zinc, cadmium, mercury, and gold. Uranyl compounds listed include the fluoride, chloride, bromide, sulphate, and nitrate. Miscellaneous compounds included comprise oxychlorides, oxybromides, uranium tetrafluoride, and uranium trioxide hydrate, uranium sulphate, and uranates.

Westrum and McCullough's article [281] summarizes many measurements on the thermodynamics of crystals.

Two translations of Russian texts provide information on work in that country. The volume by Elutin *et al.*[282] is a standard Soviet handbook and provides information on the thermodynamics of ferro-alloys, and the volume by Gonikberg [283] develops appropriate thermodynamic equations for the real states and compares the results of calculations made using the equations with experimental results.

Schexnayder's tabulation [284] presents tables of properties of some monatomic, diatomic, and polyatomic molecules which are found in many high-temperature chemical reactions, including those due to combustion, the action of shock waves, and high-energy electrical discharges.

Anderson and Wu,[285] in Bulletin 606, include the following information for 832 compounds found in coal tar: boiling temperature, the variation of boiling temperature with pressure, melting temperature, enthalpies of fusion, vaporization, and combustion, heat capacity, vapour pressure, and critical properties.

The National Bureau of Standards Monograph 68 by Schneider [286] lists the melting temperatures of 70 metal oxides published prior to January 1963. Listed are the melting temperatures as recorded and the values based on IPTS-48.

The literature up to 1959 on 65 elements, their oxides, halides, carbides,

281 E. F. Westrum, jun., and J. P. McCullough, 'Thermodynamics of Crystals' in 'Physics and Chemistry of the Organic Solid State', vol. 1, ed. D. Fox, M. M. Labes, and A. Weisberger, Interscience, New York, 1963.

282 V. P. Elutin, Yu. A. Pavlov, B. E. Levin, and E. N. Alekseev, 'Production of Ferro-alloys – Electrometallurgy', State Publishing House for Literature on Metallurgy, Moscow, 1957; translated 1961, second impression 1963, by Israel Program for Scientific Translation, Jerusalem.

283 M. G. Gonikberg, 'Chemical Equilibria and Reaction Rates at High Pressure', Publishing House of the Academy of Sciences of the U.S.S.R., Moscow, 1960; translated by Israel Program for Scientific Translations, Jerusalem, 1963.

284 C. J. Schexnayder, jun., 'Tabulated Values of Bond Dissociation Energies, Ionization Potentials and Electron Affinities for Some Molecules found in High-temperature Chemical Reactions', National Aeronautics and Space Administration, Washington, D.C., TN D-1791, 1963.

285 H. C. Anderson and W. R. K. Wu, 'Properties of Compounds in Coal-carbonization Products', U.S. Bureau of Mines, Bulletin 606, 1963.

286 S. J. Schneider, 'Compilation of the Melting Points of the Metal Oxides', National Bureau of Standards Monograph 68, 1963.

and nitrides has been surveyed by Wicks and Block.[287] Data are presented as (i) tables of enthalpy, enthalpy of formation, and free energy of formation at various temperatures as well as values for phase changes, (ii) equations representing the variation of thermodynamic functions with temperature, and (iii) graphical plots of the variation of free energy of formation with temperature. The stable form of the element at 298.15 K is used as the reference state.

Feber,[288] in his 'Heats of Dissociation of Gaseous Chlorides', treats the enthalpies of vaporization of selected elements, the thermodynamic functions for the vaporization of the chlorides of the first, second, third, and fourth long Periods, and the enthalpies of dissociation of the gaseous chlorides. Kirner has translated the Russian text by Lavrov *et al.*[289] on the thermodynamics of gasification and gas-synthesis reactions.

The data for Hultgren, Orr, Anderson, and Kelley's compilation[290] were prepared between 1955 and 1963. Information on 65 elements and 167 alloy systems is presented, and selected values for heat capacity, entropy, enthalpy, 'free energy functions', and vapour pressures of phases are given in tabular form. For alloys, the preferred values of integral, partial, and excess thermodynamic properties are listed or are presented as analytical functions. Phase diagrams and graphs are also included.

The Chemical Engineers' Handbook (1963, edited by Perry)[291] contains many tables and graphs of thermodynamic properties. Numerous temperature–enthalpy, pressure–enthalpy, enthalpy–concentration, and Mollier diagrams are provided. Tables of enthalpies and free energies of formation for inorganic and organic compounds at 298.15 K are given. Enthalpies of combustion (at 298.15 K) of organic compounds as gases and liquids are presented. Enthalpies of solution of inorganic compounds in water are listed, and tables give vapour pressures, melting and boiling temperatures, and heat capacities.

In the compilation by McBride *et al.*,[292] the thermodynamic properties to 6000 K for 210 substances, involving the first 18 elements, are given. The properties listed include: heat capacity at constant pressure, 'sensible' enthalpy and free energy, entropy, values of enthalpy changes, and the logarithms of equilibrium constants. The functions for most of the gases

[287] C. E. Wicks and F. E. Block, 'Thermodynamic Properties of 65 Elements – Their Oxides, Halides, Carbides and Nitrides', U.S. Bureau of Mines, Bulletin 605, 1963.

[288] R. C. Feber, 'Heats of Dissociation of Gaseous Chlorides', Los Alamos Scientific Laboratory, University of California, Report LA-2841, 1963.

[289] N. V. Lavrov, V. V. Korobov, and V. I. Filippova, 'Thermodynamics of Gasification and Gas-synthesis Reactions', translated by G. H. Kirner, Pergamon Press, Oxford, 1963.

[290] R. Hultgren, R. L. Orr, P. D. Anderson, and K. K. Kelley, 'Selected Values of the Thermodynamic Properties of Metals and Alloys', J. Wiley and Sons, New York, 1963.

[291] 'Chemical Engineers' Handbook', ed. J. H. Perry, McGraw-Hill, New York, 1963.

[292] J. McBride, S. Heimel, J. G. Ehlers, and S. Gordon, 'Thermodynamic Properties to 6000 °K for 210 Substances involving the First 18 Elements', National Aeronautics and Space Administration, Washington, D.C., SP-3001, 1963.

were generated from molecular data, whereas the functions for most of the condensed species are based on selected experimental data.

Feber and Herrick [293] have used experimental, and in some instances predicted, energy-level data for the calculation of the 'free energy and enthalpy functions', entropy, and heat capacity of the gaseous monatomic lanthanide elements from cerium to lutetium and the gaseous monatomic actinide elements from thorium to curium over the temperature range 100 to 6000 K.

'Phase Diagrams for Ceramists', by Levin *et al.*, [294] contains 2000 diagrams and refers to many earlier compilations.

Munson and Franklin [295] survey enthalpies of formation of gaseous ions. The data on hydrogen presented by Kubin and Presley [296] include results in the form of tables and of Mollier diagrams. Properties treated include energy, enthalpy, entropy, heat capacities, equilibrium constants for dissociation and ionization, chemical composition of equilibrium gas, density, compressibility, and speed of sound.

Gerasimov *et al.*[297] have provided a reference book on the thermodynamic properties of tungsten, molybdenum, titanium, zirconium, niobium, and tantalum, and their more important compounds, *viz.* oxides, sulphides, halides, carbides, nitrides, silicates, borides, and hydrides.

The publication of the 'JANAF' thermochemical tables [298] was a project (now terminated) for the provision of data in loose-leaf form with facilities for updating the information from time to time. These Joint Army–Navy–Air Force Tables were prepared by the Dow Chemical Company. The prime quantities tabulated are the heat capacity, entropy, 'free energy and enthalpy functions', enthalpy and free energy of formation, and the logarithm of the equilibrium constant of formation. The temperature range covered is from zero to several thousand kelvins, and includes values for 298.15 K.

Roder *et al.*,[299] in presenting the thermodynamic function of parahydrogen, have listed the following properties for selected isobars and

[293] R. C. Feber and C. C. Herrick, 'Ideal Gas Thermodynamic Functions of Lanthanide and Actinide Elements', Los Alamos Scientific Laboratory, University of California, Report LA-3184, 1964.

[294] F. M. Levin, C. R. Robbins, H. F. McMurdie, and M. K. Reser, 'Phase Diagrams for Ceramists', American Ceramic Society, Columbus, Ohio, 1964.

[295] M. S. B. Munson and J. L. Franklin, *J. Phys. Chem.*, 1964, **68**, 3191.

[296] R. F. Kubin and L. L. Presley, 'Thermodynamic Properties and Mollier Chart for Hydrogen from 300 °K to 20,000 °K', National Aeronautics and Space Administration, Washington, D.C., SP-3002, 1964.

[297] Ya. I. Gerasimov, A. N. Krestovnikov, and A. S. Shakhov, 'Chemical Thermodynamics in Non-ferrous Metallurgy', vol. III, State Publishing House for Literature on Metallurgy, Moscow, 1963; translated by Israel Program for Scientific Translations, Jerusalem, 1965.

[298] 'JANAF Thermochemical Data – 1 – Aluminium to Fluorine, 1965; 2 – Iron to Zirconium – Electron Gas – 1965; 3 – Addenda—Current Date', ed. D. R. Stull, Dow Chemical Co., Midland, Michigan.

[299] H. M. Roder, L. A. Weber, and R. D. Goodwin, 'Thermodynamic and Related Projects of Parahydrogen from the Triple Point to 100 °K at Pressures to 340 Atmospheres', National Bureau of Standards Monograph 94, 1965.

isochores: temperature, volume or pressure, the isothermal derivative $(\partial p/\partial \rho)_T$, the isochoric derivative $(\partial p/\partial T)_\rho$, internal energy, enthalpy, entropy, the heat capacity at constant volume and at constant pressure, and the velocity of sound.

Timmermans,[300] in 'Physico-chemical Constants of Pure Organic Compounds', in Volume 1 published 1950 and Volume 2 published 1965 has listed several values for a property, provided the measurements were obtained using pure samples and satisfactory measurement techniques. The properties recorded include melting and boiling temperatures, triple points, saturated vapour pressures, critical constants, enthalpies of melting, vaporization, transition, and combustion, heat capacities, and densities. Compounds treated include hydrocarbons, halogenated derivatives, aliphatic oxygen compounds, aromatic oxygen compounds, polymethylene compounds, heterocyclic oxygen compounds, mixed oxyhalogenated derivatives, aliphatic nitrogen compounds, cyclic nitrogen compounds, oxy-nitrogen derivatives, mixed halogen–nitrogen derivatives, sulphur compounds, and compounds with some other elements. The second volume is a supplement to the first and provides new data, with some corrections to Volume 1.

Schick's work[301] includes the study of borides, carbides, nitrides, and oxides of some elements in Groups IIA, IIIB, IVA, IVB, VB, VIIB, and VIII as well as selected rare earths and actinides. As far as possible, the tables have been made compatible with the 'JANAF' tables. Among the subjects treated are phase diagrams, heat capacities, enthalpies, entropies, enthalpies of phase transformation, formation, and reaction, melting temperatures, triple points, free energies of formation, vapour pressures, compositions of vapour species, ionization and appearance potentials, e.m.f. of cells, and enthalpies of solution and dilution. Volume 1 summarizes the techniques used to analyse data and cites the data analysed, and Volume 2 gives tables of values produced by this study.

Mah[302] has searched the literature up to April 1965 for thermodynamic data on vanadium and its compounds. The results are arranged in three main sections: (i) heat capacities, entropies, enthalpies, enthalpies of vaporization, free energies of vaporization, vapour pressures, temperatures of phase changes, enthalpies of phase changes and of formation, free energies of formation; (ii) enthalpies, free energies and equilibrium pressures, and constants for reactions involving vanadium; (iii) data for elements and compounds associated with metallurgical reactions of vanadium.

300 J. Timmermans, 'Physico-chemical Constants of Pure Organic Compounds', Elsevier, New York, vol. 1, 1950; vol. 2, 1965.

301 H. L. Schick, 'Thermodynamics of Certain Refractory Compounds', Academic Press, New York, vol. 1 and vol. 2, 1966.

302 A. D. Mah, 'Thermodynamic Properties of Vanadium and its Compounds', U.S. Bureau of Mines, PB 169 681, 1966.

Gerasimov *et al.*[303] have prepared four volumes on chemical thermodynamics in non-ferrous metallurgy. A volume on tungsten, molybdenum, titanium, zirconium, niobium, and tantalum and their more important compounds has been translated into English.[304]

Vedeneyev *et al.*[305] cover the literature up to the beginning of 1962 and include tables dealing with the dissociation energies of diatomic molecules, bond energies of organic and inorganic molecules and radicals, enthalpies of formation of atoms and of organic and inorganic radicals, ionization potentials, electron affinities of atoms, molecules, and radicals, and proton affinities.

Two volumes of Vol's work[306] have been translated into English; about 7000 binary systems are described, and among physical properties included are enthalpy of fusion and mixing, boiling temperature, density, and heat capacity.

'Tables of Physical and Chemical Constants'[307] contains many thermodynamic and thermal data. For example, it contains extensive tables of vapour pressure and values of ΔG°_f, ΔH°_f, S°, and C°_p for inorganic and organic compounds. Lumsden[115] presents many thermodynamic data in the text of his book.

Canjar and Manning[308] list values for thermodynamic functions of the following materials: methane, ethane, propane, n-butane, n-pentane, isobutane, n-hexane, ethylene, propene, isobutene, acetylene, benzene, ammonia, carbon dioxide, carbon monoxide, hydrogen, nitrogen, oxygen, sulphur dioxide, and water.

Kubaschewski, Evans, and Alcock[120] have tabulated many data of interest to metallurgists, including enthalpies of formation and standard entropies at 298.15 K, heat capacities, enthalpies and temperatures of transformation, melting and boiling temperatures, vapour pressures, and standard free energies of reaction. Some data on binary metallic systems are also given.

Margrave's book,[309] in addition to reviewing measurement techniques,

303 Ya. I. Gerasimov, A. N. Krestovnikov, and A. S. Shakhov, 'Chemical Thermodynamics in Non-ferrous Metallurgy', State Publishing House for Literature on Metallurgy, Moscow, vol. 1, 1960; vol. 2, 1961; vol. 3, 1962; vol. 4, 1966.

304 Ya. I. Gerasimov, A. N. Krestovnikov, and A. S. Shakhov, 'Chemical Thermodynamics in Non-ferrous Metallurgy', vol. III, State Publishing House for Literature on Metallurgy, Moscow, 1963; translated by Israel Program for Scientific Translations, Jerusalem, 1965.

305 V. I. Vedeneyev, L. V. Gurvich, V. N. Kondrat'yev, V. A. Medvedev, and Y. L. Frankevich, 'Bond Energies, Ionization Potentials and Electron Affinities', translated from the Russian by Scripta Technica Ltd., Edward Arnold, London, 1966.

306 A. E. Vol, 'Handbook of Binary Metallic Systems: Structure and Properties', State Publishing House for Physics and Mathematics, Moscow, 1959; translated by Israel Program for Scientific Translations, Jerusalem, 1966.

307 G. W. C. Kaye and T. H. Laby, 'Tables of Physical and Chemical Constants', 13th edn., Longmans, Green, and Co., London, 1966.

308 L. N. Canjar and F. S. Manning, 'Thermodynamic Properties and Reduced Correlations for Gases', Gulf Publishing Co., Houston, Texas, 1967.

309 'The Characterization of High Temperature Vapours', ed. J. L. Margrave, J. Wiley and Sons, New York and London, 1967.

studies the thermodynamic properties of high-temperature species and has an appendix on the vapour-pressure data for the elements.

A volume of the well-known Landolt–Börnstein tables [310] contains 942 pages on thermodynamic data for organic and inorganic substances in the three states of matter. Techniques used for making measurements on refractory materials are described and the data obtained are presented in the form of graphs and tables in the book entitled [311] 'Thermodynamics of Ceramic Systems'.

Two volumes edited by Lax [312] include thermodynamic data on the ideal gas state properties of a wide range of elements and compounds, including organic compounds. Other publications in 1967 giving thermodynamic data include the work by Rudman *et al.*[313] on phases in metals and alloys, by Klotz [131] on biochemical reactions, by Mueller [314] on metallurgical phenomena, and by Wendlandt and Smith [315] on the thermal properties of metal ammine complexes.

Wilson *et al.*[316a] have discussed the heat capacity, enthalpy and p, V, T properties of helium, nitrogen, ethane, neon, methane, and carbon monoxide. Zwolinski and his colleagues have prepared a compilation [316b] of the critical properties of organic substances which also includes values for oxygen, water, carbon monoxide, and carbon dioxide.

Gallant [317] has listed the following thermodynamic properties needed in manufacture and production: vapour pressure, enthalpy of vaporization, vapour and liquid heat capacity, and liquid density. These properties are plotted graphically against temperature. Tables are also given of boiling temperature, freezing temperature, and critical properties. Compounds considered include hydrocarbons (paraffins, olefins, and diolefins), halogenated hydrocarbons, alcohols, and glycols.

310 Landolt–Börnstein, Band 4, Teil 4, Bandteil a, Wärmetechnische Messverfahren und Thermodynamische Eigenschaften Homogener Stoffè, ed. W. Dienemann *et al.*, Springer-Verlag, Berlin, 1967.

311 Thermodynamics of Ceramic Systems: papers presented at a meeting held at Imperial College, London, April, 1966; Published Proceedings of the British Ceramic Society, No. 8, June 1967.

312 'Taschenbuch für Chemiker und Physiker, Band I und II', ed. L. Lax, Springer-Verlag, Heidelberg, 1967.

313 'Phase Stability in Metals and Alloys', ed. P. S. Rudman, Y. Stringer, and R. I. Jaffe, McGraw-Hill, New York, 1967.

314 'Energetics in Metallurgical Phenomena, III', ed. W. Mueller, Gordan and Breach, New York, 1967.

315 W. W. Wendlandt and J. P. Smith, 'Thermal Properties of Transition Metal Ammine Complexes', Elsevier, Amsterdam, 1967.

316a G. M. Wilson, R. G. Clark, and F. L. Hyman, 'Thermodynamic Properties of Cryogenic Fluids. Survey of Existing Data. Applied Thermodynamics', Symposium on Applied Thermodynamics sponsored by Industrial and Engineering Chemistry and American Chemical Society, American Chemical Society, Washington, D.C., 1968.

316b A. P. Kudchadker, G. H. Alani, and B. J. Zwolinski, 'The Critical Constants of Organic Substances', *Chem. Rev.*, 1968, **68**, 659.

317 R. W. Gallant, 'Physical Properties of Hydrocarbons – Vol. 1', Gulf Publishing Co., Houston, Texas, 1968.

Furukawa *et al.*[318] have presented a critical analysis of the thermodynamic properties of copper, silver, and gold from 0 to 300 K.

Haywood[319] has presented certain thermodynamic tables and diagrams in SI units.

In the third edition of Gaydon's book[320] 'Dissociation Energies and Spectra of Diatomic Molecules', the properties of 460 molecules are discussed. The results presented are more reliable than previous values.

Hamblin's abridged tables[321] include the thermodynamic properties of water and steam, Freon-12, and air; the enthalpies of some hydrocarbons and the critical constants of several inorganic and organic compounds are tabulated.

Rabinovich's collection[322] of data includes some thermodynamic properties of carbon dioxide, water, lithium, mercury, ethylene, butene, halogenated monosilanes and methanes, liquid ammonia, and hydrogen peroxide, and the densities of liquid alkali metals.

'Chemistry, Book of Data', published[323] for the Nuffield Foundation, includes the heat capacities of solid elements at 25 °C, and graphs showing the heat capacity of selected elements and water at different temperatures. Data presented also include enthalpies of fusion and vaporization of elements and compounds, enthalpies of atomization and energies of ionization of the elements, and entropies and Gibbs energies of formation of inorganic and organic compounds. In addition, some heat capacities and enthalpies of dilution of aqueous solutions are given.

The thermophysical properties of carbon dioxide presented by Vukalovich and Atunin[324] include phase equilibria, enthalpy, heat capacities, equations of state, and the thermodynamic functions for the ideal gas. The content of Sarkin's work[325] is well represented by the title 'Gas Dynamics and Thermodynamics of Solid-propellant Rockets'.

318 G. T. Furukawa, W. G. Saba, and M. L. Reilly, 'Critical Analysis of the Heat-capacity Data of the Literature and Evaluation of Thermodynamic Properties of Copper, Silver and Gold from 0 to 300 °K', National Bureau of Standards Reference Data Series; NSRDS-NBS 181.

319 R. W. Haywood, 'Thermodynamic Tables in SI (Metric) Units and Enthalpy–Entropy Diagram for Steam. Pressure–Enthalpy Diagram for Refrigerant 12', Cambridge University Press, Cambridge, 1968.

320 A. G. Gaydon, 'Dissociation Energies and Spectra of Diatomic Molecules', Chapman and Hall, London, 1968.

321 F. D. Hamblin, 'Abridged Thermodynamics and Thermochemical Tables (with charts), British Units', Pergamon Press, Oxford, 1968.

322 'Thermophysical Properties of Gases and Liquids. No. 1', ed. V. A. Rabinovich, translated from Russian by A. Moscora. Published for the U.S. Department of Commerce and National Science Foundation, Washington, D.C., by the Israel Program for Scientific Translations. U.S. Department of Commerce Clearinghouse for Federal, Scientific, and Technical Information, Springfield, Va. Originally published in Russian, 1968.

323 'Chemistry, Book of Data', published for the Nuffield Foundation by Longmans and Penguin Books, London, 1968.

324 M. P. Vukalovich and V. V. Atunin, 'Thermophysical Properties of Carbon Dioxide'; translated into English under the direction of the U.K. Atomic Energy Authority, ed. D. S. Gaunt, Colletts, London, 1968.

325 R. E. Sarkin, 'Gas Dynamics and Thermodynamics of Solid-propellant Rockets', Nauka Publishing House, Moscow, 1967; translated by Israel Program for Scientific Translations, Jerusalem, 1969.

Morton and Beckett,[326] in an appendix to their book, present bond dissociation energies and standard electrode potentials. Brown,[327] in his book, which is mainly concerned with the techniques of microcalorimetry, gives values of ΔH_f°, ΔG_f°, S°, and C_p° (for 25 °C) for many biological substances, and partial molar properties for aqueous solutions. He also presents enthalpies and Gibbs energies of formation of adenosine phosphoric acid specimens and thermodynamic quantities for some reactions in solution.

Substances considered in a compilation[328] of the thermodynamic properties of refrigerants include hydrogen, parahydrogen, helium, neon, nitrogen, air, oxygen, argon, carbon dioxide, hydrocarbons (*e.g.* methane, ethane, propane, butane, isobutane, ethylene, and propene), and fluoro- and fluoro-chloro-hydrocarbons. Properties listed include those for the liquid and saturated vapour, superheated vapour, and unsaturated vapour. In addition, pressure–enthalpy, and in some instances pressure–entropy, diagrams are provided.

Cooper and LeFevre[329] have provided twelve pages of tables on the properties of water substance, and these include lists of thermodynamic temperature against saturation pressure, heat capacity, specific volume, and specific enthalpy and entropy.

Tables edited by Allard[330] provide information on metals, including phase changes, vapour pressures, heat capacities, entropies, and enthalpies.

A joint National Standard Reference Data System–National Bureau of Standards publication[331] presents ionization potentials, appearance potentials, and enthalpies of formation of gaseous positive ions. Table 1 of this reference lists ions arranged in the order: hydrogen; the rare gases; and then the remaining elements in order of increasing atomic number. Molecules containing two or more elements are found under the element of highest atomic number, except for the rare-gas halides, which are listed under halogens. Table 2 lists enthalpies of formation of gaseous neutral species and Appendix 1 lists enthalpies of formation, mainly of radicals.

National Bureau of Standards Technical Note 270 has been issued[332] in

[326] A. S. Morton and P. J. Beckett, 'Basic Thermodynamics', Butterworths, London, 1969.

[327] H. D. Brown, 'Biochemical Microcalorimetry', Academic Press, New York, 1969.

[328] 'Thermodynamic Properties of Refrigerants', American Society of Heating, Refrigerating, and Air-conditioning Engineers, 1969.

[329] J. R. Cooper and E. J. LeFevre 'Thermophysical Properties of Water Substance (Student's Tables in SI Units)', Edward Arnold, London, 1969.

[330] 'Tables Internationales de Constantes selectionées. Métaux Données Thermiques et Mécaniques', ed. S. Allard, Pergamon Press, London, 1969.

[331] J. L. Franklin, J. G. Dillard, H. M. Rosenstock, J. T. Herron, and K. Droxl, 'Ionization Potentials, Appearance potentials, and Heats of Formation of Gaseous Positive Ions', National Standards Reference Data Series, NSRDS-NBS 26, 1969.

[332] D. D. Wagman *et al.*, 'Selected Values of Chemical Thermodynamic Properties – Part I – Tables for the first Twenty-three Elements in the Standard Order of Arrangement', National Bureau of Standards Technical Note 270-1, 1965; 'Part II – Tables for the Elements Twenty-three through Thirty-two in the Standard Order of Arrangement', Technical Note 270-2, 1966; 'Tables for the first Thirty-four Elements in the

six parts to date. Technical Note 270-3 contains all the tables of Technical Note 270-1 and 270-2, including corrections for a number of minor errors, so that part 3 supersedes parts 1 and 2. Technical Notes 270-3 to 270-6 give values for enthalpy, Gibbs energy of formation, entropy, and enthalpy, all at 298.15 K, and the enthalpy of formation at $T = 0$ for inorganic substances and organic molecules containing not more than two carbon atoms. In some instances (*e.g.* metal–organic compounds) data are given for substances in which each organic radical contains one or two carbon atoms.

The National Bureau of Standards has published a large series of Reports [333] which survey extant literature and describe experiments with results designed to provide data on the thermodynamic properties of the light elements that are of interest in high-temperature research. Data cover materials in the solid, liquid, and gaseous states in the temperature range from 0 to 6000 K and in the pressure range up to 100 atm. Roznjevic's thermodynamic tables and charts [334] present data for temperatures from 10 to 1273 K and pressures up to 100 MPa.

Gosman *et al.*[335] have tabulated, for argon, values of density, internal energy, enthalpy, and entropy of the liquid and gas for temperatures from 83.8 to 300 K at pressures from 0.01 to 1000 atm. Diagrams for heat capacities, compression factors, and entropies are included. A vapour-pressure equation covering the temperature range from the triple point to the critical point is also given.

Golden and Benson [336] have listed enthalpies of formation of free radicals and bond dissociation energies for many entities. Morton and Beckett [144] have provided useful tables of bond energies, enthalpies, and free energies of formation.

The book often referred to as the 'Rubber Handbook' [337] contains much thermodynamic information on elements, oxides, hydrocarbons, and inorganic and organic compounds (not more than two carbon atoms).

Standard Order of Arrangement', Technical Note 270-3, 1968; 'Tables for Elements Thirty-five through Fifty-three in the Standard Order of Arrangement', Technical Note 270-4, 1969; 'Tables for Elements Fifty-four through Sixty-one in the Standard Order of Arrangement', Technical Note 270-5, 1971; 'Tables for the Alkaline Earth Elements (Elements Ninety-two through Ninety-seven) in the Standard Order of Arrangement', Technical Note 270-6, 1971.

[333] 'Preliminary Reports on the Thermodynamic Properties of Selected Light Elements and Some Related Compounds', National Bureau of Standards Reports Nos. 6297, 6484, 6645, 6928, 7093, 7192, 7437, 7587, 7796, 8033, 8186, 8504, 8628, 8919, 9028, 9389, 9500, 9601, 9803, and 9905 (Period 1959—1969).

[334] K. Roznjevic, 'Thermodynamic Tables and Charts', translation by N. Boskovic, Butterworths, London, 1969.

[335] A. L. Gosman, R. D. McCarty, and J. G. Hust, 'Thermodynamic Properties of Argon from the Triple Point to 300 K at Pressures to 1000 Atmospheres', National Standard Reference Data Series–National Bureau of Standards, NSRDS-NBS 27, 1969.

[336] D. M. Golden and S. W. Benson, 'Free-radical and Molecule Thermochemistry from Studies of Gas-phase Iodine-atom Reactions', *Chem. Rev.*, 1969, **69**, 125.

[337] 'Handbook of Chemistry and Physics, 50th Edition', ed. R. C. West, The Chemical Rubber Co., Cleveland, 1969.

Many of the data relate to 298.15 K. Other information includes heat capacities, freezing and boiling temperatures, vapour pressures, and enthalpies of dilution.

Stull, Westrum, and Sinke [338] have searched the literature up to January 1966 for thermodynamic data. The heat capacity, entropy, enthalpy, and Gibbs energy of formation have been tabulated for 741 organic compounds in the ideal gas state at temperatures from 298 to 1000 K. The entropy, enthalpy, and Gibbs energy of formation are presented as fully as possible for 4400 organic compounds in the ideal gaseous and condensed states at 298 K.

Steam tables in metric units have been prepared by Keenan, Keyes, Hill, and Moore [339] by the use of an equation for the Helmholtz energy expressed in terms of density and temperature.

Stark and Wallace's 'Chemistry Data Book. SI edition' [340] gives thermodynamic data and refers to a standard temperature of 298.15 K. Among the properties listed are the heat capacities and first and second ionization energies of the elements, molar heat capacities of substances, and standard enthalpies and Gibbs energies of formation of compounds from the elements. A table lists temperatures and enthalpies of phase change, and an Ellingham diagram is given for a series of oxides. Tables give enthalpies of combustion, solution, and neutralization, enthalpies of formation of aqueous ions, hydration enthalpies of ions, and lattice enthalpies. Equilibrium data are given for three reactions and a list of standard electrode potentials is presented.

Kusuda [341] has provided very accurate values for properties of moist air, required for measuring the capacity of air-conditioning apparatus, moisture transfer analysis in cold storage machines, and problems concerned with the simultaneous transfer of heat and moisture.

Christensen and Izatt's handbook [342] consists of a compilation of enthalpies, equilibrium constants, and entropy and heat-capacity changes for reactions. Data are summarized to the end of 1969.

Vasserman and Rabinovich [343] deal with the thermodynamic properties

338 D. R. Stull, E. F. Westrum, jun., and G. C. Sinke, 'The Chemical Thermodynamics of Organic Compounds', J. Wiley and Sons, New York, 1969.

339 J. H. Keenan, F. G. Keyes, P. H. Hill, and J. G. Moore, 'Steam Tables: Thermodynamic Properties of Water including Vapour, Liquid, and Solid Phases. International Edition – Metric Units', J. Wiley and Sons, New York, 1969.

340 J. G. Stark and H. G. Wallace, 'Chemistry Data Book. SI Edition', John Murray, London, 1970.

341 T. Kusuda, 'Algorithms for Psychrometric Calculations (Skeleton Tables for the Thermodynamic Properties of Moist Air)', Building Science Series 21, National Bureau of Standards, Washington, D.C., 1970.

342 J. J. Christensen and R. M. Izatt, 'Handbook of Metal Ligand Heats', Contribution 13 from the Centre for Thermochemical Studies, Brigham Young University, Provo, Utah, Marcel Dekker, New York, 1970.

343 A. A. Vasserman and V. A. Rabinovich, 'Thermophysical Properties of Liquid Air and its Components'; translated by Israel Program for Scientific Translations; published Ann Arbor–Humphrey, London, 1970.

of liquid nitrogen, oxygen, argon, and air at pressures up to 50 MPa. The first chapter considers the form of equations of state for the liquids.

Paule and Mandel[344] have provided an intercomparison of different measurements on the vapour pressure of gold for the temperature range 1300 to 2100 K and for the pressure range 10^{-8} to 10^{-3} atm.

Dasent's book[345] on inorganic energetics contains many data on enthalpy and Gibbs energy changes for inorganic reactions, including those involving gaseous species, ionic crystals, and covalent compounds.

Mayhew and Rogers[346] have published thermodynamic and transport properties of fluids in SI units.

The United Kingdom Committee on the Properties of Steam has published 'U.K. Steam Tables in SI Units'.[347] These tables, although published in 1970, use IPTS-48. The tables include values for the specific enthalpy, entropy, volume, and heat capacity. Data are given for the saturation line for the temperature range 0.01 to 374.15 °C, pressure range 0.006 11 to 221.2 bar, and for water substances at pressures 0 to 1000 bar and temperatures 0 to 800 °C.

Moore[348] gives tables of ionization potentials expressed in electron volts, and of the ionization limits from which they have been derived, for elements of atomic number 1 to 95.

Darwent,[349] in an NBS publication, has prepared a critical compilation of bond dissociation energies of simple compounds, excluding organic compounds except for those containing one carbon atom. Values are quoted for $T = 0$ or 298 K and are referred to the gaseous state.

Ashcroft and Mortimer[350a] have surveyed the literature on the thermochemistry of transition-metal complexes up to mid-1968. A collection of numerical data is presented in the form of approximately 200 tables.

Riddick and Bunger[350b] have listed the physical properties of organic solvents and include values for the boiling temperature, vapour pressure, density, enthalpy of vaporization, critical constants, heat capacity, and cryoscopic and ebullioscopic constants.

[344] R. C. Paule and J. Mandel, 'Analysis of Interlaboratory Measurements on the Vapour Pressure of Gold', Certification of Standard Reference Material 7451, NBS Special Publication 260-19, 1970.

[345] W. E. Dasent, 'Inorganic Energetics', Penguin Library of Physical Sciences, Penguin Books, London, 1970.

[346] Y. R. Mayhew and G. F. C. Rogers, 'Thermodynamic and Transport Properties of Fluids. SI Units', 2nd edn., Blackwell, Oxford, 1970.

[347] 'U.K. Steam Tables in SI Units 1970'; published for the United Kingdom Committee on the Properties of Steam, by E. Arnold, London, 1970.

[348] C. E. Moore, 'Ionization Potentials and Ionization Limits Derived from the Analysis of Optical Spectra', National Standard Reference Data Series, NSRDS-NBS 34, 1970.

[349] B. de B. Darwent, 'Bond Dissociation Energies in Simple Molecules', National Standard Reference Data Series, NSRDS-NBS 31, 1970.

[350a] S. J. Ashcroft and C. T. Mortimer, 'Thermochemistry of Transition Metal Complexes', Academic Press, London, 1970.

[350b] J. A. Riddick and W. B. Bunger, 'Organic Solvents. Physical Properties and Methods of Purification', 'Techniques of Chemistry', vol. II, 3rd edn., Wiley-Interscience, New York and London, 1970.

Cox and Pilcher[187] have reviewed the enthalpies of formation and vaporization of organic and organometallic compounds published since 1930. The enthalpies of formation of some 3000 substances are listed, with estimates of error.

Reference must be made to continuing sources of information. Thermodynamic data continue to be published in many scientific journals but the *Journal of Chemical Thermodynamics* (published by Academic Press)[351] was set up in 1969 to publish factual papers in which there is a full discussion of experimental methods and which contain extensive tables of results.

In 1958 the IUPAC Commission on Thermodynamics and Thermochemistry inaugurated the 'Bulletin of Thermodynamics and Thermochemistry'[352] to present terse mention of work that is completed but unpublished. Numerical values are not given, but the Bulletin records the names of workers and laboratories that have completed work. Currently the Bulletin reports under five sections: Section A, thermochemical quantities; Section B, thermal properties; Section C, vaporization studies; Section D, other non-calorimetric studies; and Section E, correlation and compilation endeavours. Thus thermochemists and thermodynamicists can obtain early knowledge of data that will soon be available.

Finally, reference will be made to a source of names of organizations that make compilations. The CODATA 'International Compendium of Numerical Data Projects'[353] does not contain scientific data but lists organizations that make compilations, gives factual information on the substances covered, states the aims of the compilers, and gives names and prices of publications. Many subjects are covered, including thermodynamics.

6 Estimation of Thermodynamic Data

The systematization of the results of measurement is a normal scientific activity which leads to methods for the interpolation and extrapolation of data. For example, the Periodic Table may be used to seek relations between a thermodynamic property for compounds in a Group of elements. In the estimation of the properties of aliphatic compounds it is common practice to consider members of a homologous series, and therefore to study the change in a property produced by an increment of one CH_2 group. For this purpose it is often convenient to use the appropriate hydrocarbon homologous series as a source of reference. However, a word of caution is necessary here because some of the recorded properties of hydrocarbons have themselves been obtained by incremental methods,

351 *J. Chem. Thermodynamics*, 1969, **1**, Nos. 1—3.
352 'Bulletin of Chemical Thermodynamics', Commission on Thermodynamics and Thermochemistry, Section of Physical Chemistry, IUPAC.
353 CODATA, 'International Compendium of Numerical Data Projects', Springer-Verlag, Berlin, 1969.

and it is therefore important to avoid cyclic arguments. If estimated values of thermodynamic properties are used, it is advisable to designate clearly the source of such data. Many examples of correlation and estimation methods will be found in the text-books and data compilations recorded earlier in this article.

Engineers and chemical engineers need to know the thermodynamic properties of materials in order to plan and build new plant. Although an enormous amount of information has been collected and collated, yet technology advances so rapidly that a gap often exists between the demand for data and their experimental determination. In these circumstances it is frequently necessary to estimate values for the properties of substances. Estimation methods may be tied to basic theory (*e.g.* statistical mechanics), or may be largely empirical. The use of any method is often conditioned by the information already at hand, so that it may be necessary to find what information is available and then proceed, by various strategies, to estimate the desired quantity. The subject is a vast and rapidly expanding one, and it will only be possible to indicate some general sources of information on estimation techniques. The latest edition of Reid and Sherwood's book,[354] which has 646 pages, illustrates the size of the subject and how quickly it is developing. Thus the authors state that the present half-life of an existing correlation or estimation procedure is now about four years. Most of the new methods are empirical, because the fundamental laws describing forces between molecules are not yet clearly established and probably, even if they were, the mathematics required for a solution would be formidable and costly in time. Reid and Sherwood's book presents a review of the various estimation procedures for many thermodynamic properties, including critical properties, p, V, T relations, vapour pressures, enthalpies of phase transition, enthalpies and free energies of formation, and heat capacities. Comparisons are made between experimental and estimated values. It may be remarked that the American Institute of Chemical Engineers, with the assistance of 35 major companies, has written a large and sophisticated computer program, the AIChE Physical Property Estimation System (APPES), containing all the best estimation methods and having the ability to choose and follow the best estimation route.[355, 356]

Bondi[357] has provided a somewhat similar book to that by Reid and Sherwood, but treats substances of higher molecular weight, up to and including polymers and glasses.

354 R. C. Reid and T. K. Sherwood, 'The Properties of Gases and Liquids', McGraw-Hill, New York, 1966.
355 'Physical Property Estimation System', American Institute of Chemical Engineers, New York, 1965.
356 R. E. Heitman and G. H. Harris, *Ind. and Eng. Chem.*, 1968, **60**, 50; see also C. N. B. Martin, *Chem. Eng.*, 1970, 285.
357 H. Bondi, 'Physical Properties of Molecular Crystals, Liquids and Glasses', J. Wiley and Sons, New York, 1968.

Returning to earlier texts, reference must be made to Janz's book,[242] which has as object the provision of a collection of practical methods for the computation of thermodynamic properties based on the long recognized principle that the regularity and systematization found in the chemical properties of organic compounds applies to thermodynamic properties. The calculation of properties by statistical thermodynamics is presented, but the greater part of the book is devoted to estimation methods and the construction of correlations used for empirical estimations. Part II of the book gives numerical data in the form of 48 tables, which deal with subjects such as bond energies, Einstein functions, functions for restricted rotation, and increments for various thermodynamic properties.

Hougen *et al.*[250] discuss the estimation of thermodynamic properties from molecular structure data. Gambill,[358] in numerous articles to be found in the four volumes listed, deals with methods for the prediction of heat capacities of liquids and gases, enthalpies of vaporization, fusion and sublimation, critical temperature and pressure, and p, V, T data, including liquid densities. Dasent[359a] presents methods for the estimation of the standard Gibbs energy of formation of non-existent compounds and compounds of low stability by procedures based on the use of ionic and covalent models.

A small book by Tatevskii *et al.*[359b] summarizes many empirical methods for prediction of the properties of paraffin hydrocarbons.

A brief account of correlation methods for chemical engineers would not be complete without reference to the work of Othmer. As an example, a paper by Othmer and Chen[360] may be cited which lists 30 such contributions.

7 Uses of Thermodynamic Data

Values for many thermodynamic quantities are essential for use in chemical engineering for the design of plant. Thus, for example, a knowledge of enthalpy changes is necessary for the design of heaters, coolers, and condensers and for the removal or supply of heat to chemical reactors.

In industry, a position is now being approached where very large plant can be erected using basic data on the properties of the starting materials and products. For example, Lord Shackleton[361] has summarized the situa-

358 W. R. Gambill, 'How to Estimate Engineering Properties', *Chem. Eng.*, 1957, **64**; 1958, **65**; 1959, **66**; 1960, **67**.

359a W. E. Dasent, 'Non-existent Compounds. Compounds of Low Stability', Edward Arnold and M. Dekker, London and New York, 1965.

359b V. M. Tatevskii, V. A. Benobrskii, and S. S. Yarovoi, 'Rules and Methods for Calculating the Physico-chemical Properties of Paraffinic Hydrocarbons'; translated by M. F. Mullins, ed. B. P. Mullins, Pergamon Press, London, 1961.

360 D. F. Othmer and Hung-Tsung Chen, 'Correlating and Predicting Thermodynamic Data. Applied Thermodynamics', Symposium sponsored by Industrial and Engineering Chemistry and American Chemical Society, 1967; American Chemical Society Publications, Washington, D.C., 1968.

361 Lord Shackleton, Speech on the European Technological Community, 25 April, 1968; Ministry of Technology press notice, 26 April, 1968.

tion in the statement that 'today we can build very large plants indeed . . . purely from basic physical data'. The basic data necessary to build chemical plant include values of thermodynamic properties such as enthalpies of reaction and phase change, heat capacities, and p, V, T data for single components and mixtures. Very accurate data on mixtures are also required to carry out calculations on methods for the separation of substances from the products of reactions.

The practical importance of thermodynamic theory lies in the fact that, given certain basic data, its principles permit the exact calculation of other data and the state of equilibrium of systems under specified conditions. Thus thermodynamic theory will allow one to state the feasibility of a hypothetical reaction and the maximum yield of a desired product which is to be expected. Use of approximate expressions for this latter purpose may sometimes be justified. For example, the defining equation for standard Gibbs energy change at temperature T is the expression:

$$\Delta G^\circ = \Delta H^\circ - T\Delta S^\circ,$$

where ΔH° and ΔS° are functions of T. However, often the values of ΔH° and ΔS° (and hence ΔG°) are available only for $T = 298$ K, and then in rough calculations ΔH° and ΔS° can be treated as constants to obtain approximate values of ΔG° at other temperatures. Dodge [362] has provided a useful (if somewhat crude) criterion for studying the feasibility of reactions. If ΔG° is less than zero, the reaction is promising; if ΔG° is greater than zero but less than 2000 J mol^{-1}, the reaction is of doubtful promise but warrants further study; if ΔG° is greater than 2000 J mol^{-1}, the reaction is very unfavourable and may not warrant further study.

Many of the text-books (*e.g.* ref. 250) and sources of data (*e.g.* ref. 291) listed in earlier sections describe the applications of thermodynamic data, but there are a number of books concerned mainly with physicochemical calculations which also provide examples of the derivation of scientifically and technologically important data by means of thermodynamics. Thus Smith's book [363] 'Chemical Thermodynamics. A Problems Approach', which is of degree level, relates theory to worked examples, and Guggenheim and Prue's 'Physicochemical Calculations' is of this type.[364]

Hawes and Davies' book 'Calculations in Physical Chemistry' [365] is designed to deal with the requirements of students from the upper forms at school to the completion of the undergraduate course. A number of thermodynamic problems appear throughout the book, but Chapter VIII, of 21 pages, is entirely devoted to thermodynamic problems.

[362] B. F. Dodge. *Trans. Amer. Inst. Chem. Engineers*, 1938, **34**, 540.
[363] N. O. Smith, 'Chemical Thermodynamics. A Problems Approach', Reinhold, New York, 1967.
[364] E. A. Guggenheim and J. E. Prue, 'Physico-chemical Calculations', 2nd edn., North-Holland Publishing Co., Amsterdam, 1956.
[365] B. W. Hawes and N. H. Davies, 'Calculations in Physical Chemistry', English Universities Press, London, 1962.

Adamson's book[366] contains 312 problems grouped under the following topics: ideal-gas law, non-ideal gases, First Law, heat capacity of gases and thermochemistry, Second Law, liquids and their phase equilibria, heterogeneous and homogeneous gas equilibria, and phase diagrams.

In the English translations[367] of the book in German by Fromherz 'Physico-chemical Calculations in Science and Industry', forty pages are devoted to the properties of gases and eighty-five pages are concerned with thermochemistry and equilibria. 'Problems in Inorganic Chemistry', by Aylett and Smith,[368] places much emphasis on thermodynamic principles. A recent book 'Physical Chemistry – Problems and Solutions', by Labowitz and Arents,[369] must also be cited.

Bodsworth and Appleton have provided a book[370] entitled 'Problems in Applied Thermodynamics', which is solely concerned with thermodynamics. One hundred problems are treated, and although for most part the questions have a metallurgical bias, nevertheless the treatment is general and therefore of value to the chemist. The book has been recommended for degree students and for research workers who need to improve their knowledge of thermodynamics.

Noddings and Mullet have provided a computational aid[371] for the calculation of equilibria, which contains 587 tables of equilibrium constant against equilibrium compositions. It is claimed to cover about 90% of all chemical stoicheiometries.

Henley and Rosen's 'Material and Energy Balance Computations'[372] is an undergraduate text concerned with the performance of thermodynamic calculations on mathematically complex but conceptually straightforward systems which arise in the chemical industry. The book should be of interest chiefly to the chemical enginner because it describes the setting up of equations and their solution by manual or machine methods.

The advent of digital computers has changed the whole situation with regard to thermodynamic calculations and the use of thermodynamics in science and industry. For example, the use of digital computers for the calculation of the composition of materials that may exist in equilibrium mixtures is a rapidly developing subject, as may be seen by study of 'The

[366] A. W. Adamson, 'Understanding Physical Chemistry. Part I. Properties of Matter, Thermodynamics, Chemical Equilibria', W. A. Benjamin, New York and Amsterdam, 1964.

[367] H. Fromherz, 'Physico-chemical Calculations in Science and Industry', Butterworths, London, 1964.

[368] B. J. Aylett and B. C. Smith, 'Problems in Inorganic Chemistry', English Universities Press, London, 1965.

[369] L. C. Labowitz and J. S. Arents, 'Physical Chemistry – Problems and Solutions', Academic Press, New York, 1969.

[370] C. Bodsworth and A. S. Appleton, 'Problems in Applied Thermodynamics', Longmans, Green, and Co., London, 1965.

[371] C. R. Noddings and G. M. Mullet, 'Handbook of Compositions at Thermodynamic Equilibrium', J. Wiley and Sons, New York, 1965.

[372] E. J. Henley and E. M. Rosen, 'Material and Energy Balance Computations', J. Wiley and Sons, New York, 1969.

Computation of Chemical Equilibria' by von Zeggeren and Storey.[373] This treatise is largely concerned with equilibria in high-temperature high-pressure processes, in rocketry, in explosives, and in biological systems. However, the computational methods which are discussed are of general applicability and are classified under two headings, 'optimization' procedures, that involve the search for extrema of a function, and 'non-linear' procedures. The use of optimization techniques has been practically possible only since digital computers have become available. There is an eight-page bibliography, over 60% of whose entries are published since 1960. The last chapter discusses 'best methods' and is enlightening because it reveals the complexity of the subject and the difficulty of defining 'best' in this context. Some particular examples of the use of computers in thermodynamic calculations will now be mentioned. Prausnitz and Chueh [374] use the method of corresponding states with empirical equations to calculate high-pressure vapour–liquid equilibria. The U.K. Atomic Energy Authority [375] has published a computer program which makes for easier use of existing methods the representation of oxidizing, sooting, and carburizing tendencies of $CO + CO_2 + H_2 + H_2O + CH_4$ mixtures in contact with metals, and their oxides and carbides.

Another recent book that is concerned with thermodynamics and computers is 'Computer Calculation of Phase Diagrams with Special Reference to Refractory Metals', by Kaufman and Bernstein.[376]

The uses of thermodynamic data in technology may be illustrated by the following account of steps normally taken to bring a new chemical process to production. A research chemist, having appraised himself of the potential market for a new product, will consider possible reactions for use in its production. Some preliminary experiments will then be made, and a reaction producing some of the desired product will usually be found. At this stage it will often become apparent that the yield is disappointingly low and that side-products are being produced. Thermodynamic calculations will then be made to obtain the requisite equilibrium constants. Equilibrium compositions will then be calculated for different temperatures and pressures in order to identify the optimum temperature and pressure conditions. More experiments will then follow with the aim of getting the maximum obtainable yield of the desired product at a suitable rate and at a favourable temperature with minimum production of undesirable side-products.

373 F. von Zeggeren and S. H. Storey, 'The Computation of Chemical Equilibria', Cambridge University Press, Cambridge, 1970.

374 J. M. Prausnitz and P. L. Chueh, 'Computer Calculations for High-pressure Vapour–Liquid Equilibria', Prentice-Hall, New Jersey, 1968.

375 E. D. Hyam and S. Timperley, 'A Computer Programme for the Numerical Evaluation of Equilibrium Involving $CO–CO_2–H_2–H_2O–CH_4$ Mixtures and Non-stoichiometric Oxides and Carbides', U.K. Atomic Energy Authority, Report 1864-W.

376 L. Kaufman and H. Bernstein, 'Computer Calculation of Phase Diagrams', Academic Press, New York, 1970.

The following sources may be cited as examples of the study of equilibria by thermodynamics to ascertain the composition of equilibrium mixtures and to obtain thermodynamic data on chemical reactions.

Rossini [25] describes the calculation of equilibria, and the processes studied include isomerization, the production of toluene and of iso-octane, the syntheses of rubber and alcohols, and the transformation of graphite to diamond.

Stull, Westrum, and Sinke [338] devote a chapter to the discussion of the applications of thermodynamics to industrial problems. Subjects covered include the petroleum industry, chemicals from methane, styrene manufacture, acrylonitrile and vinyl chloride syntheses, methanol synthesis, formaldehyde production from methanol, acetic acid manufacture, the Gatterman–Koch reaction, and catalyst selection.

Cox and Pilcher [187] discuss the thermodynamic feasibility of reactions involved in the dehydrogenation of ethane, in the hydration of ethylene, and in polymerization reactions. The calculation of numerous enthalpies of reactions involving organic compounds is also considered, including enthalpy changes resulting from hydrogenation and hydrogenolysis, oxidation, hydrolysis, hydration, halogenation, isomerization, combustion, decomposition, and polymerization. Often polymerization is accompanied by a decrease in entropy and as a result there will usually be a ceiling temperature for a polymerization reaction above which depolymerization becomes dominant. Measurements of enthalpies and entropies enable the ceiling temperature to be found (see Dainton and Ivin [377]). In some instances this temperature may be quite low.

In the development of a new process, when interaction between experiment and thermodynamic calculation has established suitable reaction conditions, a large plant can be designed. Information required for these calculations includes both thermodynamic and kinetic quantities; the latter are not the subject of this article.

In industrial practice it is often necessary to separate the products obtained from the primary reaction by use of a suitable phase change. For example, products may be fractionally distilled so as to separate individual components by means of the difference in composition that often exists between a vapour and liquid mixture in equilibrium. The design of fractionating equipment depends on the use of vapour–liquid equilibrium data and it is now widely recognized that the experimental measurement of vapour–liquid equilibrium properties is difficult. Fortunately, however, experimental measurements on vapour–liquid systems are usually made in such a manner that the properties of the system are over-determined in a thermodynamic sense. For example, the activity coefficient of each component of a mixture can usually be derived from the experimental observations, and these are not independent but are linked by the Gibbs–Duhem

[377] F. S. Dainton and K. J. Ivin, *Quart. Rev.*, 1958, **12**, 61.

relation. The self-consistency of the experimental data can therefore be studied by tests based on the use of that relation. Such tests are most readily applicable to data obtained on isothermal systems, although tests are now available for isobaric data which are of importance in distillation column design. There is a vast literature on this subject, but for a short introduction to the position up to 1967 two papers by Ellis *et al.*[378, 379] may be consulted, and for information on the position from 1967 to 1969 the Proceedings of the International Symposium on Distillation, Brighton, 1969, can be studied.[380]

Thermodynamics finds some additional rather different practical applications in high-temperature chemistry and metallurgy, where the entities involved in equilibria are often atomic or small molecular species. Many of the text-books and sources of data (*e.g.* see Kubaschewski *et al.*,[120] Rand and Kubaschewski,[280] Bodsworth and Appleton[370]) discussed earlier include examples of practical application to high-temperature systems.

Topics of practical importance include studies of the impurity contents of products resulting from reaction with vapours (*e.g.* oxygen, nitrogen, carbon oxides, water) or with crucibles. Contamination of products *via* the gas phase due to the formation of volatile compounds stable at high temperatures must also be taken into account. Phase stability may be of practical interest at temperatures where equilibrium is achieved only slowly, and here measurements of thermodynamic quantities at higher temperatures may, by reasonable extrapolation procedures, enable phase diagrams to be obtained for lower temperatures. The compatibility of materials of construction for long periods of use at high temperature may also be considered by this method. Calculations of the loss of valuable metals (*e.g.* platinum) by volatilization from mixed catalysts can also be made. Such applications of thermodynamics to inorganic and metallurgical systems are additional to studies of the feasibility of inorganic reactions, which may be illustrated by U.S. Bureau of Mines Bulletin 406, where Kelley[199] examines the possibility of producing Al_2Cl_6 from $Al_2(SO_4)_3$ and NaCl or from $Al_2(SO_4)_3$ and $CaCl_2$, the recovery of elemental sulphur from low-grade copper matte, and the possibility of using Fe in the treatment of Zn + Pb sulphide ores. He also considers the action of Mn in desulphurizing steel.

Some examples of the contribution of thermodynamics to the understanding of industrial processes are provided by the Catalyst Handbook.[381]

[378] S. R. M. Ellis and J. R. Bourne in 'Proceedings of the International Symposium on Distillation', ed. P. Rottenberg, The Institution of Chemical Engineers, London, 1960.

[379] S. R. M. Ellis, C. McDermott, and J. C. L. Williams in 'Distillation – Final Report ABCM/BCPMA. Distillation Panel', ed. H. Sawistowski, The Chemical Industries Association, London, 1967.

[380] Proceedings of the International Symposium on Distillation 1969, Brighton, England, *Inst. Chem. Engineers, Symposium Series*, The Institution of Chemical Engineers, London, 1969.

[381] 'Catalyst Handbook', I.C.I. Ltd., Wolfe Scientific Books, London, 1970.

For example, in this book the reactions which take place during the use of iron oxide or zinc oxide as absorbents for hydrogen sulphide are compared. Fresh iron oxide (Fe_2O_3) is converted to Fe_3O_4 in the temperature range 340 to 400 °C when iron oxide is used as absorbent for H_2S. The reactions are thus:

$$3Fe_2O_3 + H_2 = 2Fe_3O_4 + H_2O,$$

$$Fe_3O_4 + H_2 + 3H_2S = 3FeS + 4H_2O.$$

For zinc oxide the corresponding reaction is:

$$ZnO + H_2S = ZnS + H_2O.$$

Thus in both reversible reactions the absorption of H_2S would be adversely affected by the presence of water vapour in the feed gas. Values of $\log_{10}K_p$ for the two desulphurization reactions are given, and hence the equilibrium quantities of H_2S (p.p.m. volume/volume) are tabulated, showing that ZnO is more efficient than Fe_3O_4 in the removal of H_2S. A further potential disadvantage of iron oxide (compared with zinc oxide) is associated with the relative ease of reduction of the two sulphides according to the equation:

$$MS + H_2 = M + H_2S.$$

The variation of the equilibrium constant with temperature is given for reactions of this type and a table showing the $H_2S : H_2$ ratio (p.p.m. volume/volume) for a feedstock of 100% hydrogen is presented. Thus, at 400 °C, the equilibrium $H_2S : H_2$ ratio is 19.1 for the iron sulphide but only 2×10^{-7} for the zinc sulphide. Appendix 1 of this ICI book gives examples of a calculation of sulphur absorption. Other examples of equilibrium studies discussed include ammonia synthesis, the water-gas shift reaction, and hydrocarbon-reforming catalysis. Equilibrium constants over a range of temperatures are given (in appendices) for a number of reactions, such as the ZnO + H_2S reaction, methane + steam reaction, water-gas shift reaction, methanation reaction, and ammonia synthesis. Examples of equilibrium composition calculations are given for sulphur absorption, steam-reforming reactions, and the high- and low-temperature water-gas shift reactions. In industrial practice, when catalysts are employed equilibrium may not always be achieved, and the 'Catalyst Handbook' suggests the use of a quantity ΔT, termed 'the approach to equilibrium', which is essentially the difference in temperature between that used and that which would have given the yield of product actually produced, assuming equilibrium conditions had been achieved.

So far in this section the practical applications of thermodynamics in chemistry and in industry have been considered, but an extremely important use of thermodynamics is in the systematic understanding of chemistry. There is of course no clear demarcation between the practical and theoretical interest of thermodynamics. Indeed, many estimation procedures designed to obtain values of thermodynamic quantities required for practical applications depend on the understanding, in depth, of chemistry.

Ultimately, of course, it may be expected that all quantum mechanical calculations will yield thermodynamic quantities, *e.g.* enthalpies of formation. However, this stage has not yet been reached, but, at present, relations between classical chemical structures and thermodynamic quantities are being established. For example, Cox and Pilcher's [187] book devotes some 70 pages to bond energy terms, bond energy schemes, steric effects, destabilization energies, and related matters.

Some aspects of chemistry have now been treated in text-books concerned entirely with the thermodynamic approach. For example, Johnson,[382] in 'Some Thermodynamic Aspects of Inorganic Chemistry', considers various chemical problems after they have been re-expressed in terms of the thermodynamic fundamentals involved. Most of the treatment is concerned with ionic structures, but bond energies are considered and some non-metal chemistry is studied where the bonding is predominantly covalent. The book is of final-year degree course level but to get the maximum value from the information the reader will need to use other books as well.

Dasent [345] discusses the thermodynamic principles which underlie the formation of inorganic substances, and their stabilities and their binding energies. Thus lattice energies, bond energies, ionization potentials, and electron affinities are reviewed, and thermochemical cycles of the Born–Haber type are used to examine the energy changes associated with a wide range of inorganic reactions.

For many years much of chemistry and metallurgy was taught descriptively, but as equilibrium is governed by Gibbs energy change, as the enthalpy of reaction can be calculated from the enthalpy of formation, and as entropy (and heat capacity) of gases can be calculated from molecular properties (geometry of the molecule and energy levels), it is natural that both chemistry and metallurgy should now be taught with increasing emphasis on thermodynamic considerations. For example, the extraction of metals, which is an important branch of industrial chemistry and which for many years was taught by piecemeal instruction about industrial processes, can now be unified by thermodynamic treatment. As an example may be quoted the Royal Institute of Chemistry Monograph for Teachers 'Principles of the Extraction of Metals' by Ives,[383] where the first chapter, which gives the thermodynamic background, is followed by a chapter on the thermodynamics of pyrometallurgical processes.

8 Recent Trends

Thermodynamics has long been regarded as a difficult subject but it is so fundamental to chemistry that it is now taught at all levels. For example,

[382] D. A. Johnson, 'Some Thermodynamic Aspects of Inorganic Chemistry', Cambridge University Press, Cambridge, 1968.

[383] D. J. G. Ives, 'Principles of the Extraction of Metals', Monographs for Teachers, No. 3, Royal Institute of Chemistry, London, 1960.

the Chemistry Data Book[323] for school teachers (first published 1968) contains extensive tables on enthalpy, energy, entropy, and other thermodynamic quantities. Similarly, the Chemistry Data Book by Stark and Wallace,[340] which provides a comprehensive source of data likely to be needed by sixth-form, technical college, and first-year university students, contains many tables of thermodynamic functions. As mentioned in the last section, recent text-books reflect the increasing impact of thermodynamics on the understanding of chemical principles. In this connection it is of interest to note that Unilever's booklet[384] for schools on detergency starts with a discussion of Gibbs energy. Clearly, quantitative treatments of the energetics of reactions are likely to figure to an increasing extent in all courses on chemistry as more data become available.

It might have been thought that classical thermodynamics is such a well-studied subject that the chance of development of basic new theory was slight. This is not the case; for example, new inequality relations concerning the critical state have been developed since 1963 (for example see Rowlinson[151]). Another example is provided by the Gibbs–Duhem relation, which has now been known for nearly a century and the theoretical and graphical implications of which have been discussed at length by numerous workers, yet as recently as 1968 a treatment (published under the title 'Symmetrical Area Tests') which reveals the high symmetry of the Gibbs–Duhem relation was presented for the first time (Herington[385]). This treatment is not only of theoretical interest but can readily be used to study the goodness of experimental data employed in the industrial design of fractionating columns (Herington[386]).

It might also have been thought that the scope of application of classical thermodynamics is well understood, but a recent controversy shows that areas of serious argument still exist. The controversy continues but the present position in this discussion of energetics of biological systems can be summarized briefly. Phosphorylation is one of the most fundamental processes occurring in living tissue, and Lipmann[387, 388] divided biochemically important phosphoric acid derivatives into two groups depending on whether their standard free energy of hydrolysis is large or small. Compounds in the first group were said to contain an 'energy-rich' or 'high-energy' O—P or N—P bond; those in the second group an 'energy-poor' or 'low-energy' O—P or N—P bond. These concepts have been widely applied but Gillespie, Maw, and Vernon,[389] in 1953, criticized these ideas on the grounds that 'energy-rich' phosphate bonds are irreconcilable

384 'Theory of Detergency', A Unilever Education Booklet, Advanced Series No. 7, Information Division, Unilever Ltd., London, 1969.
385 E. F. G. Herington, *J. Appl. Chem.*, 1968, **18**, 285.
386 E. F. G. Herington, *Inst. Chem. Engineers, Symposium Series*, No. 32, The Institution of Chemical Engineers, London, 1969, Section 3, p. 17.
387 F. Lipmann, *Adv. Enzymol*, 1941, **99**, 1.
388 F. Lipmann, *Fed. Proc.*, 1949, 597.
389 R. J. Gillespie, G. A. Maw, and C. A. Vernon, *Nature*, 1953, **171**, 1147.

with physicochemical principles and furthermore that the relevance of classical thermodynamics to the interpretation of detailed mechanisms of biological systems requires scrutiny. One of these authors, Vernon,[390] in an article entitled 'A Sceptical Chemist', points out that standard free energy cannot be passed on from one reaction to another and expresses the view that biochemistry has stepped into vitalism under the respectable guise of thermodynamics. These papers, however, were only the opening statements in a debate which was recently reactivated by an article entitled 'Thermodynamics and Biology' by Banks, published[391] in November 1969, which states that the discussion of biological processes is often misleading due to failure to distinguish between kinetic and thermodynamic control. This paper has given rise to a lively discussion, which has been joined by Pauling,[392] 'The Problem of Biological Energetics'; by Wilkie,[393] 'Thermodynamics and Biology'; and by Huxley,[394] 'Energetics of Muscle'. These articles have elicited replies from Banks and Vernon,[395] 'A Reply to Linus Pauling and A. F. Huxley', and from Ross and Vernon,[396] 'Biological Energetics – the Other View'. All the papers and articles referred to above, and other papers cited in them, should be consulted to appreciate the range of the arguments.

A controversy[397] in relativistic thermodynamics appears to have recently been largely resolved by Hamity.[398] This subject is perhaps of only marginal interest to chemists but concerns such interesting topics as whether the effective temperature of a rapidly moving body is lower than the temperature noted by a co-mover observer or not.

It is regrettable that, in the past, different symbols have been adopted in compilations, but it is expected that, in the future, symbols advocated by IUPAC will be employed universally and that SI will be used for the units. To secure a further unification in thermodynamic tables, the International Council of Scientific Unions (ICSU) and the Committee on Data for Science and Technology (CODATA) set up in 1968 a Task Group on Key Values for Thermodynamics. The first objective of the Task Group is to prepare a set of values of the basic thermodynamic properties of a number of chemical species, to be agreed internationally. The set is to include the elements in both standard and monatomic gaseous states, aqueous ions, and simple compounds.[399]

When these values or slight modifications to them have been internationally agreed, it is hoped that they will be applied generally, so that

[390] C. Vernon, 'Science News 50', Penguin Books, 1958, p. 44.
[391] B. E. C. Banks, *Chem. in Britain*, 1969, **5**, 514.
[392] L. Pauling, *Chem. in Britain*, 1970, **6**, 468.
[393] D. Wilkie, *Chem. in Britain*, 1970, **6**, 472.
[394] A. F. Huxley, *Chem. in Britain*, 1970, **6**, 477.
[395] B. E. C. Banks and C. A. Vernon, *Chem. in Britian*, 1970, **6**, 541.
[396] R. A. Ross and C. A. Vernon, *Chem. in Britain*, 1970, **6**, 539.
[397] 'Relativistic Thermodynamics – Controversy Settled', *Nature*, 1970, **226**, 497.
[398] V. T. Hamity, *Phys. Rev.*, 1969, **187**, 1745.
[399] *J. Chem. Thermodynamics*, 1971, **3**, 1; see also CODATA, Bulletin 2, November, 1970; Bulletin 6, December, 1971.

tabulated critically selected thermodynamic data from all countries (*e.g.* from the U.S.A. and the U.S.S.R.) will be based on the same key values.

Statistical thermodynamics has made large advances, and since 1960 the theory of liquids has progressed considerably. More recently, attention has been paid to mixtures (see ref. 151 and 'New Light on Solutions'[400]) and further advances are expected which will enable thermodynamic quantities to be calculated.

There is little doubt that rapid automatic methods for the measurement of thermodynamic quantities, coupled with computer analysis of the observations and manipulation of the results, will enable large chemical plants for the production of new products to be quickly designed, and adequately controlled in operation.

400 'New Light on Solutions', *Nature*, 1970, **228**, 1260.

3
Combustion and Reaction Calorimetry

BY A. J. HEAD

1 Introduction

Reaction calorimetry is the experimental determination of the enthalpy changes accompanying chemical reactions by direct methods using calorimeters. It is the principal means by which enthalpies of formation of pure chemical compounds are determined. With the exception of certain binary compounds, chiefly oxides, it is impractical to measure the enthalpy of formation of a compound from its elements directly, and it is necessary to determine the enthalpy of a reaction involving the compound in which the enthalpies of the other reactants and products are all known, and then to apply Hess's law. Occasionally, enthalpies of formation can be derived from the study of equilibria (as measured by the e.m.f.s. of electrochemical cells, dissociation pressures, *etc.*) by means of second- or third-law methods, or from electron impact experiments, but such indirect approaches are outside the scope of the present review, which is confined to the discussion of experimental procedures used in direct calorimetric methods.

Not all reaction calorimetry is undertaken with the object of determining enthalpies of formation of pure compounds. The enthalpy change accompanying a process can sometimes be used to yield information about the mechanism of that process, especially when results are available for a series of related reactions; it can be used as the basis for analytical methods and may also be required for technological reasons. However, since this review forms part of a report on chemical thermodynamics, the main emphasis will be on the use of reaction calorimetry to determine enthalpies of formation.

Because chemical thermodynamic properties are usually referred to a temperature of 298.15 K it is convenient to work with systems with the initial and final temperatures close to this value (even though the temperature of the reacting mixture may be considerably different) and thus avoid the necessity for accurate heat capacities from 298.15 K to the reaction temperature. Where the thermodynamic quantities are principally of interest in connection with processes occurring at elevated temperatures, *e.g.* in the fields of metallurgy, refractory materials, fused salts, flames, *etc.*, reaction calorimetry is frequently carried out at temperatures remote from 298.15 K, but such work will not be discussed in this chapter, which deals only with methods used to obtain values of ΔH_f° at 298.15 K. Thus, for example, hydrogenation of organic compounds at 250 °C is mentioned whereas the solution calorimetry of alloys in liquid metals (at temperatures from 240 °C upwards) is not.

The methods used to determine standard enthalpies of formation (ΔH_f°) involve calorimetric observations on both constant pressure and constant volume systems, yielding respectively values of the change in enthalpy (ΔH) or internal energy (ΔU), differing by the volume work term $\Delta(pV)$. This term is significant only if one or more of the substances involved in the reaction is gaseous, when it may amount to several kilojoules per mole, and is readily calculated.

2 General Classification of Calorimeters

Although the reactions carried out in calorimeters are extremely diverse, the instruments used may be classified into a few main types.

Adiabatic and Isoperibol Calorimeters.—Most calorimeters used in combustion and reaction calorimetry undergo a change of temperature when reaction takes place. If the calorimeter is surrounded by a jacket, the temperature of which is controlled to be the same as that of the calorimeter, no heat-exchange occurs between the surroundings and the calorimeter, which is then described as *adiabatic*. However, if the temperature of the environment is maintained constant (in a type of calorimeter conveniently described as *isoperibol* and sometimes, incorrectly, as isothermal) some heat-exchange occurs between the calorimeter and its surroundings, but may be accurately determined by analysis of the temperature–time curves before and after reaction takes place, provided the reaction is of short duration (say not exceeding 15 min). With slower processes, isoperibol calorimeters are less useful, and the adiabatic principle is easier to effect and yields more accurate results.

Reaction is carried out in a convenient vessel which is completely surrounded by a calorimeter so that temperature changes are rapidly equalized throughout the system. The calorimeter is commonly a metal vessel filled with a well-stirred liquid (usually water) or it may be a metal block (aneroid calorimeter); alternatively, the reaction vessel itself may serve as calorimeter also. Because heat-exchange between a calorimeter and its surroundings is governed by its surface temperature, the ideal calorimeter is bounded by a perfectly conducting envelope, which is at all times isothermal and at the same temperature as the interior. In practice, the measured 'temperature of the calorimeter' corresponds to the temperature neither throughout the whole calorimeter nor at its surface. However, reaction calorimeters are not used to measure energy or enthalpy absolutely, but to compare the energy or enthalpy of the process under investigation with that of some calibration process (either chemical reaction or Joule heating). The more nearly the measurement and calibration experiments are matched the less stringent are the requirements for the calorimeter to behave ideally.

In isoperibol and adiabatic calorimeters the quantity actually measured is the temperature change of a calorimeter. Although a detailed discussion of thermometry would be inappropriate, it is useful to survey briefly the main methods used in reaction calorimetry.

Very few modern calorimeters employ mercury-in-glass thermometers. The limit of accuracy of the most accurate instrument of this type, the Beckmann thermometer, is about 0.001 K; it is easily broken, and subject to errors caused by exposed stem, pressure, sticking of the mercury column, and drift in calibration.

Thermocouples offer a number of advantages although they tend to have been used rather more in high-temperature calorimeters, where other methods are less applicable, than in instruments operating close to ambient temperature. The small size of thermocouple junctions enables several to be connected in series to increase the sensitivity and to obtain a mean temperature from measurements at several positions in a calorimeter, although this advantage tends to be offset by the corresponding increased thermal leakage to the surroundings (a property used to advantage, however, in heat-flow calorimeters, see below). Thermocouples are particularly useful for the measurement of temperature *differences* between a calorimeter and its environment and, provided that the temperature of the isothermal jacket remains constant, they can be successfully used in isoperibol calorimeters.[1] The chief disadvantage of thermocouples lies in errors due to spurious e.m.f.s. arising from temperature gradients at inhomogeneities in all but the purest of metal wires serving as leads to the potentiometer or other measuring instrument. It is convenient for multi-junction thermopiles to be permanently built into calorimeters so that calibration has to be carried out *in situ*. Modern developments in the construction of digital voltmeters, which include the use of integrated circuits, should tend to increase the application of thermocouples to calorimetry of moderate precision by making available comparatively inexpensive instruments, even at a resolution level of 1 μV, which are conveniently coupled to automatic data-recording systems.

For the highest accuracy and stability platinum resistance thermometers are still unsurpassed. When they are used with four leads in conjunction with resistance bridges of the Smith's difference or Mueller types, errors due to lead resistance are eliminated. The more recent development of a.c. bridges based on inductive dividers has overcome interference by thermoelectric e.m.f.s. Automatic versions of the a.c. bridge are now commercially available with direct digital display of the ratio $R_1/(R_1 + R_2)$, where R_1 and R_2 are the resistances of the thermometer and of a reference standard respectively. Automatic a.c. bridges have many advantages over manual d.c. bridges and are satisfactory for all measurements save those of the highest precision (better than 0.0005 K with a $R_0 = 25\ \Omega$ thermometer). With both thermocouples and resistance thermometers the variation

[1] V. P. Vasil'ev and G. A. Lobanov, *Russ. J. Inorg. Chem.*, 1966, **11**, 381.

of the response with temperature is nearly linear over the temperature range used in reaction calorimetry (usually < 3 K). The chief disadvantages of platinum resistance thermometry are the relatively large size of the sensing elements (compared with thermistors, see below) and the expensive nature of the equipment. Some of the advantages of resistance thermometry are retained when resistance elements, *e.g.* of copper,[2, 3] are wound on the surface of aneroid calorimeters, and the benefit of integrating the temperature measurements over a large part of the calorimeter is gained. However, some of the stability and accuracy associated with a carefully constructed strain-free platinum resistance thermometer element may be lost and calibration has to be carried out *in situ.*

Thermistors have proved useful in reaction calorimetry except for work of the very highest precision. Because of their high resistance and large (negative) temperature coefficient the temperature sensitivity attained is comparable with that of resistance thermometry, but their use requires much less expensive auxiliary equipment. Lead resistance is very small compared with that of the thermistor itself so fairly simple Wheatstone bridge circuits can be used. Their small size produces rapid response and they are particularly useful in small calorimeters. Care must be taken to avoid excessive self-heating by limiting the current through the thermistor. They may also be somewhat sensitive to light, shock, and pressure. Whilst the relation between resistance and temperature, *viz.* $R = \exp\{A + B/(T + \theta)\}$, is inconvenient for desk computation, the less exact expression

$$(T_1 - T_2) = - C\log(R_1/R_2),$$

is sufficiently accurate for most calorimetric purposes. The limits of accuracy of various approximate methods of calculating the corrected temperature rise in isoperibol calorimetry from values of thermistor resistance have been discussed recently by Gunn.[3a] The main disadvantage of thermistors is in their slight tendency for the resistance to drift, although modern manufacturing methods now result in products considerably improved over earlier thermistors in this respect. Because of this risk of drift, thermistors find their most usual application in calorimeters where the relation between change in resistance and energy is frequently determined, as for example in instruments where the calorimeter is also the reaction vessel and is electrically calibrated in every experiment. The apparent very small drift in the measured *absolute* temperature is normally insignificant in reaction calorimetry.

An important development in recent years has been the advent of quartz crystal thermometry. When a quartz crystal is cut at a certain orientation to the axes of the crystal lattice the temperature dependence of the resonant frequency is large and nearly linear, and the Hewlett Packard Company has

[2] W. A. Keith and H. Mackle, *Trans. Faraday Soc.*, 1958, **54**, 353.
[3] M. A. Frisch and H. Mackle, *J. Sci. Instr.*, 1965, **42**, 186.
[3a] S. R. Gunn, *J. Chem. Thermodynamics*, 1971, **3**, 19.

developed a direct-reading digital instrument based on this principle[4]. The sensing element is small and robust with quick response. Because the instrumentation basically consists of a counter, the degree of resolution depends on the period of observation, thus for example temperature differences of 0.0001 K can be discriminated over a period of 10 s. Measurements of temperature can in effect be made throughout the greater part of the duration of a calorimetric experiment, and recorded values of temperature are integrated mean values rather than instantaneous readings; this leads to greater accuracy in the determination of areas under temperature–time curves which forms the basis for calculation of the heat-exchange correction in isoperibol calorimetry. Although comparable in price with instrumentation for the best resistance thermometry using d.c. bridges, the associated equipment is much more compact and the advantage of digital output, with its ease of connection to data-recording systems, is considerable. Moreover, two sensing elements can be used to measure temperature differences without extra instrumentation. Several modern high-precision calorimeters now use quartz thermometry and, significantly, some workers of high repute have recently substituted quartz thermometers in calorimeters where platinum resistance thermometers were formerly used.[5]

Heat-flow Calorimeters.—In calorimeters of the adiabatic or isoperibol types, heat-exchange between the calorimeter and its surroundings is either eliminated or is restricted to a small, accurately determined amount. An alternative method is to transfer the heat of reaction completely to a heat-sink, so that both the calorimeter and the heat-sink remain essentially isothermal and the calorimetric determination consists of measuring the heat transferred. Two main types have been employed.

In the first type the heat-sink is in good thermal contact with the reaction vessel and consists of two phases in equilibrium, the amount of phase-change being a measure of the change in energy, as in the well-known Bunsen ice calorimeter where the amount of phase-change is determined by dilatometry. Diphenyl ether has proved particularly useful in reaction calorimetry since its melting temperature (26.9 °C) is conveniently close to 25 °C, it may be readily purified by zone-refining, and exhibits a large volume change on melting. The volume change is conveniently determined by weighing the amount of mercury displaced from the two-phase region by melting of the solid ether. The energy equivalent of the calorimeter may thus be expressed as energy per unit mass of mercury and should be the same for all calorimeters employing diphenyl ether. Some of the most reliable values for this quantity have been summarized by Peters and Tappe.[6]

[4] D. L. Hammond, C. A. Adams, and P. Schmidt, *Trans. Instrument Soc. America*, 1965, **4**, 349.

[5] E. Greenberg and W. N. Hubbard, *J. Phys. Chem.*, 1968, **72**, 222.

[6] H. Peters and E. Tappe, *Monatsberichte*, 1967, **9**, 901.

Although isothermal phase-change calorimeters operate on the heat-flow principle, the use of the term 'heat-flow' is usually restricted to calorimeters of the second type where thermometry is used to measure the small temperature differences which arise. In the most widely used form of heat-flow calorimeter, a thermopile provides the main thermal conduction path between the reaction vessel and the heat-sink and is also used to measure the small temperature difference between them. The enthalpy change is calculated from the area under the temperature–time curve, or thermogram.

Calorimeters based on the heat-flow principle are most useful for the measurement of small quantities of heat absorbed or evolved over long periods, but have also been used for the study of rapid reactions.

The application of calorimeters of these main classes to the measurement of enthalpies of reaction will be described in subsequent sections. For many of the techniques described much greater detail is available in the two excellent volumes on 'Experimental Thermochemistry', prepared under the auspices of the I.U.P.A.C., which no new worker in the field can afford not to read.[7, 8] A very useful review entitled 'Recent Developments in Calorimetry' has been published by Wilhoit[9] in which he includes references to many commercially available calorimeters.

3 Combustion Calorimetry

Combustion in Oxygen.—*Static-bomb calorimetry.* More effort has been devoted to the determination of energies of combustion in oxygen than to any other thermochemical measurement. Initially, there was a need to know the 'calorific values' (specific energies of combustion) of various fuels and, later, interest in the enthalpies and Gibbs energies of formation of pure chemical substances developed. Every organic compound containing only the elements C, H, and O can be burnt in oxygen to carbon dioxide and water so the number of substances amenable to the method is enormous, since the enthalpies of formation of these two products are well established. The presence of nitrogen in the molecule results in the formation of elemental nitrogen and some nitric acid, which must be determined after each combustion but does not otherwise limit the applicability of the method.

The combustion bomb has changed little in design since it was introduced by Berthelot in 1881. It consists essentially of a thick-walled stainless steel vessel fitted with a head or lid which is clamped to the bomb body by means of a threaded ring, a suitable O-ring or gasket providing a seal tested to withstand 300 bar. The bomb head is fitted with preferably two valves (although sometimes only one is provided) for the admission and discharge of gases, and carries a support for the crucible containing the sample. Two

[7] 'Experimental Thermochemistry', ed. F. D. Rossini, Interscience, New York, 1956.
[8] 'Experimental Thermochemistry', Vol. II, ed. H. A. Skinner, Interscience, London, 1962.
[9] R. C. Wilhoit, *J. Chem. Educ.*, 1967, **44**, A571, A629, A685.

electrodes are provided, one insulated from the bomb, between which a piece of fine platinum wire can be supported and electrically heated to ignite the sample, either directly or *via* a fuse of cotton or plastic such as polyethylene. The bomb is charged to a pressure of usually 30 bar with oxygen, which has been freed from combustible impurities by passage over heated copper oxide, and, most commonly, is contained in an isoperibol calorimeter of the stirred-water type.

It is usual to calibrate combustion calorimeters by burning in them benzoic acid, for which the specific energy of combustion has been certified by a national standardizing laboratory. Benzoic acid has been accepted as the primary standard for combustion calorimetry since 1934 [10] and its specific energy of combustion has been determined several times using electrically-calibrated calorimeters. Electrical calibration is complicated by the fact that the heating takes place at a different situation in the calorimeter and with a different dependence on time from that due to the combustion. Considerable care has to be taken in calorimeter design and in the application of corrections to produce results of the precision and accuracy of those reported from three different laboratories during 1968–9 for the specific energy of combustion of standard reference sample 39i issued by the National Bureau of Standards;[11–13] very good agreement was obtained.

Because the magnitude of the energy of combustion of an organic compound is large compared with that of its enthalpy of formation, high accuracy in the measurement of energies of combustion is required to yield enthalpies of formation of moderate accuracy. Thus for an organic compound of molecular weight in the region of 100, the uncertainty in the enthalpy of formation is approximately 0.2 per cent when calculated from energies of combustion known to $\pm$ 0.02 per cent, a typical uncertainty for combustions of simple organic substances. Frequently, the accuracy of an energy of combustion determination is limited more by the purity of the compound or the completeness of reaction than by the measurement technique. Even with the simplest compounds containing only C, H, and O it is desirable to determine the carbon dioxide and test for the absence of carbon monoxide in the products to ensure that combustion has been complete. Particular care must be taken to exclude water from substances which are to any degree hygroscopic. Indeed, elaborate calorimetric methods are superfluous unless the criteria of purity of sample and completeness of combustion have been met.

The energy of the process occurring in a combustion bomb is not identical to the energy of combustion required for thermodynamic calculations, in

[10] Premier Rapport de la Commission Permanente de Thermochimie, Union Internationale de Chimie, Paris, 1934.

[11] K. L. Churney and G. T. Armstrong, *J. Res. Nat. Bur. Stand.*, 1968, **72A**, 453.

[12] C. Mosselman and H. Dekker, *Rec. Trav. chim.*, 1969, **88**, 161.

[13] H. A. Gundry, D. Harrop, A. J. Head, and G. B. Lewis, *J. Chem. Thermodynamics*, 1969, **1**, 321.

which the reaction is required to proceed isothermally with the reactants and products in their standard states. It was Washburn who first applied the elaborate procedure for corrections to standard states which now often bears his name.[14] This calculation is simplified if the bomb is saturated with water-vapour before and after combustion and for this reason (and to absorb oxides of nitrogen in the products) water (usually 1 cm^3) is added to the bomb before every combustion.

Although the majority of static-bomb combustion calorimeters are of the stirred-water isoperibol type a number of aneroid instruments have been used. These overcome the disadvantages of having to weigh the water in the calorimeter before every determination and having to guard against evaporation, but the thermal equilibration of the calorimeter is generally less rapid. Thus the calorimeter described by Pilcher and Sutton [15] requires a main reaction period of 1 h. However, Meetham and Nicholls [16] showed that, when the bomb was constructed from Sterling silver, temperature equilibrium could be attained as rapidly with an aneroid calorimeter as with a stirred-water calorimeter; moreover, the space between the calorimeter and the isothermal jacket could be easily evacuated to reduce heat transfer.

One of the disadvantages of conventional bomb calorimeters is the relatively large amount of material (1 to 2 g) which is consumed in each determination, so several semi-microcalorimeters have been designed where the mass of sample is in the range 10 to 100 mg. Although the conventional stirred-water calorimeter may be suitably scaled down,[17] the most successful semi-microcalorimeters have been of the aneroid type. Precisions of $\pm$ 0.05 per cent are claimed for the designs of Quitzsch, Schaffernicht, and Geiseler [18] and of Mackle and O'Hare.[19] The latter calorimeter permits the use of a thin-walled bomb by the ingenious device of pressurizing both the bomb and the interspace between the bomb and the jacket. A semi-micro bomb has been constructed for use with the Tian–Calvet calorimeter (see below) which operates on the heat-flow principle. It was necessary to reduce the thermal diffusivity by coating the interior bomb walls with sintered alumina to prevent incomplete combustion (by cooling the flame on the walls) and to give a thermogram from which the energy of combustion could conveniently be calculated.[20] Tachoire showed how the Washburn correction could be simplified by omitting the addition of water to this bomb (since all the water formed from the combustion of 5 to 10 mg samples is in the gas phase), and by working at an oxygen pressure of only

[14] E. W. Washburn, *J. Res. Nat. Bur. Stand.*, 1933, **10**, 525.
[15] G. Pilcher and L. E. Sutton, *Phil. Trans.*, 1955, **248A**, 23.
[16] A. R. Meetham and J. A. Nicholls, *Proc. Roy. Soc.*, 1960, **A256**, 384.
[17] E. A. Miroshnichenko, V. P. Leiko, and Y. A. Lebedev, *Russ. J. Phys. Chem.*, 1964, **38**, 582.
[18] von K. Quitzsch, H. Schaffernicht, and G. Geiseler, *Z. phys. Chem.* (*Leipzig*), 1963, **223**, 200.
[19] H. Mackle and P. A. G. O'Hare, *Trans. Faraday Soc.*, 1963, **59**, 2693.
[20] E. Calvet, P. Chovin, H. Moureu ,and H. Tachoire, *J. Chim. phys.*, 1960, **57**, 593.

1 bar.[21] A semi-micro bomb has been used in conjunction with an isothermal phase-change calorimeter (diphenyl ether) and a standard error of $\pm$ 0.01 per cent has been obtained in the combustion of 150 mg samples of benzoic acid.[22]

Despite the large amount of combustion calorimetry which has been carried out, there are still many relatively simple organic compounds for which the enthalpies of formation have not been reliably established. Some examples of compounds for which definitive values have only recently been obtained include aliphatic tertiary amines,[23] aliphatic ketones,[24, 25] and the isomeric pentanes,[26] and much work in static-bomb oxygen calorimetry remains to be carried out. Nevertheless, the scope of the method is limited since few compounds (outside the large class of substances containing only C, H, O, and N) burn to yield a final state which is uniform and both chemically and thermodynamically well defined. Metals which form only one oxide can be handled satisfactorily; no water is added to the bomb and the completeness of reaction can be checked by weighing: thus the enthalpies of formation of HfO_2 [27] and PuO_2 [28] have been obtained. Some carbides, *e.g.* of tantalum,[29] nitrides, *e.g.* of plutonium,[28] and lower oxides, *e.g.* Ce_2O_3 burning to CeO_2,[30] are amenable to static-bomb calorimetry, and the enthalpies of formation of several tin alkyls have also been determined by this method.[31] Of the organic halogen compounds, only those containing iodine give rise to a single halogen-containing product and are suitable for burning in a static bomb protected against corrosion. Elemental iodine is formed virtually in its standard reference (crystalline) state; the corrections for dissolution in water and for vaporization are small, and no other products have been detected.[32]

Many other compounds have been burnt in static-bomb calorimeters with some success, but the knowledge concerning the final state is frequently inadequate and the result therefore uncertain. Thus, for example, boron trialkyls [33] and carboranes [34] can be burnt to yield CO_2, H_2O, and H_3BO_3, with only traces of side reactions or of incomplete combustion, but it has to be assumed that a saturated aqueous solution of boric acid is produced

[21] H. Tachoire, *Compt. rend.*, 1962, **254**, 477.
[22] H. Peters and E. Tappe, *Monatsberichte*, 1967, **9**, 828.
[23] N. D. Lebedeva, *Russ. J. Phys. Chem.*, 1966, **40**, 1465.
[24] P. Sellers, *J. Chem. Thermodynamics*, 1970, **2**, 211.
[25] D. Harrop, A. J. Head, and G. B. Lewis, *J. Chem. Thermodynamics*, 1970, **2**, 203.
[26] W. D. Good, *J. Chem. Thermodynamics*, 1970, **2**, 237.
[27] E. J. Huber and C. E. Holley, *J. Chem. and Eng. Data*, 1968, **13**, 252.
[28] G. K. Johnson, E. H. Van Deventer, O. L. Kruger, and W. N. Hubbard, *J. Chem. Thermodynamics*, 1969, **1**, 89.
[29] A. N. Kornilov, I. D. Zaikin, S. M. Skuratov, L. B. Dubrovskya, and G. P. Shveikin, *Russ. J. Phys. Chem.*, 1967, **41**, 172.
[30] F. B. Baker and C. E. Holley, *J. Chem. and Eng. Data*, 1968, **13**, 405.
[31] J. V. Davies, A. E. Pope, and H. A. Skinner, *Trans. Faraday Soc.*, 1963, **59**, 2233.
[32] W. A. Roth, *Ber.*, 1944, **77**, 535.
[33] G. L. Gal'chenko and N. S. Zaugol'nikova, *Russ. J. Phys. Chem.*, 1967, **41**, 538.
[34] G. L. Gal'chenko, L. N. Martynovskaya, and V. I. Stanko, *Proc. Acad. Sci.* (*U.S.S.R.*), (*Chem.*), 1969, **186**, 497.

during the main reaction period. Some organophosphorus compounds appear to burn completely to CO_2, H_2O, and a solution of orthophosphoric acid [35] but the tests applied for other phosphorus-containing products were not very sensitive. Moving-bomb methods, discussed below, frequently enable the only condensed-phase product of a combustion to be a homogeneous solution, and wherever this can be achieved the method is inherently more reliable than when a mixture of solid phases, or of solid and liquid phases is obtained by static-bomb methods.

Moving-bomb calorimetry. The moving-bomb calorimeter includes provision for thoroughly mixing the products of combustion with a liquid reagent initially added to the bomb. In the stirred-water form developed at the U.S. Bureau of Mines, Bartlesville,[36] and at the University of Lund, Sweden,[37] the bomb may be continuously rotated end-over-end and also about its cylindrical axis so that every part of the interior surface is thoroughly washed by the added reagent. The bomb is usually platinum-lined to withstand corrosion and is often fired in the inverted position to prevent the flame from attacking the gas inlets and electrodes, although in Kolesov's design [38] a platinum shield serves this purpose and the bomb is fired in the upright position.

It is somewhat easier to construct a moving-bomb mechanism when the calorimeter is of the aneroid type. Keith and Mackle built a calorimeter from a block of high-conductivity copper surrounded by a copper jacket which oscillated through 360° about its cylindrical axis and rocked to and fro through 90° about a second axis at right angles to the first.[2] This calorimeter operated in an air thermostat, whereas an otherwise similar instrument was immersed in a water thermostat.[39] Adams, Carson, and Laye have developed the design further: they evacuated the interspace between the copper calorimeter and the jacket and devised an ingenious method whereby the apparatus could be continuously rotated in two planes simultaneously without twisting or straining either the electrical leads or the vacuum connection.[40] The aneroid calorimeters so far mentioned use samples in the range 0.2 to 0.5 g, rather less than the stirred-water instruments. A semi-micro aneroid moving-bomb calorimeter, in which a cylindrical cavity in a copper sphere accommodates a nickel bomb of interior volume only 8 cm^3, has been described by Ponomarev and Alekseeva.[41] The copper calorimeter is surrounded by a spherical water-filled thermostat jacket and can be rocked through 180°. Precisions of about $\pm$ 0.1 per cent have been claimed for results obtained using this

35 A. F. Bedford and C. T. Mortimer, *J. Chem. Soc.*, 1960, 1622.
36 W. N. Hubbard, C. Katz, and G. Waddington, *J. Phys. Chem.*, 1954, **58**, 142.
37 L. Bjellerup, *Acta Chem. Scand.*, 1959, **13**, 1511.
38 V. P. Kolesov, E. M. Tomareva, S. M. Skuratov, and S. P. Alekhin, *Russ. J. Phys. Chem.*, 1967, **41**, 817.
39 R. H. Boyd, R. L. Christensen, and R. Pua, *J. Amer. Chem. Soc.*, 1965, **87**, 3554.
40 G. P. Adams, A. S. Carson, and P. G. Laye, *Trans. Faraday Soc.*, 1969, **65**, 113.
41 V. V. Ponomarev and T. A. Alekseeva, *Russ. J. Phys. Chem.*, 1961, **35**, 800.

apparatus. An interesting attempt to build a moving-bomb calorimeter which incorporates the best features of stirred-water and aneroid types has recently been reported by Hajiev and Kerinov.[42] The combustion bomb is in good thermal contact with two hollow copper hemispheres which surround it and which are filled with carbon tetrachloride, hermetically sealed, and fitted with magnetic stirrers. The resulting spherical calorimeter not only has the advantages of an aneroid system in that the calorimetric liquid does not have to be weighed for each experiment, there is no danger of loss by evaporation, the system has low heat capacity, and is easily assembled; but also it possesses the benfits of high conductivity and rapid temperature equilibration associated with stirred-liquid calorimeters.

One of the first uses of moving-bomb calorimetry was in the determination of energies of combustion of organic compounds containing sulphur. When these compounds are burnt in a static bomb initially containing no water, as in the method due to Huffman and Ellis,[43] the liquid product is a mist of concentrated sulphuric acid of varying concentration throughout the bomb and the final state is impossible to define. However, if 10 cm^3 of water is added to a moving-bomb a final state with a uniform concentration of moderately dilute sulphuric acid is obtained.[36] To ensure that all the sulphur is oxidized to the hexavalent state it is necessary for sufficient oxides of nitrogen to be present after combustion. This is achieved by deliberately introducing nitrogen into the bomb (frequently it is sufficient merely to allow the atmospheric air to remain before admitting oxygen),[44] or by burning urea as an auxiliary substance.[45]

When organic compounds containing chlorine or bromine are burnt in oxygen the halogen is present in the products partly as hydrogen halide and partly as the element. If a suitable reducing agent is allowed to react with the products of such combustions in a moving-bomb calorimeter then a solution of halogen hydracid is the only halogen-containing compound in the final state. Both arsenious oxide and hydrazine hydrochloride have been used as reductants, the potential advantage of the latter being its greater solubility in water; combustions of some highly-chlorinated compounds have required the addition of as much as 60 cm^3 of 0.08 mol dm^{-3} arsenious oxide solution to a 350 cm^3 bomb.[46] However, the molar energy of oxidation for hydrazine hydrochloride is twice that for arsenious oxide and the extent of decomposition of hydrazine hydrochloride to ammonium chloride varies widely. Moreover, the oxidation of hydrazine hydrochloride by oxygen is catalysed by platinum and it is necessary to use a tantalum-lined rather than a platinum-lined bomb.[47] Several workers have investigated the stability of arsenious oxide solutions to oxygen in platinum-lined

[42] S. N. Hajiev and K. K. Kerimov, *Russ. J. Phys. Chem.*, 1969, **43**, 1513.
[43] H. M. Huffman and E. L. Ellis, *J. Amer. Chem. Soc.*, 1935, **57**, 41.
[44] W. D. Good, J. L. Lacina, and J. P. McCullough, *J. Phys. Chem.*, 1961, **65**, 2229.
[45] M. Månsson and S. Sunner, *Acta Chem. Scand.*, 1963, **17**, 723.
[46] A. T. Hu and G. C. Sinke, *J. Chem. Thermodynamics*, 1969, **1**, 507.
[47] N. K. Smith, D. W. Scott, and J. P. McCullough, *J. Phys. Chem.*, 1964, **68**, 934.

bombs, but there has been only one report of catalytic oxidation having occurred.[48] The absence of any serious systematic error in the combustion calorimetry of chlorine and bromine compounds has recently been demonstrated by Laynez, Ringnér, and Sunner, who measured the energy of combustion of tris(hydroxymethyl)aminomethane (TRIS), its hydrochloride, and its hydrobromide, and compared the results thus calculated for the enthalpies of the reactions

$$\text{TRIS(c)} + \text{HX(g)} = \text{TRIS.HX(c)}, \quad (\text{X} = \text{Cl, Br}),$$

with the values obtained by reaction calorimetry: no significant differences were found.[49]

The combustion of organic fluorine compounds presents different problems. Where the atomic ratio $n(\text{F}) / n(\text{H})$ in the compound is less than unity, all the fluorine appears in the products as hydrofluoric acid, so that by adding sufficient water to the bomb a moderately dilute homogeneous solution of the acid can be obtained by moving-bomb calorimetry. The more highly-fluorinated compounds are less easy to burn completely and it is frequently necessary to add an auxiliary substance, such as a hydrocarbon oil, and to raise the oxygen pressure, say to 40 bar.[50] The products contain carbon tetrafluoride, but mass spectrographic analysis has demonstrated the absence of any other fluorine-containing species in significant quantity. Mass spectrographic analysis is unsuitable for quantitative analysis of carbon tetrafluoride, but it is possible to obtain the amount formed in the combustion by gravimetric determination of the carbon dioxide produced and attributing the deficiency of carbon in the products compared with carbon in the sample to formation of carbon tetrafluoride.[50] Some workers have preferred to determine the hydrofluoric acid in the products and calculate the amount of carbon tetrafluoride from the fluorine balance rather than from the carbon balance.[51] The energies of combustion of a number of gaseous organic fluorine compounds have been measured using a moving-bomb calorimeter. Here it is necessary to obtain by product analysis the mass of sample burnt but sometimes, even when the absence of carbon tetrafluoride has been demonstrated by mass spectroscopy, such determinations based on carbon dioxide and hydrofluoric acid analyses are discrepant.[52] It seems probable that carbon dioxide analysis is the more reliable technique and that the carbon balance method for the determination of carbon tetrafluoride is to be preferred. One useful consequence of the addition of hydrogen-containing material to promote the combustion of highly-fluorinated compounds is that formation of carbon tetrafluoride is reduced and any uncertainty in its determination becomes less important.

48 W. H. Johnson, Reported at the 25th Calorimetry Conference, Washington, D.C., 1970.
49 J. Laynez, B. Ringnér, and S. Sunner, *J. Chem. Thermodynamics*, 1970, **2**, 603.
50 J. D. Cox, H. A. Gundry, and A. J. Head, *Trans. Faraday Soc.*, 1964, **60**, 653.
51 W. D. Good, D. W. Scott, and G. Waddington, *J. Phys. Chem.*, 1956, **60**, 1080.
52 V. P. Kolesov, S. N. Shtekher, A. M. Martynov, and S. M. Skuratov, *Russ. J. Phys. Chem.*, 1968, **42**, 975.

Good and Månsson have overcome the difficulty of obtaining a well-defined final state from the combustion of organoboron compounds by mixing the compound with a fluorine-containing combustion promoter, and introducing hydrofluoric acid into the bomb.[53] The liquid in the bomb after combustion was an aqueous solution containing hydrofluoric and fluoroboric (HBF_4) acids and when the bomb was opened as soon as possible after ignition there was no trace of solid oxidation products, even from the combustion of elemental boron. A vinylidene fluoride polymer was used as the auxiliary substance.

A similar method has been used for organo-silicon compounds (and elemental silicon) by Good and co-workers.[54] These compounds are notoriously difficult to burn completely because of the protective coating of silica which is formed, but the admixture of a fluorine-containing promoter leads to complete combustion. If aqueous hydrofluoric acid is added to the bomb initially, a final solution containing fluorosilicic acid (H_2SiF_6) in excess hydrofluoric acid is obtained by the moving-bomb method. Both vinylidene fluoride polymer and benzotrifluoride have been used as the auxiliary substance. A rotating-bomb method for organic compounds containing silicon and chlorine has been described by Agarunov and Hajiev.[55] The compound is contained in a glass ampoule which, together with a glass-covered heater and platinum ignition spiral, is enclosed in a glass sphere. The heater causes the ampoule to break and the compound to volatilize and mix with oxygen. The gaseous mixture is ignited and reacts explosively, shattering the sphere and another ampoule containing hydrazine hydrochloride for the reduction of chlorine to hydrogen chloride. The success of the method depends on judicious control of the explosive force. Despite the ingenuity of the technique for obtaining complete combustion the method suffers from the disadvantage of producing a two-phase final state containing hydrated amorphous silica.

Relatively few organometallic compounds have been burnt in moving-bomb calorimeters although many are particularly amenable to this technique. Combustion under static-bomb conditions frequently gives rise to oxides or mixtures of oxides, possibly partially hydrated, difficult to characterize, and contaminated with unburnt material or products of side-reactions so that the completeness of combustion is difficult to assess. Germanium forms allotropic oxides which differ in enthalpy of formation by more than 14 kJ mol^{-1}, hence its organic compounds are best investigated by moving-bomb calorimetry, although static-bomb techniques have been used.[56] The amphoteric nature of the oxides makes possible two techniques. The oxidation products of germanium tetraethyl, with the exception of some unburnt carbon, have been dissolved in hydrofluoric acid

[53] W. D. Good and M. Månsson, *J. Phys. Chem.*, 1966, **70**, 97.
[54] W. D. Good, J. L. Lacina, B. L. De Prater, and J. P. McCullough, *J. Phys. Chem.*, 1964, **68**, 579.
[55] M. J. Agarunov and S. N. Hajiev, *J. Organometallic Chem.*, 1968, **11**, 415.
[56] A. E. Pope and H. A. Skinner, *Trans. Faraday Soc.*, 1964, **60**, 1404.

to yield fluorogermanic acid,[57] whereas the oxide obtained from the combustion of germanium tetraphenyl was dissolved in aqueous potassium hydroxide.[40] Nitric acid has been used to dissolve oxides to give aqueous solutions of metal nitrates when lead tetramethyl [58] and manganese decacarbonyl [59] were burnt. The former compound was mixed with hydrocarbon oil to moderate the reaction and prevent explosive combustion, and a small amount of arsenious acid was added to the liquid in the bomb to ensure that all the lead was present in the divalent state. Although complete combustion of manganese decacarbonyl could be obtained by adding a hydrocarbon promoter and working with an oxygen pressure of 30 bar, the product fused to an insoluble refractory mass. It proved better to reduce the oxygen pressure to 5 bar, which gave a product readily soluble in nitric acid (containing a small amount of added hydrogen peroxide to reduce any manganese in higher oxidation states to Mn^{2+}) with the exception of a carbonaceous residue, whose energy of combustion could be determined in separate experiments. The techniques described for these two compounds show how hydrocarbon oil can be used either as a promoter or as a moderator of combustion reactions in suitable cases. A further example of the application of moving-bomb calorimetry to organometallic compounds is provided by the combustion of triphenylarsine, where sodium hydroxide solution was used in the bomb to give a final liquid state containing sodium arsenite, sodium arsenate, sodium carbonate, and unchanged sodium hydroxide.[60]

Moving-bomb calorimetry has not yet been applied to measurements on organophosphorus compounds, although the enthalpy of formation of aqueous orthophosphoric acid has recently been determined by this method.[61] White phosphorus was burnt on a gold dish, at an oxygen pressure restricted to 10 bar to prevent the dish from being melted, and the oxidation products hydrolysed to orthophosphoric acid by the action of 60 per cent by mass perchloric acid at 50 °C. This temperature was chosen because careful paper-chromatographic analysis of the products obtained after hydrolysis under milder conditions revealed the presence of other phosphorus-containing species unless the hydrolysis was continued for a period excessive for satisfactory isoperibol calorimetry.

One of the chief difficulties associated with moving-bomb methods is that many of the physical properties necessary for rigorous evaluation of the Washburn correction to standard states are unknown. The relatively large volume of solution used in the bomb (10 to 60 cm^3) means that the energy of solution of carbon dioxide is appreciable, thus in the combustion of organofluorine compounds in the presence of 10 cm^3 of water it is typically

[57] J. L. Bills and F. A. Cotton, *J. Phys. Chem.*, 1964, **68**, 806.
[58] W. D. Good, D. W. Scott, J. L. Lacina, and J. P. McCullough, *J. Phys. Chem.*, 1959, **63**, 1139.
[59] W. D. Good, D. M. Fairbrother, and G. Waddington, *J. Phys. Chem.*, 1958, **62**, 853.
[60] C. T. Mortimer and P. W. Sellers, *J. Chem. Soc.*, 1964, 1965.
[61] A. J. Head and G. B. Lewis, *J. Chem. Thermodynamics*, 1970, **2**, 701.

25 J, or *ca.* 0.2 per cent of the energy due to the combustion of a highly-fluorinated substance. Cox and Head determined the solubility of carbon dioxide in aqueous hydrofluoric acid in the concentration range encountered during the combustion of organic fluorine compounds and calculated the energy of solution from the temperature variation;[62] and Robb and Zimmer have carried out similar measurements for solutions containing As_2O_3, As_2O_5, and HCl.[63] However, for a large number of solutions obtained in moving-bomb combustion calorimetry such data are not available. Fortunately, the necessity for this and other information required in the calculation of the Washburn correction can be avoided by making 'comparison experiments'. In these experiments, instead of carrying out the determination of the energy equivalent of the calorimeter in the conventional way, a mixture of benzoic acid with another substance for which the energy of combustion has been accurately measured is burnt in the presence of a suitable liquid reagent, so that both the temperature rise and the composition of the final state are the same as in combustion experiments with the compound under investigation. The other substance should differ from benzoic acid in the energy of combustion for a given carbon content so that a composition of the mixture can be devised for which both the energy and the carbon dioxide produced in comparison experiments are the same as when the substance under study is burnt in combustion experiments. By using the 'effective energy equivalent' obtained in comparison experiments to calculate the energy liberated in combustion experiments, any errors in the calculation of the Washburn correction for the latter experiments are largely eliminated. Although the composition of the final state in moving-bomb combustion experiments is accurately determined, its enthalpy of formation is not always known. In these cases it is necessary to measure the energy of mixing of substances having known enthalpies of formation to yield a final solution having the same composition as that obtained in combustion experiments, *e.g.* to measure the energy of solution of orthoboric acid in hydrofluoric acid or silica (quartz) in hydrofluoric acid. Alternatively, this determination may be combined with the measurement of other auxiliary thermal quantities: thus, for example, benzoic acid may be burnt in an experiment in which also crystalline (hexagonal) germanium dioxide is allowed to dissolve in aqueous alkali placed in the bomb.[40] The combined correction for the energy of solution in alkali of germanium dioxide and of the carbon dioxide produced in the combustion of germanium tetraphenyl may therefore be obtained from one experiment, and the energy of the reaction:

$$Ge(C_6H_5)_4 + 30O_2 = GeO_2 + 24CO_2 + 10H_2O,$$

readily calculated from the results of combustion experiments.

[62] J. D. Cox and A. J. Head, *Trans. Faraday Soc.*, 1962, **58**, 1839.
[63] R. A. Robb and M. F. Zimmer, *J. Chem. and Eng. Data*, 1968, **13**, 200.

Special Techniques in Combustion-bomb Calorimetry. In the previous section some of the ingenious methods devised for the successful combustion of compounds of different types have been indicated. Writing about organometallic compounds, Good and Scott said, 'Control of the combustion reaction is more of an art than a science. The more tricks the thermochemist has up his sleeve, and the more he has the ingenuity to devise, the better he is equipped to deal with a fickle and perverse Nature'.[64] This statement is equally true of the method used to contain the samples for combustion. It is the exception to work with a material which can be tableted, weighed, and burnt in a platinum crucible without any complications. If a substance is at all volatile, hygroscopic, or reactive with oxygen at ordinary temperatures it must be enclosed in a container which is either inert, yet opens to allow the substance to burn to completion, or itself burns completely. Thin-walled, soft-glass, roughly spherical ampoules are the traditional containers for liquids. They must be completely filled and flattened on opposite sides to withstand the oxygen pressure in a combustion bomb without premature fracture.[65] Alternatively, stronger, rigid glass ampoules can be used for substances with boiling temperatures below room temperature.[44] When the combustion products attack soft glass, as with sulphur compounds and, particularly, fluorine compounds, borosilicate glass or quartz ampoules can be used,[51] but it is preferable to use a plastic container. Flat bags[66] or satchels[67] can be made from polyester film ('Mylar', 'Melinex', *etc.*), which is impermeable to many organic liquids and conveniently heat-sealable, but suffers from the disadvantage that it is permeable to water-vapour and its mass depends on the ambient relative humidity. Polyethylene does not have these disadvantageous properties but it is permeable to many non-polar liquids; it has been used to make bags,[68] capsules,[69] and comparatively thick-walled cylindrical ampoules.[70] The last have been useful where auxiliary hydrocarbon material was required in the combustion, and for protecting a material which reacts spontaneously with oxygen, such as white phosphorus.[61] Considerable ingenuity has been demonstrated in the use of plastic bags. Two solids have been separately weighed into a polyester bag which has then been sealed with a 'bubble' of air intentionally included to facilitate the subsequent thorough mixing of the two solids. The air was then expelled through a pin-prick and the bag rolled up and compacted in a pellet press without loss of mass.[53] The same

[64] W. D. Good and D. W. Scott, 'Combustions in a bomb of organometallic compounds', Chapter 4 in ref. 8, p. 57.

[65] G. B. Guthrie, D. W. Scott, W. N. Hubbard, C. Katz, J. P. McCullough, M. E. Gross, K. D. Williamson, and G. Waddington, *J. Amer. Chem. Soc.*, 1952, **74**, 4662.

[66] W. D. Good, D. R. Douslin, D. W. Scott, A. George, J. L. Lacina, J. P. Dawson, and G. Waddington, *J. Phys. Chem.*, 1959, **63**, 1133.

[67] M. E. Butwill and J. D. Rockenfeller, *Thermochim. Acta*, 1970, **1**, 289.

[68] R. J. L. Andon, D. P. Biddiscombe, J. D. Cox, R. Handley, D. Harrop, E. F. G. Herington, and J. F. Martin, *J. Chem. Soc.*, 1960, 5246.

[69] H. Mackle and R. G. Mayrick, *J. Sci. Instr.*, 1961, **38**, 218.

[70] J. D. Cox, H. A. Gundry, D. Harrop, and A. J. Head, *J. Chem. Thermodynamics*, 1969, **1**, 77.

technique has been applied to PTFE bags and the resulting pellet protected from the action of the gas in the bomb by surrounding it with powdered PTFE and again compacting it in a pellet press.[71] Weighed quantities of two liquids have been mixed by sealing one into a polyester bag which also contained a smaller sealed bag filled with the second component. The inner bag also contained a small platinum 'knife' which could be manipulated to open the inner bag after the outer one was sealed.[54] A novel technique for handling volatile liquids has been described by Mosselman and Dekker.[72] The sample is weighed into a platinum crucible which is closed by a hinged lid, sealed with vaseline, but into which oxygen can enter *via* a capillary. When ignition takes place the vaseline melts and a counter-weight causes the lid to open and the sample and vaseline are able to burn.

The problem of preventing premature oxidation of samples by the gas in the bomb where these react spontaneously has been solved in other ways than by using special sample containers. Workers at the Argonne National Laboratory arranged for a gas-reservoir to surround the bomb. When it was required to ignite the sample the valve of the reservoir was opened by operation of the bomb rotation mechanism and gas entered the bomb.[73] This system has been improved by the construction of a two-compartment bomb which permits higher gas pressures to be used.[74] Although this arrangement has been employed for oxygen combustion calorimetry, *e.g.* for plutonium nitride,[28] it finds its most widespread application to combustion in fluorine (see below) in which most substances ignite spontaneously, a property which limits the available construction materials. Recent improvements in the design of the valve between the two compartments include the elimination of the PTFE sealing button which occasionally ignited in fluorine.[75] Another design of a two-compartment bomb for combustion in fluorine makes use of a titanium foil cover to seal the sample chamber while fluorine is admitted to the other compartment.[76]

The choice of the best conditions for a combustion can sometimes be made easier by preliminary observation of the reaction in a bomb fitted with a window. This is particularly true of the combustion of rigid inorganic materials (in oxygen or fluorine) where there exists considerable scope for variations in the size and shape of the sample and its method of support. Nuttall, Frisch, and Hubbard have described an inexpensive combustion vessel of similar dimensions to a normal bomb but made from thick-walled Pyrex pipe.[77] A more elaborate window-bomb, in which the combustion of metals, carbides, and borides has been observed, contains a

[71] E. S. Domalski and G. T. Armstrong, *J. Res. Nat. Bur. Stand.*, 1967, **71A**, 105.
[72] C. Mosselman and H. Dekker, *Rec. Trav. chim.*, 1969, **88**, 257.
[73] R. L. Nuttall, S. Wise, and W. N. Hubbard, *Rev. Sci. Instr.*, 1961, **32**, 1402.
[74] J. L. Settle, E. Greenberg, and W. N. Hubbard, *Rev. Sci. Instr.*, 1967, **38**, 1805.
[75] T. L. Denst, E. Greenberg, J. L. Settle, and W. N. Hubbard, *Rev. Sci. Instr.*, 1970, **41**, 588.
[76] B. D. Kybett and J. L. Margrave, *Rev. Sci. Instr.*, 1966, **37**, 675.
[77] R. L. Nuttall, M. A. Frisch, and W. N. Hubbard, *Rev. Sci. Instr.*, 1960, **31**, 461.

lamp for illumination of the assembly before combustion and is suitable for high-speed cinephotography.[78] The influences of the mass of sample, size and shape of crucible, and oxygen pressure on the duration of burning, flame temperature, and soot formation in the combustion of benzoic acid have been studied using a window-bomb by Peters and co-workers.[79]

Flame Calorimetry. The enthalpies of combustion of substances which are gaseous or sufficiently volatile at ambient temperature may be measured by means of a flame calorimeter. The gas to be burnt and a supply of oxygen are fed at constant flow rates into a reaction chamber where combustion takes place at a jet. A spark between two platinum electrodes is continued for several seconds and serves to ignite the mixture which burns with a quiet flame. The apparatus, constructed from borosilicate glass except for the silica jet, is placed in a calorimeter containing water, the remainder of the equipment being the same as for a static-bomb calorimeter of the stirred-water, isoperibol type. Most of the water formed during the combustion collects in a vessel below the reaction chamber, from which an exit tube leads to a heat-exchanger to ensure that the gases have reached temperature equilibrium with the calorimeter before they pass to an analysis train. The ignition energy is determined in combustion experiments which are allowed to proceed only until even burning is established; this results in enthalpy of combustion measurements being based only on that portion of the experiment when the flame is in a steady state. The apparatus and method are basically the same as were used by Rossini to determine the enthalpy of formation of liquid water by the combustion of hydrogen in excess oxygen and of oxygen in excess hydrogen, using an electrically calibrated calorimeter.[80] This reaction is recommended as the primary standard for the calibration of flame calorimeters [10] and Rossini's value obtained in 1931 is still employed.

When hydrocarbons or other organic compounds are burnt in a flame calorimeter it is often necessary to pre-mix the gas with oxygen to increase flame stability and obtain complete reaction. However, if the optimum proportion of oxygen is exceeded the flame temperature becomes too high and thermal decomposition with deposition of carbon occurs below the jet. Since the combustion takes place at constant pressure close to 1 bar, the enthalpy of combustion is measured under conditions near to those of the standard states. Flame calorimetry has the advantage that enthalpies of combustion are obtained for the gaseous state without the necessity of measuring enthalpies of vaporization in separate experiments and, moreover, completeness of combustion can be established by determination of both the water and the carbon dioxide produced.

In spite of these advantages, flame-calorimetry has largely been used only for C_2 to C_5 hydrocarbons and before 1963 the only organic oxygen

[78] D. Pavone and C. E. Holley, *Rev. Sci. Instr.*, 1965, **36**, 102.
[79] H. Peters, E. Tappe, and M. Urbanczik, *Monatsberichte*, 1966, **8**, 720.
[80] F. D. Rossini, *J. Res. Nat. Bur. Stand.*, 1931, **6**, 1.

compounds to have been studied were methanol, ethanol, ethylene oxide, propan-2-one, and butan-2-one. However, more recently, Pilcher and co-workers [81] have used flame calorimetry to obtain definitive enthalpies of combustion for a number of oxygen-containing compounds, particularly low-boiling ethers, substances which because of their high volatility are not easy to handle by static-bomb calorimetry. Previously, only one attempt had been made to burn a halogen compound in a flame calorimeter in a study for which rather low accuracy ($\pm$ 8 kJ mol^{-1}) was sought,[82] but Pilcher has recently extended his high-precision work to this class of compound.[83]

The application of flame calorimetry is limited to moderately volatile substances. Pell and Pilcher [84] have used it for a substance boiling at 81 °C (tetrahydropyran) and by raising both the temperature at which the carrier gas (argon) is saturated with organic vapour and the operating temperature of the calorimeter above 25 °C it should be possible to increase the scope of the method.

Combustion in Fluorine.—In 1961 Hubbard and his co-workers at the Argonne National Laboratory published their first paper on the determination of energies of combustion in fluorine.[85] Since that time the technique has begun to offer to inorganic thermochemists the advantages that organic thermochemists have long derived from oxygen combustion calorimetry. At the outset there existed few accurate values for the enthalpies of formation of inorganic fluorides and the early work was confined to measurement of the energies of combustion of many elements in fluorine. More recently, it has been possible to extend fluorine combustion calorimetry to the determination of enthalpies of formation of a wide range of inorganic refractory materials including oxides, nitrides, borides, carbides, phosphides, and sulphides, which were difficult, if not impossible, to handle by other thermochemical methods, now that the enthalpies of formation of the products of the combustion of such materials are known. Of the types of compound named, with the exception of oxides and nitrides which yield oxygen and nitrogen respectively, the fluorination products are BF_3, CF_4, PF_5, or SF_6, together with the metal fluoride (in the highest valence state).

The high reactivity of fluorine which is responsible for its usefulness in inorganic thermochemistry is also the cause of the principal difficulties associated with the technique: problems in selecting suitable materials for the construction of apparatus, the need for elaborate safety precautions, and difficulties in the analysis of volatile products in the presence of excess fluorine. Bombs of the conventional Berthelot type are used but are

81 G. Pilcher, H. A. Skinner, A. S. Pell, and A. E. Pope, *Trans. Faraday Soc.*, 1963, **59**, 316.

82 D. W. H. Casey and S. Fordham, *J. Chem. Soc.*, 1951, 2513.

83 R. A. Fletcher and G. Pilcher, *Trans. Faraday Soc.*, 1971, **67**, 3191.

84 A. S. Pell and G. Pilcher, *Trans. Faraday Soc.*, 1965, **61**, 71.

85 E. Greenberg, J. L. Settle, H. M. Feder, and W. N. Hubbard, *J. Phys. Chem.*, 1961, **65**, 1168.

constructed from nickel or Monel, which are fairly resistant to attack by dry fluorine by virtue of a protective fluoride film. However, it is necessary to pre-fluorinate the interior surface of the bomb by reaction at 200 °C and also to allow the coated walls to be exposed to the volatile products of a particular reaction. Sealing gaskets may be constructed from aluminium, lead, gold, or PTFE; and valves are packed with PTFE. The fluorine should be fractionated at liquid nitrogen temperature to give a product of better than 99.9 per cent purity. Because of the considerable danger that would arise from an escape of fluorine, all operations involving the gas, except for the actual calorimetric determination, must be conducted in a well-ventilated fume cupboard and behind an efficient safety screen, through which pass the extended controls of the gas valves. It is evident, therefore, that expensive preparations are necessary before fluorine-combustion calorimetry can be undertaken and it is understandable why few laboratories are equipped to use this very valuable technique.

Nevertheless, once set up, the procedure for measurements is fairly straightforward; the rest of the apparatus is the same as for conventional oxygen static-bomb calorimetry and the calorimeter is calibrated by the combustion in oxygen of benzoic acid. It is in the method used to support the sample in the bomb that the greatest variations in technique occur. Some materials, *e.g.* zirconium, are burnt in the form of a vertical rod clamped at the bottom in a massive nickel holder.[85] The rod is ignited at the top and burns downwards, but eventually the flame is extinguished as heat is conducted away from the base. The tendency for droplets of molten metal to fall to the bottom of the bomb can be controlled by diluting the fluorine with helium. In other cases it is preferable to suspend a sheet of the metal by a wire of the same material from a thick nickel rod and ignite the sample from the bottom: this technique resulted in 99.9 per cent combustion of a molybdenum sample.[86] Sometimes the walls of the bomb are protected from sputtering metal, *e.g.* magnesium,[87] aluminium,[88] or nickel,[89] by lining them with the combustion product: powdered metal fluoride is tamped down into a nickel cup fitting the interior of the bomb. The combustion of nickel was sustained and the amount of metal falling on to the bomb liner minimized by use of a complicated arrangement of the sample in which strips of nickel were contained in a suspended basket made of perforated nickel foil, and more nickel foil was stacked under the basket.[89] Domalski and Armstrong at the National Bureau of Standards overcame the problem of sputtering in the combustion of aluminium by admixture of the powdered metal with powdered PTFE. A pellet prepared from the mixed solids was burnt essentially to completion on a stainless-

86 J. L. Settle, H. M. Feder, and W. N. Hubbard, *J. Phys. Chem.*, 1961, **65**, 1337.
87 E. Rudzitis, H. M. Feder, and W. N. Hubbard, *J. Phys. Chem.*, 1964, **68**, 2978.
88 E. Rudzitis, H. M. Feder, and W. N. Hubbard, *Inorg. Chem.*, 1967, **6**, 1716.
89 E. Rudzitis, E. H. Van Deventer, and W. N. Hubbard, *J. Chem. and Eng. Data*, 1967, **12**, 133.

steel or Monel plate.[90] The technique of preparing pellets after mixing solids in PTFE bags has been described (p. 110) and was used for the combustion of beryllium[91] and graphite[71] in fluorine. PTFE made ignition easier because of its low thermal conductivity; it acted as a moderator and maintained a temperature conducive to burning. Occasionally the sample is burnt on a thin shell of calcium fluoride supported by a nickel crucible as in the combustion of boron,[92] or is supported on its own combustion product, *e.g.* cadmium on cadmium fluoride.[93] Some substances which react spontaneously with fluorine and for which it is necessary to use a two-compartment bomb (see above) can be burnt in a nickel crucible, *e.g.* boron nitride.[94] The choice of the best method for burning a sample is only made possible by the use of the window-bomb already described (p. 111).

Because of the nature of the products, determination of the amount of reaction is usually made by weighing the sample after combustion. This involves recovery of unburnt material from various parts of the bomb and its determination in the presence of combustion products from which it may have to be mechanically separated, although more convenient physical methods (*e.g.* magnetic separation of nickel from nickel fluoride[89]) and chemical methods (*e.g.* determination of the hydrogen produced by the action of acid on cadmium[93] or aluminium[88]) are sometimes applicable.

The pressures of fluorine in the bomb calorimetry described above have varied from a few bars to pressures comparable with those used in oxygen bomb calorimetry. Gross *et al.* demonstrated the use of a comparatively simple technique for substances which are spontaneously inflammable and burn to completion at pressures of about 5 to 8 bar. They used a simple glass vessel divided into two compartments by a break-seal and containing fluorine in one section and the sample, supported on alumina, in the other. The reaction vessel was contained in a stirred-water isoperibol calorimeter. Their results for the enthalpies of formation of germanium tetrafluoride[95] and phosphorus pentafluoride[96] are in good agreement with those subsequently obtained using a combustion bomb.

Although there have been various early attempts to use fluorine as the oxidant in flame calorimetry, the problem of corrosion restricted these efforts to the use of fluorine as a minor component of the reaction mixture. More recent work at the National Bureau of Standards has overcome many of the difficulties and Armstrong and Jessup have described a calorimeter in which volatile substances can be burnt in excess fluorine.[97] Because the

[90] E. S. Domalski and G. T. Armstrong, *J. Res. Nat. Bur. Stand.*, 1965, **69A**, 137.
[91] K. L. Churney and G. T. Armstrong, *J. Res. Nat. Bur. Stand.*, 1969, **73A**, 281.
[92] S. S. Wise, J. L. Margrave, H. M. Feder, and W. N. Hubbard, *J. Phys. Chem.*, 1961, **65**, 2157.
[93] E. Rudzitis, H. M. Feder, and W. N. Hubbard, *J. Phys. Chem.*, 1963, **67**, 2388.
[94] S. S. Wise, J. L. Margrave, H. M. Feder, and W. N. Hubbard, *J. Phys. Chem.*, 1966, **70**, 7.
[95] P. Gross, C. Hayman, and J. T. Bingham, *Trans. Faraday Soc.*, 1966, **62**, 2388.
[96] P. Gross, C. Hayman, and M. C. Stuart, *Trans. Faraday Soc.*, 1966, **62**, 2716.
[97] G. T. Armstrong and R. S. Jessup, *J. Res. Nat. Bur. Stand.*, 1960, **64A**, 49.

reaction between fluorine and many substances is spontaneous, the components cannot be pre-mixed, and when hydrocarbons are burnt it is necessary to separate the diffusion flame from the copper jet by means of helium to prevent deposition of carbon. The reaction vessel is immersed in a stirred-water isoperibol calorimeter which is calibrated electrically. This calorimeter was used to determine the enthalpy of the reaction:

$$2NH_3(g) + 3F_2(g) = N_2(g) + 6HF(g).$$

The amount of ammonia consumed was obtained from the change in mass of a reservoir and the amount of hydrogen fluoride formed was determined after it had been absorbed by solid sodium fluoride. The accuracy of the result was limited principally by the uncertainty in the correction applied for the non-ideality of hydrogen fluoride. An improved calorimeter has been described in which the hydrogen fluoride formed is absorbed in water contained in a PTFE-lined vessel below the combustion chamber, so that the final product is an aqueous solution of hydrofluoric acid.[98] This calorimeter has been used to study the reactions:

$$H_2(g) + F_2(g) + 100H_2O = 2(HF.50H_2O),$$
$$OF_2(g) + 2H_2(g) + 99H_2O = 2(HF.50H_2O).$$

Combustion in Other Gases.—Although the great majority of combustion calorimetry experiments have been carried out using oxygen or fluorine as the oxidant, a few reactions involving other gases have been studied. A simple two-compartment glass reaction vessel, similar to that described above for the fluorination of germanium and phosphorus, has been used for chlorination. The pressure of chlorine is, of course, limited to its saturation vapour pressure (*ca.* 7 bar at 25 °C) but the energy of combustion of titanium, which ignites spontaneously under these conditions, has been measured[99] and this element has been used to initiate the combustion of zirconium, vanadium, hafnium, and, recently, silicon.[100] Combustion experiments of this type are accompanied by blank experiments in which the enthalpy due to the evaporation of chlorine into the evacuated sample chamber and the interaction (dissolution or wetting) between liquid chlorine and the reaction product is measured. The same technique has been used to determine the energy of combustion of niobium and tantalum in bromine, except that, because of the slow rate of reaction at the saturation vapour pressure of bromine at 25 °C, it was necessary to raise the reaction vessel temperature to 110 °C and use paraffin as the stirred liquid in the calorimeter.[101]

Ludwig and Cooper[102] showed that boron burned to 95 per cent completion in excess nitrogen trifluoride (2 to 9 bar pressure), according to the

[98] R. C. King and G. T. Armstrong, *J. Res. Nat. Bur. Stand.*, 1968, **72A**, 113.
[99] P. Gross, C. Hayman, and D. L. Levi, *Trans. Faraday Soc.*, 1955, **51**, 626.
[100] P. Gross, C. Hayman, and S. Mwroka, *Trans. Faraday Soc.*, 1969, **65**, 2856.
[101] P. Gross, C. Hayman, D. L. Levi, and G. L. Wilson, *Trans. Faraday Soc.*, 1962, **58**, 890.
[102] J. R. Ludwig and W. J. Cooper, *J. Chem. and Eng. Data*, 1963, **8**, 76.

equation:

$$2B(c) + 2NF_3(g) = 2BF_3(g) + N_2(g),$$

and derived a value for $\Delta H_f^\circ(NF_3, g)$ by carrying out this reaction in a static-bomb calorimeter using a Monel bomb. The use of nitrogen trifluoride as an oxidant in combustion calorimetry has been further studied by Sinke and Walker, who showed that it could conveniently replace fluorine in some reactions, but with the advantage that it is relatively inert, so that it did not need to be separated from the sample prior to combustion and required no special handling techniques. It is necessary, however, to determine fluorine in the combustion products arising from the decomposition of excess nitrogen trifluoride. The reaction between oxidizable gases and nitrogen trifluoride is explosive when initiated by the discharge of a capacitor through a platinum fuse wire; under these conditions the excess nitrogen trifluoride is completely decomposed. Sinke determined the energy of the reaction of nitrogen trifluoride with hydrogen using a 10 per cent and a 100 per cent excess of oxidant to obtain the energy of decomposition and hence the enthalpy of formation of nitrogen trifluoride.[103] The result was in good agreement with that calculated from the energy of combustion of sulphur in nitrogen trifluoride (5 bar pressure), initiated by combustion of molybdenum, when only a small fraction of the excess oxidant decomposed.[104] Explosion with nitrogen trifluoride has been applied to hexafluoroethane, and recently to acetonitrile to yield an enthalpy of formation in good agreement with the value obtained by an equilibrium method.[105]

Combustion in hydrogen has been used to determine the enthalpy of formation of the chlorine fluorides. Because the reaction is spontaneous, chlorine pentafluoride was contained in a Monel ampoule and released by means of the rupture of a disc through a mechanism triggered by the fusion of a nichrome wire. Large energy corrections were necessary because of the non-ideality of the hydrogen fluoride produced. Fair agreement was obtained with a value derived from the energy of combustion of chlorine pentafluoride in ammonia:[106]

$$ClF_5(l) + 8NH_3(g) = NH_4Cl(s) + 5NH_4F(s) + N_2(g).$$

Barberi *et al.* have measured the energy of combustion of ClF_5, ClF_3, and ClF in hydrogen using a bomb ignited by a high-frequency spark at 25 °C and also using a two-compartment bomb operated at 130 °C.[107] The results

[103] G. C. Sinke, *J. Phys. Chem.*, 1967, **71**, 359.
[104] L. C. Walker, *J. Phys. Chem.*, 1967, **71**, 361.
[105] L. C. Walker, G. C. Sinke, D. J. Perettie, and G. J. Janz, *J. Amer. Chem. Soc.*, 1970, **92**, 4525.
[106] W. R. Bisbee, J. V. Hamilton, J. M. Gerhauser, and R. Bushworth, *J. Chem. and Eng. Data*, 1968, **13**, 382.
[107] P. Barberi, J. Caton, J. Guillin, and O. Hartmanshenn, C. E. A. Report (France), 1969, CEA-R-3761.

obtained from combustions of the chlorine fluorides were considerably less precise than those normally achieved in combustion calorimetry.

Finally, an unusual reaction to be studied by combustion calorimetry was the following reaction between nitric oxide and excess carbon monoxide:

$$2NO(g) + 2CO(g) = 2CO_2(g) + N_2(g),$$

which proceeded cleanly and gave reproducible results.[108] Apart from the reactants the combustion was usual in that it had to be initiated with an iridium fuse; the melting temperature of platinum was too low to produce the threshold concentration of dissociated molecules needed for rapid combustion and the reaction proceeded only slowly if platinum was used.

4 Hot-zone Calorimetry and Explosion Calorimetry

Before the techniques available for studying reactions other than combustion are described it is convenient to refer to the calorimetry of reactions which are explosive or are investigated using a hot-zone calorimeter, since the majority of these processes are studied using modified bombs rather than the more usual forms of reaction calorimeters.

The term 'hot-zone calorimeter' is used to describe an instrument in which, in order to bring about a reaction which proceeds at a suitable rate only at elevated temperatures, the reaction zone is heated locally in preference to raising the operating temperature of the whole calorimeter. These two approaches are illustrated by the methods which have been used to determine the energy of combustion of relatively unreactive metals in chlorine. Niobium has been chlorinated in an adiabatic calorimeter in which the reaction vessel, calorimeter, and jacket were preheated to 680 K before chlorine was passed over the metal to initiate reaction.[109] Using the other technique the energies of combustion in chlorine of zirconium, hafnium, and tantalum have been measured in a conventional bomb calorimeter in which the quartz dish containing the metal could be heated to 450 °C in 1.5 min, but the operating temperature of the calorimeter remained at 25 °C.[110] The first method has the advantage that all the measured energy is due to the reaction under investigation, but difficulties arise because thermostats have to be operated at elevated temperatures and accurate heat capacities are necessary for correction of the result to 25 °C. The advantages of operation at 25 °C in the second method tend to outweigh the loss of accuracy to be expected from the large contribution to the measured temperature change made by the energy used to heat the reaction zone, which may be considerably in excess of that liberated by the reaction yet measured with reasonable accuracy. In the determination of

108 M. A. Frisch and J. L. Margrave, *J. Phys. Chem.*, 1965, **69**, 3863.
109 L. A. Reznitskii, *Russ. J. Phys. Chem.*, 1967, **41**, 787.
110 G. L. Gal'chenko, D. A. Gedakyan, and B. I. Timofeev, *Russ. J. Inorg. Chem.*, 1968, **13**, 159.

the energy of decomposition of diborane and pentaborane[111] the electrical energy used to heat the quartz reaction vessel to 600 °C was about 160 times the energy of decomposition, which was determined with a standard error of 1 kJ mol^{-1} (3 per cent).

Some workers have used the energy of a normal calorimetric combustion reaction to heat the reaction zone. Thus the energy of decomposition of sodium and potassium chlorates (to chlorides and oxygen) was determined by using a combustion bomb in which the crucible containing the chlorate was heated by a charge of burning benzoic acid placed under it.[112] More recently, the combustion of hydrogen in oxygen has been used to heat the silica reaction vessel of a modified flame calorimeter, in which the reactions of copper and of cuprous oxide with oxygen and the reductions of the oxides of copper and lead by hydrogen have been investigated.[113,114] In the latter apparatus less than 10 per cent of the total energy measured was derived from the reaction under study. Other investigators have modified combustion calorimeters by incorporating electrically-heated reaction vessels into the bomb, as already described for combustion in chlorine. Hajiev has measured the energy of formation of the selenides of tin and lead by carrying out the synthesis from the elements in an evacuated quartz ampoule heated to 1100 °C by a furnace. The bomb also contained an electrically-driven device for shaking the reaction vessel but could still be used for the combustion of benzoic acid in order to calibrate the calorimeter.[115] A combustion bomb containing an electrically-heated oven has been used to measure the energy of decomposition of aluminium hydride.[116]

An electrically-calibrated calorimeter in which the reaction vessel is still essentially a bomb has been used to determine the energy of reaction between barium and hydrogen. The low heat-capacity quartz furnace could be heated to 300 to 400 °C within a few seconds and was turned off when the reaction started. The electrical energy was limited to about 40 per cent of the total energy and between 80 and 90 per cent of the metal reacted.[117] The enthalpies of other reactions between gases and solids, such as those of beryllium with chlorine and with ammonia (to form the nitride), have conveniently been measured using a hot-zone calorimeter through which the gas was continuously flowed; reaction was initiated by heating the reaction zone for the required period.[118]

111 E. J. Prosen, W. H. Johnson, and F. Y. Pergiel, *J. Res. Nat. Bur. Stand.*, 1958, **61**, 247.
112 A. A. Gilliland and D. D. Wagman, *J. Res. Nat. Bur. Stand.*, 1965, **69A**, 1.
113 L. Nuñez, G. Pilcher, and H. A. Skinner, *J. Chem. Thermodynamics*, 1969, **1**, 31.
114 L. Espada, G. Pilcher, and H. A. Skinner, *J. Chem. Thermodynamics*, 1970, **2**, 647.
115 S. N. Hajiev, *J. Chem. Thermodynamics*, 1970, **2**, 765.
116 G. C. Sinke, L. C. Walker, F. L. Oetting, and D. R. Stull, *J. Chem. Phys.*, 1967, **47**, 2759.
117 A. F. Vorob'ev, A. S. Monaenkova, and S. M. Skuratov, *Proc. Acad. Sci.* (*U.S.S.R.*), (*Chem.*), 1968, **179**, 250.
118 P. Gross, C. Hayman, P. D. Greene, and J. T. Bingham, *Trans. Faraday Soc.*, 1966, **62**, 2719.

Several gases decompose completely but explosively when reaction is initiated by an electrically-heated platinum ignition wire, and the energies of such processes can be measured using a bomb calorimeter. Thus bis-(fluoroxy)perfluoromethane decomposes mainly according to the equation:

$$2(OF)_2CF_2 = 2COF_2 + 2F_2 + O_2,$$

with the formation of small quantities of CF_3OF and CF_3COCF_3 and traces of CF_3OOF. The energy of this reaction has been determined using a nickel bomb pre-treated with fluorine.[119] A similar study has been carried out on various perfluoroamines, *e.g.* $F_2C(NF_2)_2$ and $C(NF_2)_4$, which decompose quantitatively to nitrogen, fluorine, and carbon tetrafluoride.[120]

Gunn and his co-workers have investigated the explosive decomposition of several hydrides, including those of B, Si, Ge, P, As, Sb, and Te. Some hydrides decompose completely to their elements under the influence of a heated platinum wire whereas others such as arsine, which are stable alone, decompose completely in the presence of stibine. Earlier work was carried out using glass reaction cells, into which platinum ignition wires were sealed, in a calorimeter of the isothermal heat-flow type originally designed for radiation measurements.[121] In later work[122] an isoperibol aneroid (copper block) calorimeter was employed into which the glass cells were inserted and which could be operated at temperatures up to 150 °C where necessary (*e.g.* for $B_{10}H_{14}$ + SbH_3 mixtures).

Two more reactions may conveniently be included in this section for, although strictly involving neither hot-zone nor explosion calorimetry, they have features in common with the processes studied using these methods. The reaction

$$3PbF_2 + 2Al = 2AlF_3 + 3Pb,$$

proceeds spontaneously and to completion when a current is passed through a tungsten wire embedded in a briquette made from the two mixed powdered solids. The reaction was carried out in an aluminium crucible in a brass bomb filled with argon and contained in a stirred-water calorimeter.[123] The ignition energy required amounted to only about 0.4 per cent of the total energy change. On the other hand, the electrical energy dissipated during a measurement of the energy of reaction between gaseous perhalogeno-hydrocarbons and sodium metal, *e.g.*

$$CF_2Cl\cdot CF_2Cl(g) + 6Na(s) = 2C(s) + 4NaF(s) + 2NaCl(s),$$

was in the range 10 to 40 per cent of the total measured energy and thus

119 G. D. Foss and D. A. Pitt, *J. Phys. Chem.*, 1968, **72**, 3512.
120 G. C. Sinke, C. J. Thompson, R. E. Jostad, L. C. Walker, A. C. Swanson, and D. R. Stull, *J. Chem. Phys.*, 1967, **47**, 1852.
121 S. R. Gunn, W. L. Jolly, and LeR. G. Green, *J. Phys. Chem.*, 1960, **64**, 1334.
122 S. R. Gunn, *Rev. Sci. Instr.*, 1964, **35**, 183.
123 P. Gross, C. Hayman, and D. L. Levi, *Trans. Faraday Soc.*, 1954, **50**, 477.

comparable with the heating energy in some hot-zone calorimetry.[124] Excess sodium was contained in a tantalum dish in a combustion bomb containing the gaseous reactant, and an arc struck between a tungsten electrode and the dish. The reaction usually proceeded to at least 99.9 per cent completion as estimated by the residual pressure in the bomb. After allowance for the quantity of sodium halides occluded in the carbon produced (and only recoverable after combustion of this carbon) the results based on analysis of fluoride ion or chloride ion, or on the mass of sample taken, showed no systematic differences and a precision (standard deviation of the mean) of about 0.1 per cent. In order to obtain reliable enthalpies of formation from the energy measurements, the energy of combustion of each sample of amorphous carbon produced was determined in separate experiments.

5 Reaction Calorimetry

It is customary to exclude combustion reactions from those covered by the term 'reaction calorimetry', but this still leaves an enormous field of study. In the discussion of the techniques available in combustion calorimetry it was appropriate to consider them according to the chemical nature of the compound and the combustion process. In the much wider field of reaction calorimetry very similar apparatus and techniques may be employed for the study of quite different reactions and it is therefore more helpful to classify methods according to the physical state of the reactants, the magnitude and sign of the enthalpy change, and the rate of reaction, but, inevitably, the distinctions are not always clearly defined. Although some general-purpose reaction calorimeters are commercially available, most have been built for a specific investigation and rarely are any two exactly alike. From the considerable amount of published work it is possible to draw only relatively few examples to illustrate the main types of instrument available to the experimental thermochemist.

Solid + Liquid and Liquid + Liquid Reactions.—*Macrocalorimeters for Rapid Reactions.* In this section are described techniques which are available for measuring the energy or enthalpy of reactions which proceed sufficiently rapidly for isoperibol calorimeters to be used, *i.e.* the reaction is complete in about 15 min, and for which quantities of reactant between 100 mg and several grams are available for each determination. Such reactions constitute the most widely studied group.

In its essentials, the calorimeter consists of a reaction vessel of about 50 to 200 cm^3 capacity which contains one liquid reactant and is equipped with a thermometer, a stirrer, and the means of introducing the second (solid or liquid) reactant with minimum disturbance of the system. Commonly the second substance is contained in a frangible ampoule which is attached to the stirrer and broken by being depressed against a spike. Sometimes the

[124] V. P. Kolesov, O. G. Talakin, and S. M. Skuratov, *Russ. J. Phys. Chem.*, 1968, **42**, 1617.

reaction vessel is immersed in a stirred-water calorimeter,[125] but more commonly it also serves as the calorimeter. In this case, the reaction vessel contains an electrical heater and is surrounded by an isothermal jacket, often a submarine vessel immersed in a water thermostat. Much useful calorimetry has been carried out using Dewar flasks as calorimeters (*e.g.* Vanderzee has reported results of good precision [126]) but in general they are slow to reach temperature equilibrium. Sunner and Wadsö have investigated the efficiency of various designs of reaction calorimeter, including Dewar vessels, and describe one with an equilibration time of less than 2 min which yields results having accuracy of ± 0.1 per cent.[127] Accuracy of this order or better is obtainable from the best design of reaction calorimeter, but it must always be borne in mind that considerably less reliable results may arise from the chemical reaction being ill-defined or not proceeding to completion. Because enthalpies of reaction are in general much less than enthalpies of combustion, since fewer chemical bonds are broken, they need not be known as accurately to yield enthalpies of formation of accuracy comparable with those obtained by combustion calorimetry.

Because the overall accuracy of reaction calorimeters is seldom better than ± 0.1 per cent they may be calibrated electrically without the necessity of the high accuracy of measurement demanded by combustion calorimetry. Electrical calibration is widely used for reaction calorimeters and there is no standard chemical reaction internationally recognized for calibration purposes. However, it is important that the performance of reaction calorimeters be checked by carrying out test reactions. The dissolution of potassium chloride in water has been used for this purpose but Gunn has demonstrated that variations of up to 0.2 per cent in the result can occur according to the pre-treatment of the sample.[128] Instead of this reaction, which unlike the majority of processes studied is endothermic, it is now more usual to employ the exothermic neutralization of tris(hydroxymethyl)-aminomethane (usually referred to as TRIS or THAM), first suggested by Wadsö and Irving.[129] However, recent work has indicated that the influence of experimental conditions on the enthalpy of this process is not completely understood, and the reaction may not fulfil all the requirements of a test reaction.[129a] The reaction between solutions of sulphuric acid and sodium hydroxide is also suitable.[128] Gunn and his co-workers have recently demonstrated the absence of serious systematic errors in the values obtained for these two test reactions by a scheme of measurements which involved determination of the energy of the H_2SO_4 + NaOH reaction using a

[125] G. A. Nash, H. A. Skinner, and W. F. Stack, *Trans. Faraday Soc.*, 1965, **61**, 640.
[126] C. E. Vanderzee and R. A. Myers, *J. Phys. Chem.*, 1961, **65**, 153.
[127] S. Sunner and I. Wadsö, *Acta Chem. Scand.*, 1959, **13**, 97.
[128] S. R. Gunn, *J. Phys. Chem.*, 1965, **69**, 2902.
[129] I. Wadsö and R. J. Irving, *Acta Chem. Scand.*, 1964, **18**, 195.
[129a] M. V. Kilday and E. J. Prosen, reported at the 26th Calorimetry Conference, Orono, Maine, U.S.A., 1971.

rotating-bomb calorimeter calibrated with benzoic acid, and thus obtained a result independent of the electrical calibration of the reaction calorimeter.[130, 131] Where the reaction vessel serves also as the calorimeter there is danger of errors due to 'hot spots' on the surface caused by inadequacies in the heater design or by local concentrations of reactants.[128]

Although isoperibol reaction calorimeters are usually operated close to ambient temperatures, the method is not restricted to this condition and one Russian instrument has an operating temperature range stated to be from − 20 to + 120 °C.[132] Whilst glass is frequently used for the construction of the reaction vessel several calorimeters designed for reactions involving hydrofluoric acid employ noble metals, and Robie has described a technique for plating a copper calorimeter with an impermeable coating of gold, which is resistant to hot concentrated hydrochloric, nitric, or hydrofluoric acids.[133] Heat-exchange between a calorimeter and its environment can be considerably reduced, and hence longer reaction periods (and slower reactions) tolerated, if the interspace between the calorimeter and jacket is thoroughly evacuated (10^{-3} Pa).[132, 133]

The reaction vessels used in the calorimeters so far outlined are not sealed and are unsuitable if one of the products is gaseous because evolution of gas is likely to result in evaporation of an unknown amount of liquid phase, and therefore introduce a thermal error. Gunn has described a sealed copper bomb which may be rocked to and fro in an evacuated submarine enclosure; increasing the angle of oscillation causes the ampoule-breaking hammer to be released. The rocking motion produces efficient stirring and uniform distribution of temperature over the calorimeter surface with low generation of thermal power.[134] In a recent application of this calorimeter the energies of reaction of rubidium and caesium and their hydrides with water were determined.[135] In another design the rocking calorimeter is sealed by means of a flexible diaphragm attached to the ampoule-breaking mechanism which can therefore be operated from outside the bomb.[136] The rotating combustion bomb can also be used as a sealed reaction calorimeter and used *e.g.* for metal + acid reactions.[137]

Auxiliary measurements of enthalpies of dilution are necessary to obtain the enthalpy of reaction between two substances from measurements of the change in enthalpy when solutions of those substances are mixed. This additional labour may often be avoided and greater accuracy obtained by the use of a differential calorimeter in which the dilution process takes place

130 S. R. Gunn, *J. Chem. Thermodynamics*, 1970, **2**, 535.
131 S. R. Gunn, J. A. Watson, H. Mackle, H. A. Gundry, A. J. Head, M. Månsson, and S. Sunner, *J. Chem. Thermodynamics*, 1970, **2**, 549.
132 V. K. Abrosimov and G. A. Krestov, *Russ. J. Phys. Chem.*, 1967, **41**, 1699.
133 R. A. Robie, *Rev. Sci. Instr.*, 1965, **36**, 484.
134 S. R. Gunn, *Rev. Sci. Instr.*, 1958, **29**, 377.
135 S. R. Gunn, *J. Phys. Chem.*, 1967, **71**, 1386.
136 A. F. Vorob'ev, A. F. Broier, and S. M. Skuratov, *Russ. J. Phys. Chem.*, 1967, **41**, 487.
137 C. J. Thompson, G. C. Sinke, and D. R. Stull, *J. Chem. and Eng. Data*, 1962, **7**, 380.

in a twin vessel situated in the same temperature environment as the reaction vessel, and by observation of changes in the *difference* in temperature between the two vessels. Thermometric methods where temperature differences can be measured directly are particularly suitable for use in a calorimetric system of this kind; thermopiles, thermistors in a bridge network, and, more recently, quartz thermometers [138] have been used in this way.

Macrocalorimeters for Slow Reactions. Isoperibol calorimeters become unsuitable when the main reaction period exceeds about 30 min as the correction for heat exchange is then generally large and inaccurately determinable. This difficulty may be overcome by controlling the temperature of the jacket to be always the same as that of the calorimeter so that no heat-exchange occurs. Since there is energy continuously dissipated in the calorimeter by the stirrer, and by Joule heating of the thermometer when thermistors or resistance thermometers are used, it is preferable to control the jacket temperature to be slightly below that of the calorimeter. Fitzsimmons and Kirkbride have recently reported the use of a calorimeter basically similar to the best of the designs investigated by Sunner and Wadsö [127] but in which the jacket could be controlled either to be isothermal or to follow the calorimeter temperature, using thermistors (in adjacent arms of a Wheatstone bridge circuit) in the calorimeter and in the jacket to sense the temperature difference between them.[139] Very satisfactory results were obtained for the enthalpy of dissolution of silica in hydrofluoric acid; good agreement was obtained between the value determined using the isoperibol mode of operation with material of small particle size (0.5 h reaction time) and the value determined using the adiabatic mode with coarser material (3 h reaction time). In this calorimeter, and in general, the completely adiabatic condition is never perfectly realized but the correction for heat-exchange becomes very small and large proportionate errors in its determination can be tolerated.

Rather better adiabatic control can be achieved using a low heat-capacity shield which responds more quickly than a water bath to temperature changes sensed by the detector. A recent Russian calorimeter designed for reactions lasting 30 to 40 min employs both a metal shield and a water jacket to achieve the adiabatic condition.[140] Heat-exchange between calorimeter and jacket is also reduced by evacuation of the space between them and by using highly polished surfaces to limit radiation.

As with other reaction calorimeters, it is sometimes necessary that adiabatic calorimeters should be sealed. Palkin *et al.* used a rubber membrane in the lid which provided a seal but permitted manipulation of the ampoule breaker attached to it; the contents were stirred magnetically.[140]

[138] D. Barnes, P. J. Laye, and L. D. Pettit, *J. Chem. Soc.* (*A*), 1969, 2073.
[139] C. P. Fitzsimmons and B. J. Kirkbride, *J. Chem. Thermodynamics*, 1970, **2**, 265.
[140] V. A. Palkin, V. E. Gorbunov, and T. A. Kapitonova, *Russ. J. Phys. Chem.*, 1969, **43**, 914.

Benjamin has described a sealed metal calorimeter contained in a partially evacuated submarine vessel, which could be rocked through 180° in a temperature-controlled water-bath.[141] The automatic adiabatic control had to be supplemented by manual assistance during the first 10 to 20 s of a reaction.

An alternative method to reduce the magnitude of the heat-exchange correction is to maintain the calorimeter at a steady temperature in a constant-temperature environment throughout the determination. This is most easily achieved in an isothermal phase-change calorimeter. In a typical apparatus described by Tischer [142] a platinum-lined copper reaction vessel is positioned in good thermal contact with a stainless-steel well at the interior of the calorimeter. The well is provided with fins for efficient heat-exchange and surrounded by a mantle of solid diphenyl ether. The rest of the calorimeter is constructed from glass and surrounded by a vacuum jacket, the whole assembly being placed in a thermostat. A small controlled heat-leak is deliberately arranged so that the meniscus of the mercury dilatometer is continuously moving in one direction and is not liable to stick. The calorimeter is calibrated electrically and has been tested by determination of the enthalpy of solution of zinc oxide (contained in gelatin capsules) in hydrofluoric + nitric acid mixtures. The standard error in measurements of *ca.* 4000 J liberated over 90 min was ± 0.2 per cent. Although diphenyl ether is the most commonly used calorimetric substance, ice was employed in a calorimeter to determine the enthalpy of reaction between hex-1-ene and sulphuric acid, a two-phase system in which reaction was brought about by vigorous stirring and emulsification.[143] A naphthalene calorimeter has been used to study the slow oxidation of solids at 80 °C.[144]

Isothermal phase-change calorimeters based on liquid + vapour equilibria have been used for the determination of energies of polymerization. The monomer was sealed into an ampoule immersed in the calorimetric liquid (*e.g.* carbon tetrachloride, benzene, toluene) contained in a tube, which was suspended from a balance and surrounded by vapour of the same refluxing liquid. As the polymerization proceeded the heat liberated by the reaction caused liquid to vaporize from the tube and was measured by the consequent loss in mass.[145]

Kanbour and Joncich have described a reaction calorimeter in which the isothermal condition is maintained by balancing constant Peltier cooling supplied by a thermoelectric module against Joule heating.[146] A steady-state condition in which the calorimeter temperature changed by less than 0.001 K was attained before the initiation of reaction. The Joule heating

141 L. Benjamin, *Canad. J. Chem.*, 1963, **41**, 2210.
142 R. E. Tischer, *Rev. Sci. Instr.*, 1966, **37**, 431.
143 K. L. Butcher and G. M. Nickson, *Trans. Faraday Soc.*, 1958, **54**, 1195.
144 A. Thomas, *Trans. Faraday Soc.*, 1951, **47**, 569.
145 L. K. J. Tong and W. O. Kenyon, *J. Amer. Chem. Soc.*, 1945, **67**, 1278.
146 F. Kanbour and M. J. Joncich, *Rev. Sci. Instr.*, 1967, **38**, 913.

was then decreased or increased according to whether the process was exo- or endo-thermic and the steady temperature maintained to within ± 0.005 K; this method yielded satisfactory results for relatively rapid reactions. The extension of the technique, which is applicable to both exo- and endo-thermic reactions, to slow processes has not been described, but a titration calorimeter based on this principle has been reported.[147] In this apparatus small heat inputs were produced by successive additions of small quantities of reactant over a period comparable with the duration of a slow reaction, and the temperature maintained constant to within ± 0.0002 K. A somewhat similar method of making isothermal calorimetric measurements on endothermic reactions is to use the twin differential principle and supply a measured amount of Joule heating to the calorimeter in which reaction takes place.[148]

Microcalorimeters. The term 'microcalorimeter' is generally reserved for those instruments used for measuring energy changes of less than about 1 J. For the chemist the main advantage of microcalorimetry is that it enables measurements to be made on substances which are available only in small quantities, although the results are generally less accurate than those attainable by macro-methods. Sometimes, however, it is desirable to obtain calorimetric results on very dilute solutions, *e.g.* to avoid uncertainties in the extrapolation to infinite dilution of results obtained for electrochemical systems.[149] Calorimeters used for measurement of small energy changes on mixing dilute aqueous solutions are particularly suitable for investigation of the energetics of biochemical reactions. The inclusion of a series of lectures on 'Calorimetry in Biology' at the 1970 American Calorimetry Conference is indicative of the growth of interest in this subject which will doubtless be reflected by future developments in microcalorimetry.

Successful microcalorimeters for the study of rapid reactions have been made by scaling down the conventional isoperibol instrument and taking special precautions to reduce heat-leakage. Westrum and Eyring have described a calorimeter with a tantalum reaction vessel only 8 cm^3 in volume containing a platinum stirrer on a thin quartz shaft and a copper resistor, with only two fine external leads, which served both as heater and thermometer.[150] Results accurate to a few tenths of 1 per cent were obtained in the measurement of energy changes of about 0.8 J. A rather similar instrument designed, like that of Westrum and Eyring, for work on actinides uses a thermistor instead of a resistance thermometer.[151] Because fluctuations in jacket temperature, heat of stirring, thermistor current, *etc.*, have a much larger effect relative to the energy being measured in micro- than

[147] J. J. Christensen, H. D. Johnston, and R. M. Izatt, *Rev. Sci. Instr.*, 1968, **39**, 1356.
[148] A. Buzzell and J. M. Sturtevant, *J. Amer. Chem. Soc.*, 1951, **73**, 2454.
[149] W. J. Canady, H. M. Papée, and K. J. Laidler, *Trans. Faraday Soc.*, 1958, **54**, 502.
[150] E. F. Westrum jun. and LeR. Eyring, *J. Amer. Chem. Soc.*, 1952, **74**, 2045.
[151] G. R. Argue, E. E. Mercer, and J. W. Cobble, *J. Phys. Chem.*, 1961, **65**, 2041.

in macro-calorimetry, considerable advantage accrues from the use of the twin differential principle. This has been employed in a calorimeter designed for the measurement of energies not less than 0.1 J with an accuracy of ± 2 per cent in which the volumes of liquid reactants are 3 cm^3 and 1 cm^3.[152] The twin vessels were connected by a thermopile and surrounded by an isothermal block; corrections for heat-exchange were made according to a computer programme based on observations of the thermopile e.m.f. following inputs of known quantities of electrical energy.

The basic principles of the heat-flow calorimeter have already been outlined. Calvet and his co-workers and successors at Marseilles have over many years developed the original design of Tian into a sophisticated differential microcalorimeter with many uses.[153] The twin reaction cells are in the form of narrow cylinders and as far as possible are completely surrounded by a regular array of thermocouple junctions (as many as 1000) attached to very small silver plates to give good thermal contact. The reference junctions of the thermopiles are at the temperature of the massive metal block in which the cells are embedded and which is carefully thermostatted; suitable instruments can be operated up to 1000 K. The thermopiles around the two cells are connected in opposition so that only temperature *differences* between the two cells are recorded and the effect of any fluctuations in the temperature of the heat-sink is eliminated; slow processes generating as little as 0.004 J h^{-1} can be studied. It is not possible to stir the contents of the reaction cell, which is a disadvantage for solid + liquid reactions[154] although errors due to temperature inequalities throughout the cells are avoided by the arrangement of thermocouples, which causes the temperature measurements to be integrated over the whole surface of the cell, except the top and bottom. Provision is made for compensation of the temperature change produced in the reaction cell by Peltier cooling or Joule heating; for precise thermochemical measurements approximately 90 to 95 per cent of the total energy change is compensated and measured with high precision; only the uncompensated energy gives rise to a thermogram from which the energy can be calculated with an accuracy of about 1 per cent. The Calvet calorimeter has been used more for investigation of biological, adsorption, and hydration processes (where the kinetic information provided by an uncompensated thermogram has been of as much interest as the energy change) than for determination of the energies of well-defined chemical reactions. However, recent examples of the latter category include determinations of the energy of polymerization of propiolactone,[154] the energy of hydrolysis of ethyl acetate,[155] and the enthalpy of formation of metal carboxylates.[156]

[152] R. L. Berger, Yu-Bing Fok Chick, and N. Davids, *Rev. Sci. Instr.*, 1968, **39**, 362.
[153] E. Calvet and H. Prat, 'Recent Progress in Microcalorimetry', Pergamon, Oxford, 1963.
[154] B. Börjesson, Y. Nakase, and S. Sunner, *Acta Chem. Scand.*, 1966, **20**, 803.
[155] C. Zahra and L. Lagarde, *Bull. Soc. chim. France*, 1969, 1092.
[156] M. Le Van and G. Perinet, *Bull. Soc. chim. France*, 1969, 2681.

Benzinger has designed a heat-flow calorimeter which has many similarities to the Calvet instrument, but without provision for compensation of the energy.[157] The thermopile (10 000 junctions) was constructed by the ingenious method of removing the insulation from every alternate half-turn of a constantan coil and plating the exposed metal with copper. The coil was itself coiled around the thin shell containing the annular-shaped twin reaction vessels which were thus almost entirely surrounded by the 'coiled-coil' thermopiles, the junctions being tangentially adjacent to either the reaction vessels or the copper-block heat sink. The Benzinger calorimeter is rather more convenient than the Calvet instrument for the study of (non-volatile) liquid–liquid reactions, which are simply initiated by rotation of the whole calorimeter when mixing of the contents of the two-compartment reaction cells takes place. The emphasis in design has been on the rapid transfer of energy across the thermopile and the name 'heat-burst calorimeter' indicates that it is intended more for rapid reactions (typically involving 10 to 200 mJ) than for slow reactions, although it is suitable for investigating continuous processes of about 40 $\mu J\ s^{-1}$. Stack and Skinner used a commercial model to determine the energies of complexing of various metal ions with an accuracy of about 1 per cent,[158] but the Benzinger calorimeter has so far not been widely used for reaction calorimetry of other than biochemical systems.

Wadsö has described a 'batch microcalorimeter' basically similar to Benzinger's but principally intended for measurement of enthalpies of reaction involving small sample volumes (less than 5 cm^3 of each component). Electrical calibration experiments indicated that the precision attainable was better than 1 per cent for rapid processes involving 4 mJ or slow (1 h) processes of 60 mJ, but better than 0.05 per cent for the rapid production of *ca.* 0.5 J.[159]

There are advantages in measuring the enthalpy of reaction between two liquids by means of a flow system. This is simple to operate and little energy is generated by the mixing process, which can take place in a vessel without vapour space, thus avoiding errors due to condensation and vaporization effects which complicate the use of volatile solvents and even of some aqueous solutions in microcalorimetry. Surface effects are eliminated since measurements are not made until a steady state has been reached. Stoesser and Gill first described a flow microcalorimeter, which is basically a twin differential adiabatic calorimeter in which the two thermally-equilibrated liquids are pumped into a mixing cell and allowed to react.[160] The mixed liquids are then pumped to a second, reference cell, the two cells being maintained at the same temperature by Joule heating. For exothermic processes, where the compensating energy is supplied to the reference cell

157 T. H. Benzinger and C. Kitzinger, 'Temperature: Its Measurement and Control in Science and Industry', Reinhold, New York, 1963, Vol. 3, Part 3, p. 43.

158 W. F. Stack and H. A. Skinner, *Trans. Faraday Soc.*, 1967, **63**, 1136.

159 I. Wadsö, *Acta Chem. Scand.*, 1968, **22**, 927.

160 P. R. Stoesser and S. J. Gill, *Rev. Sci. Instr.*, 1967, **38**, 422.

it is necessary to know the relative heat capacities of the two cells. Wadsö built a somewhat similar flow calorimeter but based it on his design of 'batch microcalorimeter' (see above) in which the heat-flow principle was used.[161] Although flow calorimeters can be operated using a pulse technique, where intermittent flow is employed, the results are far more satisfactory when both liquids flow at constant rates and a steady-state condition is attained. However, this does limit their use to reactions which are relatively rapid and complete well within the retention time of liquid in the mixing chamber. Both the flow calorimeters described enable measurements to be made with 1 per cent precision on heat effects of about 40 $\mu J\ s^{-1}$, but Monk and Wadsö[161] claim 0.1 per cent for heat effects greater than 400 $\mu J\ s^{-1}$.

Two flow calorimeters of novel design have recently been reported by Picker, Jolicoeur, and Desnoyers.[162] In the first the thermostatted liquids are pumped along concentric tubes into a PTFE mixing tube, and then fed to a second, reference, mixing tube of the differential calorimeter. Both mixing-tubes are situated in an evacuated chamber containing thermal shields and the system is essentially adiabatic, although some variation of the calibration constant (determined electrically) with liquid flow rate was demonstrated. Overall uncertainties in the results obtained (including the uncertainty associated with determination of the specific heat capacity of the liquid product) amounted to about 0.5 per cent. Where important changes in properties occur during mixing or where the heat capacity of the flowing system is small compared with that of the reaction vessel, the second, isothermal, calorimeter is to be preferred. This operates on the heat-flow principle but the heat-sink is provided by liquid flowing through a counter-current heat-exchanger, and the heat flux is measured by the temperature change of this liquid. Clearly, a minimum flow rate of heat-exchange liquid is required in order to ensure that the calorimeter operates isothermally; if the flow rate is reduced to the level where the outlet of the heat-exchange liquid has come to temperature equilibrium with the mixing cell then the conditions are adiabatic. Both calorimeters have the feature that steady-state conditions are reached within 1 min. Because of this, the composition or volume ratio of the reacting liquids can be continuously varied and curves relating energy of reaction to composition ratio readily obtained.

Reactions Involving Gases.—Rapid reactions between gases and liquids are generally studied using isoperibol calorimeters similar to those already described for liquid + liquid or liquid + solid reactions, but which yield results of rather lower precision (typically about 1 per cent). Reaction may be initiated either by substitution of gaseous reactant for inert gas passing through the reaction vessel containing the liquid component, as in the

[161] P. Monk and I. Wadsö, *Acta Chem. Scand.*, 1968, **22**, 1842.
[162] P. Picker, C. Jolicoeur, and J. E. Desnoyers, *J. Chem. Thermodynamics*, 1969, **1**, 469.

reaction between aqueous thallous hydroxide and cyclopentadiene,[163] or by breaking an ampoule of liquid in a reaction vessel filled with gas and connected to a gas burette for determination of the quantity reacted; the latter technique was used to measure the enthalpies of addition of alkanols to 1,1-dichloro-1,2-difluoroethylene, where twin differential calorimeters were employed.[164] Frisch and Mackle[3] have described a semi-micro reaction calorimeter of the aneroid type which is suitable for gas–liquid reactions involving energy changes of the order of 40 J. It has been used for the determination of the energy of addition of diborane and of boron trifluoride to organic bases dissolved in benzene. The standard error of such measurements was in the region of 1 to 2 per cent.[165]

Thermochemical measurements on reactions between gases and solids, other than combustions, are uncommon. Gross and his co-workers measured the enthalpy of reaction between alkali-metal fluorides MF (M = Li or Na) and BF_3 to form MBF_4 using an isoperibol calorimeter operated at 110 °C.[166] The stream of argon flowing over the heated fluoride was replaced by BF_3 for a period of 10 min. Duus and Mykytiuk used a flow method to determine the enthalpy of the reaction

$$2PCl_3 + 3CaF_2 = 2PF_3 + 3CaCl_2,$$

which was carried out in a vessel thermostatted by boiling mercury.[167] The solid bed was completely fluidized and thus in thermal equilibrium with the exit gases, but the value obtained for $\Delta H_f^\circ(PF_3, g)$ was in very poor agreement (12 kJ mol^{-1}) with that recently determined using the far more reliable method of fluorine-combustion calorimetry.[168]

Reactions between gases in static systems require only simple reaction vessels in conventional calorimeters. The gases can be introduced into the vessel by simple Joule expansion, when account has to be taken of the energy of the expansion process,[169] or they may be contained in a two-compartment glass reaction vessel and separated by a break-seal, a technique employed by Gunn for the chlorination of B_2Cl_4 using a heat-flow calorimeter suitable for the study of processes of an hour's duration.[170]

Lacher and his co-workers have used a flow system operated at temperatures up to 250 °C to measure the enthalpies of the catalytic addition of hydrogen chloride to olefins and the catalytic hydrogenation of alkyl halides.[171, 172] The calorimeter is maintained at constant temperature by

163 H. Hull and A. G. Turnbull, *Inorg. Chem.*, 1967, **6**, 2020.
164 M. B. Kennedy, J. R. Lacher, and J. D. Park, *Trans. Faraday Soc.*, 1969, **65**, 1435.
165 J. McAllister and H. Mackle, *Trans. Faraday Soc.*, 1969, **65**, 1734.
166 P. Gross, C. Hayman, and H. A. Joël, *Trans. Faraday Soc.*, 1968, **64**, 317.
167 H. C. Duus and D. P. Mykytiuk, *J. Chem. and Eng. Data*, 1964, **9**, 585.
168 E. Rudzitis, E. H. Van Deventer, and W. N. Hubbard, *J. Chem. Thermodynamics*, 1970, **2**, 221.
169 J. D. Ray and R. A. Ogg, *J. Phys. Chem.*, 1957, **61**, 1087.
170 S. R. Gunn, L. G. Green, and A. I. Von Egidy, *J. Phys. Chem.*, 1959, **63**, 1787.
171 J. R. Lacher, H. B. Gottlieb, and J. D. Park, *Trans. Faraday Soc.*, 1962, **58**, 2348.
172 P. Fowell, J. R. Lacher, and J. D. Park, *Trans. Faraday Soc.*, 1965, **61**, 1324.

compensating with electrical energy the continuous cooling produced by bubbling hydrogen at a constant rate through a volatile liquid. After the flow of reactants is started the electrical power is adjusted until a new steady-state condition is attained at the same temperature, the difference in electrical energy being a measure of the enthalpy of reaction. In a few cases, significant discrepancies occur between enthalpies of formation obtained in this way and those determined by combustion calorimetric methods, which are much less likely to be subject to systematic errors. The isothermal flow calorimeter of Picker *et al.* described previously [162] is well suited to the study of gaseous reactions, where the heat capacity of the reactants is small compared with that of the reaction vessel.

Thermal Decomposition Reactions.—When a substance undergoes thermal decomposition, the enthalpy of the process can often be determined by an indirect calorimetric method based on measurements of enthalpy differences. If the difference is determined between the enthalpy of a substance at a temperature where it is essentially stable and the enthalpy at a temperature where it is completely decomposed, this will differ from the enthalpy change predicted by extrapolation of the heat capacity–temperature curve in the thermally stable region by an amount equal to the enthalpy of the decomposition process. Gunn used this principle to obtain the energy of decomposition of KrF_2 by comparison of the temperature change of an aneroid calorimeter (initially at 93 °C) when steel containers of the gas (initially at 40 °C) were dropped into it with the corresponding temperature change observed when the mixture of krypton and fluorine, which resulted from decomposition of the sample, was used under otherwise identical conditions.[173]

Several designs of dynamic scanning calorimeter have been described. They have in common continuous change of the sample temperature with time and the simultaneous recording of some property related to change in enthalpy. Scanning calorimetry yields results for enthalpies of decomposition processes of lower precision and accuracy than most of the methods of combustion and reaction calorimetry described in this review (typical values are in the 1 to 5 per cent region) but where the accuracy is sufficient the method is convenient and rapid, and can also serve as a useful preliminary investigation to a subsequent thermochemical determination undertaken with more refined methods. The simplest and least accurate instrument is a modified differential thermal analysis apparatus in which twin calorimeters are enclosed in a jacket which is heated at a constant rate, the temperature difference between the two calorimeters being recorded.[174] In a more accurate twin differential scanning calorimeter the jacket is maintained at constant temperature and the reference calorimeter heated at a constant rate; the differential electrical power needed to keep

[173] S. R. Gunn, *J. Phys. Chem.*, 1967, **71**, 2934.
[174] K. A. Sherwin, *J. Sci. Instr.*, 1964, **41**, 7.

the sample calorimeter at the same temperature as the reference calorimeter is measured.[175] The area under the power–time curve is proportional to the enthalpy of any process occurring during the temperature scan made by the instrument, which is usually calibrated by means of the known enthalpy of fusion of a standard substance which melts at a convenient temperature. Amongst many examples of the use of instruments of this type is the determination of the enthalpy of decomposition of complexes of the general formula CoL_4X_2 where L = pyridine, picoline, aniline, *etc.*, and X = Cl, Br, or I, at temperatures up to 350 °C.[176] Two forms of adiabatic scanning calorimeter have been described. In the first, constant power is supplied to the sample and the variation of temperature with time is recorded,[177] whereas in the second the jacket is heated at a constant rate and the power supplied to the calorimeter to cause its temperature to follow that of the jacket is determined. The calibration of an adiabatic calorimeter of the second type, using the enthalpy of fusion of tin, and the testing of its performance in measurements of the enthalpies of dehydration of $CuSO_4.5H_2O$ and $BaCl_2.2H_2O$ have been reported.[178] Scanning calorimeters are easy to automate and commercial models of all four types are available. In general, they provide useful information about the energetics of thermally induced processes, but rarely enable enthalpies of formation of high accuracy to be obtained.

[175] E. S. Watson, M. J. O'Neill, J. Justin, and N. Brenner, *Analyt. Chem.*, 1964, **36**, 1233.
[176] G. Beech, C. T. Mortimer, and E. G. Tyler, *J. Chem. Soc. (A)*, 1967, 925.
[177] C. Solomons and J. P. Cummings, *Rev. Sci. Instr.*, 1964, **35**, 307.
[178] E. L. Dosch and W. W. Wendlandt, *Thermochim. Acta*, 1970, **1**, 181.

4
The Heat Capacities of Organic Compounds

BY J. F. MARTIN

1 Introduction

The equilibrium constant for an organic reaction in the gas phase can be determined directly for a few favourable reactions but it can be calculated if sufficient thermodynamic data on all the reactants and products are available. If the standard enthalpy and entropy of each substance taking part in the reaction are known, then the equilibrium constant for the reaction may be calculated:

$$-RT\ln K = \Delta G^{\ominus} = \Delta H^{\ominus} - T\Delta S^{\ominus}.$$

Standard entropies at 298.15 K are evaluated from measurements of heat capacities at constant pressure and enthalpies of phase changes from low temperatures to 298.15 K, assuming the third law of thermodynamics is applicable to the substance. The entropy at 298.15 K is given by an expression such as

$$S^{\ominus}(298.15\ \mathrm{K}) = \int_0^{T_\mathrm{t}}(C_p/T)\,\mathrm{d}T + \Delta H_\mathrm{t}/T_\mathrm{t} + \int_{T_\mathrm{t}}^{T_\mathrm{m}}(C_p/T)\,\mathrm{d}T + \Delta H_\mathrm{m}/T_\mathrm{m} + \int_{T_\mathrm{m}}^{298.15\ \mathrm{K}}(C_p/T)\,\mathrm{d}T,$$

in which ΔH_t and T_t are the enthalpy and temperature of an isothermal phase transition and ΔH_m and T_m are the enthalpy and temperature of fusion.

The entropy of the compound in the ideal gas state at 298.15 K and the standard pressure (101.325 kPa) is calculated from the entropy at 298.15 K of the condensed phase, the enthalpy of vaporization at 298.15 K, and vapour pressure and gas imperfection data, using the equation:

$$S^{\ominus}(\mathrm{g}, 298.15\ \mathrm{K}) = S^{\ominus}(\mathrm{l\ or\ c}, 298.15\ \mathrm{K}) + \Delta H_\mathrm{v}(298.15\ \mathrm{K})/298.15\ \mathrm{K} + R\ln(p/101.325\ \mathrm{kPa}) + (\mathrm{d}B/\mathrm{d}T)p + (\mathrm{d}C'/\mathrm{d}T)p^2/2,$$

where B and C' are virial coefficients in the equation of state, $pV = RT + Bp + C'p^2$.

In practice, heat capacity measurements are usually made from about 10 K. Values of C_p below the lowest temperature of the measurements are obtained by extrapolation. For organic compounds with melting temperatures below the highest temperature of the C_p measurements, values for the enthalpy and entropy of fusion are obtained in the course of the measurements and it is usual to calculate the equilibrium temperature (T_m) in the melting temperature region as a function of the fraction (F) of the sample melted. The mole fraction x of impurity in the sample is then given by the expression:

$$x = (\Delta H_m / RT_m^2)\, F\{T_m(\text{pure}) - T_m\}.$$

As indicated above, many organic compounds exist in more than one crystalline form and the transition between crystalline forms is frequently manifested by an anomaly in the heat capacity. The enthalpy and entropy of the transition are obtained from measurements in the transition region. The experimental apparatus for low-temperature heat capacity measurements is often usable for measurements up to about 400 K. Heat capacities of organic compounds above 298 K are required for chemical engineering computations, *e.g.* for calculations using Kirchhoff's equation, of enthalpies of formation and reaction at temperatures other than those at which the thermodynamic data are available.

Several recently published text books cover various aspects of the measurement and theory of heat capacities of organic compounds. In particular, the reader is referred to 'Experimental Thermodynamics,' Vol. I, 'Calorimetry of Non-reacting Systems,' edited by J. P. McCullough and D. W. Scott (Butterworths, London, 1968), which gives excellent accounts of calorimetric methods and related topics. 'Physics and Chemistry of the Organic Solid State,' Vol. I, edited by D. Fox, M. M. Labes, and A. Weissberger (Interscience, New York, 1963) and 'Melting and Crystal Structure' by A. R. Ubbelohde (Clarendon Press, London, 1965) contain a great deal of information on phase transitions. 'The Chemical Thermodynamics of Organic Compounds' by D. R. Stull, E. F. Westrum jun., and G. C. Sinke (John Wiley, New York, 1969) covers basic principles, experimental methods, and applications of chemical thermodynamics and is especially valuable for the tabulation of thermodynamic properties (published before 1965), including heat capacities, of several thousand organic compounds.

The present review describes recent developments in the experimental techniques for low-temperature (10 to 400 K) calorimetry and in the measurement of heat capacities of organic compounds during the past ten years. Data on heat capacities, enthalpies of phase changes, and entropies of organic compounds published before 1960 can be found in Landolt-Börnstein, 'Zahlenwerte und Funktionen,' II. Band, 4. Teil (Berlin, Springer–Verlag, 1961). See also Chapter 2.

2 Experimental Techniques

The most frequently used experimental methods for the determination of heat capacities in the temperature range 10 to 400 K require the measurement of electrical energy introduced into a sample of the compound in a suitable container and the measurement of the initial and the final equilibrium temperatures of the sample. The measurements are made in an adiabatic calorimeter in which heat leakage is reduced to a minimum by control of the temperature of an adiabatic shield. Corrections are made for any residual heat leakage.

Precise measurements of heat capacity may also be made in a calorimeter with a shield at constant temperature (the Nernst calorimeter). The heat lost or gained during each measurement is determined accurately.

Detailed reviews have been published on calorimetric design,[1] on isothermal calorimetry,[2] and on adiabatic calorimetry.[3] The emphasis in the present account is on automatic methods for measurements of heat capacity and enthalpies of phase changes by adiabatic calorimetry.

Automatic Calorimetry.—Before the development of modern electronic techniques, precise measurements of heat capacities were made by manual methods both for control of adiabatic conditions and for measurements of temperatures and electrical energies. Typical examples of such measurements have been described by Southard and Brickwedde [4] and by Ruhrwein and Huffman.[5] To obtain the highest possible accuracy, the measurement required two operators, and, to cover the temperature range (10 to 300 K), a large number of potentiometric observations were necessary. It is now usual to have automatic control of the adiabatic shield.[3] Measurements of temperature and energy with a manually-operated potentiometer can be simplified by the use of constant current supplies for the potentiometer and thermometer circuits.

Calorimetric apparatus in which the experimental observations were recorded on charts has been described by Dauphinee *et al.*,[6] by Stull,[7] and by Suga and Seki.[8] Stull [7] and Suga and Seki [8] used automatic recorders for resistance thermometry of the type developed by Stull.[9] Several laboratories now have manually-operated potentiometers or resistance bridges, the dials of which are equipped with extra banks of contacts for encoding and recording in print and on punched tape. One of the earliest descriptions of apparatus of this type was by Blythe *et al.*[10] who modified a commercial two-dial potentiometer to enable its dial settings to be recorded by a tape punch. Systems of this type require an operator to balance the potentiometer and initiate the recording of the dial readings at a particular time which is also recorded. Other semi-automatic systems record small temperature changes by use of a manually-operated potentiometer or d.c. resistance bridge coupled to a very stable d.c. amplifier and a digital voltmeter. The potentiometer is balanced to within 10 μV and the residual

1 G. K. White, 'Experimental Techniques in Low-temperature Physics,' Clarendon Press, Oxford, 1959.
2 J. W. Stout in 'Experimental Thermodynamics, Vol. I, Calorimetry of Non-reacting Systems,' ed. J. P. McCullough and D. W. Scott, Butterworths, London, 1968, p. 215.
3 E. F. Westrum jun., G. T. Furukawa, and J. P. McCullough in ref. 2, p. 133.
4 J. C. Southard and E. G. Brickwedde, *J. Amer. Chem. Soc.*, 1933, **55**, 4378.
5 R. A. Ruhrwein and H. M. Huffman, *J. Amer. Chem. Soc.*, 1943, **65**, 1920.
6 T. M. Dauphinee, D. K. C. MacDonald, and H. Preston-Thomas, *Proc. Roy. Soc.*, 1954, **A 221**, 167.
7 D. R. Stull, *Analyt. Chim. Acta*, 1957, **17**, 133.
8 H. Suga and S. Seki, *Bull. Chem. Soc. Japan*, 1965, **38**, 1000.
9 D. R. Stull, *Rev. Sci. Instr.*, 1945, **16**, 318.
10 H. J. Blythe, T. J. Harvey, F. E. Hoare, and D. E. Moody, *Cryogenics*, 1964, **4**, 28.

10 μV or less is amplified and displayed on a digital voltmeter. If the potentiometer or bridge is fitted with a set of coding contacts, the readings of the instrument and those of the digital voltmeter can be recorded automatically on tape and in print.

A fully-automatic Mueller bridge has been described [11] and an instrument of this type has been used at the National Bureau of Standards for heat capacity measurements.[12]

Resistance Thermometry with Alternating Current Instruments.—D.c. instruments (potentiometers or resistance bridges) in conjunction with a platinum resistance thermometer were in general use for precise calorimetry and for accurate temperature measurements generally until about ten years ago, when the first publications on the application of a.c. transformer-ratio arm bridges to precision temperature measurement appeared.[13, 14] Hill and Miller [14] described the use of an a.c. double bridge with inductively coupled ratio arms [15] for precision platinum resistance thermometry. The bridge network is illustrated in Figure 1, which also gives equations for balance of the bridge.

In the Figure, R_t and R_s represent the resistances of a four-lead platinum resistance thermometer and a four-lead standard resistor, z_1 to z_5 represent the impedances of the connecting wires, and Z_0 and Z_1 the impedances of the inductive ratio arms. The exact equation (1) of balance can be simplified to equation (2) which is accurate to 1 part in 10^7 when the impedances of the connecting wires are less than 1 Ω.[14] If accurate values of temperature changes less than 5 K only are required and n_0 lies between 0.2 and 0.8, equation (3) may be used. The resistance bridge described by Hill and Miller [14] was manually operated but a self-balancing a.c. instrument was developed commercially* in 1963, firstly as a six-decade ratio bridge and later as a seven-decade instrument. Instruments of this type have been in use at NPL for temperature measurements since 1964.[16] The seven-decade bridge utilizes inductive voltage dividers arranged as a Kelvin double bridge. The bridge is energized at about 400 Hz and the bridge out-of-balance signal is detected by an integrating phase-locked detector, which in turn operates a digital servo to control the setting of the decade switches of the voltage dividers. Reactive components in the servo signal are balanced by a separate servo. Changes in resistance are thus automatically balanced and the ratio $R_t/(R_t + R_s)$ is displayed and is also available in binary-coded decimal form. The smallest indicated change in ratio is 1 part in 10^7 but the sensitivity of the instrument depends upon internal and external electrical noise. Typically, the sensitivity is 2×10^{-7} and the speed of balance, which depends on the setting of the detector integration period,

11 A. J. Williams, jun. and G. C. Mergner, *I.E.E.E. Trans. Instr. Meas.*, 1966, **IM-15**, 121.
12 G. T. Furukawa and M. L. Reilly, *J. Res. Nat. Bur. Stand.*, 1970, **74A**, 617.
13 W. H. P. Leslie, J. J. Hunter, and D. Robb, *Research*, 1960, **13**, 250.
14 J. J. Hill and A. P. Miller, *Proc. Inst. Elec. Engrs.*, 1963, **110**, 453.
15 J. J. Hill and A. P. Miller, *Proc. Inst. Elec. Engrs.*, 1962, **109**, 157.
16 J. F. Martin, *Proc. Brit. Ceram. Soc.*, 1967, **8**, 1.

* By Automatic Systems Laboratories Ltd., Leighton Buzzard, Beds., England.

is about 10 s for a large change of ratio and almost instantaneous for a small change. For maximum sensitivity, the value of the standard resistance (R_s) should be equal to the resistance of the thermometer (R_t) but in practice a sensitivity of at least 2×10^{-4} K can be achieved for the temperature range 10 to 400 K from a seven-decade bridge in conjunction with a 25 Ω (at 273 K) platinum resistance thermometer and 1, 10, 25, and 40 Ω

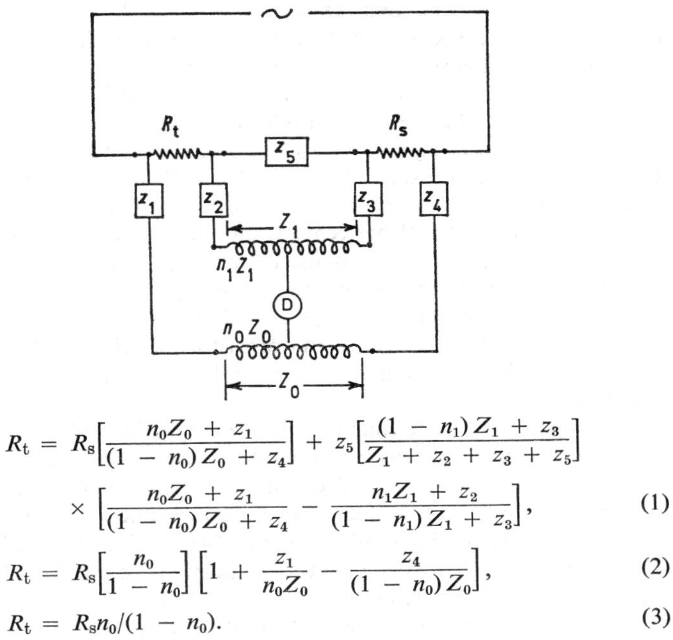

$$R_t = R_s\left[\frac{n_0Z_0 + z_1}{(1 - n_0)Z_0 + z_4}\right] + z_5\left[\frac{(1 - n_1)Z_1 + z_3}{Z_1 + z_2 + z_3 + z_5}\right]$$

$$\times \left[\frac{n_0Z_0 + z_1}{(1 - n_0)Z_0 + z_4} - \frac{n_1Z_1 + z_2}{(1 - n_1)Z_1 + z_3}\right], \quad (1)$$

$$R_t = R_s\left[\frac{n_0}{1 - n_0}\right]\left[1 + \frac{z_1}{n_0Z_0} - \frac{z_4}{(1 - n_0)Z_0}\right], \quad (2)$$

$$R_t = R_sn_0/(1 - n_0). \quad (3)$$

Figure 1 *Alternating current resistance bridge*

standard resistors. Standard resistors specially designed for a.c. measurements have been described [14, 17] and are available commercially.

The automatic a.c. bridge is comparable in accuracy and sensitivity to precision d.c. potentiometers or bridges but has two additional advantages. Calibration during use is not necessary because the accuracy of the a.c. instrument depends on the turns-ratio of the voltage dividers which, unlike the ohmic value of resistance coils, will not change with time. Measurements are simplified because thermal e.m.f.s. in the measuring instruments and external circuits are automatically compensated by current reversals arising from the use of alternating current for energizing the bridge.

The problems of energy measurement in adiabatic calorimetry have been discussed (see ref. 3). Storage batteries previously used as a source of power for the calorimeter heater can now be replaced by very stable

[17] D. L. H. Gibbings, *Proc. Inst. Elec. Engrs.*, 1963, **110**, 335.

electronically regulated power supplies, which may be either constant-potential or constant-current supplies. The use of constant-current supplies for calorimeter heaters has been discussed in a recent publication.[18] The potentiometer and voltage divider usually used for measurements of calorimeter heater currents and potentials can be replaced by a digital voltmeter.

Automatic Measurement and Data Collection.—A description of an automatic data acquisition system for calorimetry has recently been published.[19] The instrumentation for low-temperature calorimetry now used in the Division of Chemical Standards, N.P.L., is illustrated by Figure 2. It is used in conjunction with cryostats and sample containers similar to published designs (Figures 2 and 9, respectively, of ref. 3).

Temperatures are measured by a 25 Ω (at 273 K) platinum resistance thermometer and a seven-decade a.c. bridge of the type described above. A digital voltmeter of 0.001 per cent maximum accuracy and with an input impedance of $> 2.5 \times 10^{10}$ Ω is used for energy measurements and the duration of the power input is determined by a programme unit which also controls the recording of the bridge and digital voltmeter readings. The programme unit, constructed from transistorized circuit blocks, counts 1 Hz pulses derived from a quartz crystal oscillator and performs the following operations at pre-set counts.

(a) It supplies a print command pulse to the bridge serializer during the equilibration period and to the digital voltmeter serializer during the heating period. The frequency of recording the reading of either instrument can be pre-set, by decade switches, in the range once in 10 to 99 s. The same counter also supplies a pulse at the half-time of each recording period to operate reed switches to change the connections of the digital voltmeter from voltage to current readings. The digital voltmeter is sampled continuously at about 1 s intervals and is supplied with a hold signal from the serializer during print-out. The bridge reading is also frozen by a signal from its serializer during print-out. The counter is reset at each print-out and the reset pulses are counted on a 'divide-by-five' counter which initiates a carriage return-line feed signal to the teleprinter.

(b) It controls the equilibration and heating periods by means of a four-decade counter with two pre-setting switches. Equilibration and heating periods are thus separately adjustable up to a total time of cycle of 9999 s. After the elapse of the time set up on the first switch, *i.e.* at the end of the equilibration period, a pulse operates reed switches to change current from a ballast resistor to the calorimeter heater. When the counter reaches the exact time pre-set on the second switch, the current through the calorimeter heater is stopped, the counter is reset, and the next equilibration period begins.

A constant-current power supply with potential-limiting circuits is used for supplying current to the calorimeter heater which is switched into

[18] S. S. Chang, *Rev. Sci. Instr.*, 1969, **40**, 822.

[19] D. L. Martin and R. L. Snowden, *Rev. Sci. Instr.*, 1970, **41**, 1869.

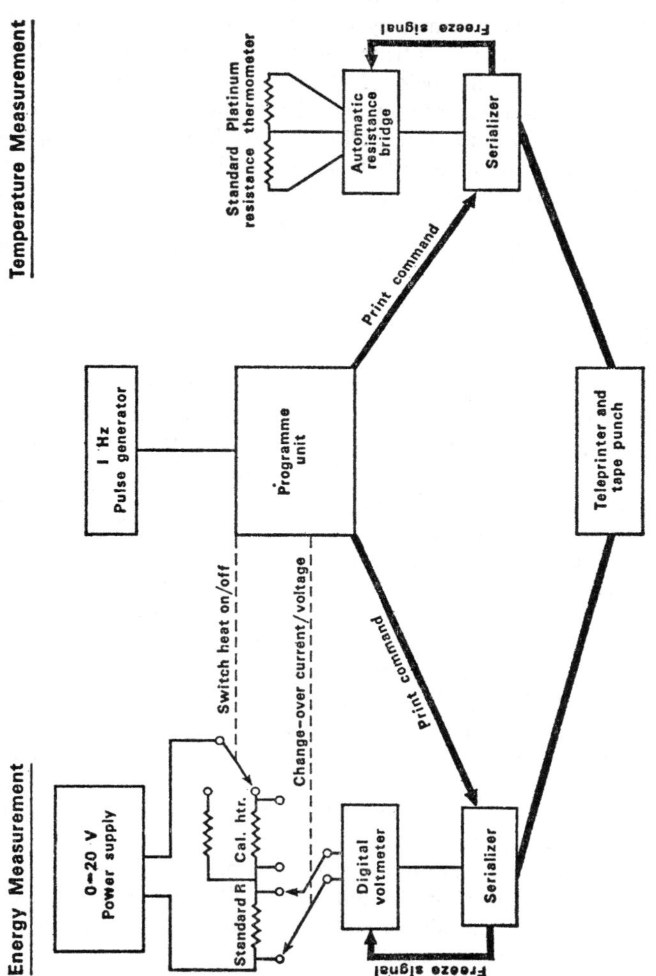

Figure 2 *Measurement and data collection*

circuit by break-before-make switches, *viz.* the reed switches referred to above. The duration of heating is always an exact integral number of seconds; the accuracy of the time interval depends on the performance of the quartz crystal oscillator from which the 1 Hz pulses are derived.

The total time of a measurement cycle (heating plus equilibration periods) for an organic compound varies from about four minutes to several hours. For example, at temperatures below 50 K, heat capacities are small and the sample usually equilibrates rapidly, so that a heating period of 120 s followed by a similar equilibration period is typical. In contrast, equilibration may require many hours for crystals in the region of a phase transition.

Automatic controls are used to maintain the sample under adiabatic or near-adiabatic conditions. It is necessary to keep the environment of the sample container at the same temperature as the sample container both when the temperature is rising during a heat input and when thermal equilibrium is being established during the fore and after periods. Methods for controlling the adiabatic shield automatically have been described.[3] In the equipment used in the Division of Chemical Standards, N.P.L., the adiabatic shield has separate controls for the top, side, and bottom. Differential thermocouples sense any temperature difference between the sample container and each section of the shield and the signals from the differential thermocouples are amplified by low-noise microvolt amplifiers. For the main control on the side of the shield, the amplified d.c. output operates a three-way controller (proportional, differential, and integral) followed by a d.c. power amplifier with the side heaters as load. Proportional control only is applied to the top and bottom of the shield but otherwise the controls are similar to those used for the side. Proportional control is also used for the guard ring, which minimizes heat loss by conduction through the electrical leads. The four d.c. power amplifiers are similar to a published design,[20] modified to give a higher output.

The equipment described is capable of unattended operation for periods of many hours. At temperatures above 50 K, only occasional manual adjustments to the shield controls and to the level of the power input to the sample are necessary, but at lower temperatures frequent adjustments of the controls are usually required.

The experimental observations, recorded in print and on paper tape, consist of alternate series of bridge and digital voltmeter readings taken at regular time intervals. The digital voltmeter readings, distinguished by a negative sign, are alternate readings of potentials across a standard resistance and the calorimeter heater, from which the power input to the calorimeter is calculated. The exact duration of the power input is obtained from the number of digital voltmeter readings. The initial and final temperatures for each energy input are obtained by extrapolation to the mid-point of the heating period of the bridge readings recorded when the sample has reached its equilibrium state. Typical recordings of observations, and computations from them of mean temperatures and heat capacities of the

[20] T. M. Gayle and W. T. Berg, *Rev. Sci. Instr.*, 1966, **37**, 1740.

sample and container are shown below (Tables 1a and 1b). The experimental values of temperature and heat capacity are stored on magnetic tape, from which graphs of the experimental values may be obtained at any time and values of the molar heat capacity of the compound are finally computed. A detailed discussion of the corrections necessary for the computations has been published.[3]

The molar heat capacities of the compound are smoothed graphically or by computer. For computer smoothing Chebyshev polynomials in temperature, usually up to order 15 or 20, are fitted to the experimental values, and the order which best represents the data is selected as follows:

(i) The curve of the sum of the squares of the residuals,

$$\sum[C_p(\text{obs}) - C_p(\text{calc})]^2,$$

plotted against the order of the polynomial has regions which are almost horizontal. The order chosen is the lowest in one of these regions.

(ii) First differences of the calculated heat capacities for temperature increments of 2.5 K below 50 K and 5 K above 50 K are examined for smoothness.

(iii) Residuals $[C_p(\text{obs}) - C_p(\text{calc})]$ are examined for randomness.

(iv) Residuals $[C_p(\text{obs}) - C_p(\text{calc})]$ are compared with expected errors at the temperature of measurement.

By use of these criteria, an order of the polynomial can usually be selected to represent the experimental data for the compound in a particular condensed state. If it is necessary to divide the data for one state into more than one temperature range, the fitting is calculated to make the heat capacity and its first differential the same at the common point.

The measured heat capacities of organic compounds can usually be extrapolated to $T = 0$ by fitting of Debye heat capacity functions (or a combination of Debye and Einstein functions) to the measured values below 20 K. Smoothed heat capacities from the experimental results and from the Debye functions are used to obtain the thermodynamic properties $S^{\ominus}(T)$, $H^{\ominus}(T) - H^{\ominus}(0)$, and $[G^{\ominus}(T) - H^{\ominus}(0)]/T$ of the crystals and liquids.

3 Results of Measurements on Organic Compounds

Table 2, containing results derived from measurements of heat capacities of some pure organic compounds published since 1961, gives values of temperatures, enthalpies, and entropies of phase transitions, and also of standard entropies. In the Table, (t) refers to crystal–crystal transitions; (m) to crystal–liquid transitions; (c), (l), and (g) to crystal, liquid, and gas; G to glass transition; and m, s to metastable, stable crystal.

In addition to thermodynamic studies of phase transformations, other methods, such as crystal structure determinations, dielectric constant measurements, dilatometry, n.m.r. and i.r. spectrometry, and neutron scattering techniques, have been used to obtain a more complete understanding of the phenomena. Several reviews of recent work on solid-state

Table 1a *Record of experimental observations. Negative readings are digital voltmeter measurements. Other readings are from resistance bridge. First five numbers are calorimeter, cryostat, sample, standard resistor, and time interval between readings*

```
4,1,4,3,50,
4058707,4058707,
-052578,2,-104738,2,-052574,2,-104739,2,-052569,2,
-104740,2,-052565,2,-104743,2,-052560,2,-104744,2,
-052555,2,-104748,2,4096020,4095841,4095832,
4095828,4095828,4095827,4095826,4095825,
4095825,4095825,4095825,4095824,4095824,
4095824,4095824,4095824,4095823,4095823,
4095823,4095823,4095823,4095822,4095823,
4095822,-052552,2,-104750,2,-052547,2,-104753,2,
-052544,2,-104755,2,-052539,2,-104755,2,-052534,2,
-104759,2,-052530,2,-104759,2,4132501,4132325,
4132316,4132313,4132311,4132311,4132309,
4132309,4132308,4132308,4132308,4132308,
4132307,4132308,4132307,4132307,4132307,
4132306,4132306,4132306,4132306,4132305,
4132306,4132306,-052525,2,-104760,2,-052521,2,
-104764,2,-052517,2,-104767,2,-052513,2,-104770,2,
-052509,2,-104771,2,-052505,2,-104774,2,4168369,
4168199,4168191,4168188,4168186,4168184,
4168184,4168184,4168184,4168183,4168183,
4168183,4168182,4168182,4168182,4168182,
4168181,4168181,4168181,4168181,4168180,
4168180,4168181,4168179,-052501,2,-104775,2,
-052497,2,-104777,2,-052493,2,-104780,2,-052489,2,
-104783,2,-052485,2,-104783,2,-052480,2,-104782,2,
4203650,4203483,4203476,4203472,4203471,
4203470,4203469,4203468,4203468,4203468,
4203467,4203466,4203466,4203466,4203466,
4203466,4203465,4203465,4203465,4203464,
4203465,4203465,4203465,4203464,-052476,2,
-104786,2,-052471,2,-104786,2,-052468,2,-104790,2,
-052465,2,-104792,2,-052461,2,-104794,2,-052458,2,
-104794,2,4238300,4238136,4238130,423817,
4238126,4238123,4238124,4238123,4238122,
4238122,4238121,4238121,4238121,4238121,
4238120,4238121,4238121,4238120,4238120,
4238119,4238120,4238120,4238119,4238119,
-052453,2,-104796,2,-052449,2,-104798,2,-052446,2,
-104799,2,-052441,2,-104801,2,-052437,2,-104802,2,
-052433,2,-104802,2,4272450,4272293,4272284,
4272282,4272281,4272280,4272278,4272278,
4272278,4272278,4272277,4272277,4272276,
4272276,4272275,4272275,4272276,4272276,
4272275,4272274,4272274,4272274,4272274,
4272274,-052430,2,-104809,2,-052427,2,-104810,2,
-052422,2,-104811,2,-052420,2,-104810,2,-052416,2,
-104815,2,-052412,2,-104816,2,4306064,4305906,
4305900,4305898,4305897,4305895,4305895,
4305894,4305894,4305893,4305893,4305892,
4305893,4305892,4305891,4305891,4305891,
4305890,4305890,4305890,4305890,4305889,
4305889,4305889,-052408,2,-104815,2,-052404,2,
-104818,2,-052401,2,-104819,2,-052397,2,-104822,2,
-052394,2,-104822,2,-052391,2,-104825,2,4339140,
4338998,4338991,4338989,4338987,4338986,
```

Table 1b *Computer output. Point nos.* 31 *to* 38 *inclusive are from observations recorded in Table* 1*a. The first column is the total number of bridge readings of which the numbers in the second column were used to obtain the extrapolated values (columns* 3 *and* 4*) and initial* (TI) *and final* (TF) *temperatures. Energies are in joules.* TM *is the mean temperature and* CP *is the heat capacity of sample plus sample container*

CALORIMETER NO 4 BRIDGE READINGS		EXTRAPOLATED VALUES				CALCULATED RESULTS				
TOTAL	USED	INITIAL	FINAL	TI	TF	ENERGY	DELTA T	POINT NO.	TM	CP
24	21	4058707	4095830	188.0908	190.5693	330.3243	2.4785	31	189.3300	133.2733
24	21	4095820	4132314	190.5687	193.0384	330.2035	2.4697	32	191.8035	133.7024
24	21	4132303	4168189	193.0376	195.4989	330.0794	2.4612	33	194.2683	134.1116
24	21	4168177	4203473	195.4981	197.9511	329.9664	2.4530	34	196.7246	134.5162
24	20	4203461	4238127	197.9502	200.3910	329.8459	2.4408	35	199.1706	135.1376
24	21	4238117	4272284	200.3903	202.8273	329.7286	2.4370	36	201.6088	135.3019
24	22	4272271	4305901	202.8264	205.2560	329.6286	2.4296	37	204.0412	135.6716
24	21	4305885	4338991	205.2549	207.6770	329.5164	2.4221	38	206.4659	136.0431
24	21	4338977	4371572	207.6760	210.0909	329.4231	2.4149	39	208.8834	136.4137
24	21	4371556	4403657	210.0897	212.4976	329.3162	2.4079	40	211.2937	136.7634
24	21	4403638	4435256	212.4962	214.8972	329.2192	2.4010	41	213.6967	137.1169
24	22	4435236	4466385	214.8957	217.2900	329.1342	2.3943	42	216.0928	137.4641
24	21	4466362	4497047	217.2882	219.6754	329.0325	2.3872	43	218.4818	137.8294
24	19	4497020	4527251	219.6733	222.0534	328.9485	2.3801	44	220.8634	138.2103

Table 2

Formula	*Compound*	$T(t)$ / K	$\Delta H(t)$ / J mol^{-1}	$\Delta S(t)$ / J K^{-1} mol^{-1}	$T(m)$ / K	$\Delta H(m)$ / J mol^{-1}	$\Delta S(m)$ / J K^{-1} mol^{-1}	$S^{\ominus}(298.15\text{ K})$ / J K^{-1} mol^{-1} c or l	$S^{\ominus}(298.15\text{ K})$ / J K^{-1} mol^{-1} g	*Reference*
CH_4	Methane	20.49		1.46	90.675	9284	10.24		188.7	21
CH_3D	Deuteriomethane	16 to 26								22
CH_2D_2	Dideuteriomethane	16 to 26								22
CHD_3	Trideuteriomethane	16 to 26								22
CD_4	Tetradeuteriomethane	22.2		2.04	89.784	9104	10.13			21
		27.0								
C_5H_8	1,2-Pentadiene				135.89	7559	55.62	245.0	334.8	23
C_5H_8	1,*cis*-3-Pentadiene				132.35	5639	42.61	233.2	322.8	23
C_5H_8	1,*trans*-3-Pentadiene				185.71	7144	38.47	227.1	315.6	23
C_5H_8	1,4-Pentadiene				124.91	6115	48.96	248.9	334.0	23
C_5H_8	2,3-Pentadiene				147.52	6628	44.93	237.3	329.1	23
C_5H_8	3-Methyl-1,2-butadiene				159.53	7956	49.87	231.8	321.2	23
C_5H_8	2-Methyl-1,3-butadiene				127.27	4925	38.70	228.3	314.8	23
C_5H_{12}	n-Pentane				143.47	8401	58.56	263.5		24
C_5H_{12}	Isopentane	65 (G)			113.36	5140	45.34	260.7		25
C_5H_{12}	Neopentane	140	2630	18.70	256.76	3096	12.05		298.1[a]	26
C_6H_{14}	n-Hexane				177.83	13080	73.55	296.1		24, 28
C_7H_8	Toluene				178.15	6636	37.25	221.0	321.2	27
C_7H_{16}	n-Heptane				182.55	14037	76.90	328.6	427.9	24
C_7H_{16}	2-Methylhexane	71.5			154.90	9184	59.29	323.3	420.0	28
C_7H_{16}	3-Ethylpentane				154.58	9548	61.77	314.6	411.5	28
C_7H_{16}	2,2-Dimethylpentane	83.2			149.43	5825	38.98	300.3	392.9	28
C_7H_{16}	2,4-Dimethylpentane				153.97	6845	44.46	303.2	396.7	28
C_7H_{16}	2,2,3-Trimethylbutane	86.8								
		108.0								
		121.4	2451	20.19	248.57	2261	9.096	292.3	383.3	28
C_8H_{12}	Bicyclo[2.2.2]octene	110.50	188	3.26	389.75	5381	13.81	210.5		29
		176.47	5648	32.05						
C_8H_{14}	Bicyclo[2.2.2]octane	164.25	4586	27.9	447.48	8347	18.7	210.0		29

C_9H_{12}	n-Propylbenzene				s 173.59	9268	53.39	287.8	397.9	31
					m 171.6	8498	49.52			
C_9H_{18}	n-Butylcyclopentane				165.18	11314	68.49	343.8	453.8	30
C_9H_{18}	n-Propylcyclohexane				178.25	10372	58.19	311.9	419.9	31, 32
C_9H_{20}	n-Nonane	217.18	6280	28.91	219.65	15468	70.42	393.6		24
$C_{10}H_{14}$	n-Butylbenzene				s 185.30	11221	60.56	321.2	437.9	31
					m 185.14	11259	60.81			
$C_{10}H_{20}$	n-Butylcyclohexane				198.42	14159	71.36	345.0	459.8	31, 32
$C_{10}H_{22}$	n-Decane				243.50	28715	117.9	425.9		24
$C_{11}H_{24}$	n-Undecane	236.60	6858	28.98	247.58	22179	89.58	458.1		24
$C_{12}H_{18}$	Hexamethylbenzene	116.48	1124	9.65						33
$C_{12}H_{26}$	n-Dodecane				263.58	36836	139.8	490.8		24
$C_{13}H_{28}$	n-Tridecane	255.00	7661	30.04	267.78	28501	106.4	522.9		24
$C_{14}H_{10}$	Anthracene				488.97	29360	60.04	207.1		34
$C_{14}H_{10}$	Phenanthrene	345	1590							35
$C_{14}H_{30}$	n-Tetradecane				279.02	45 070	161.5	555.4		24
$C_{15}H_{30}$	n-Decylcyclopentane				251.02	33 125	132.0	538.5	688.1	30
$C_{15}H_{32}$	n-Pentadecane	270.90	9167	33.84	283.10	34 593	122.2	587.5		24

[21] J. H. Colwell, E. K. Gill, and J. A. Morrison, *J. Chem. Phys.*, 1963, **39**, 635; 1964, **40**, 2041.
[22] J. H. Colwell, E. K. Gill, and J. A. Morrison, *J. Chem. Phys.*, 1965, **42**, 3144.
[23] J. F. Messerly, S. S. Todd, and G. B. Guthrie, *J. Chem. Eng. and Data*, 1970, **15**, 227.
[24] J. F. Messerly, G. B. Guthrie, S. S. Todd, and H. L. Finke, *J. Chem. Eng. and Data*, 1967, **12**, 338.
[25] M. Sugisaki, K. Adachi, H. Suga, and S. Seki, *Bull. Chem. Soc. Japan*, 1968, **41**, 593.
[26] H. Enokido, T. Shinoda, and Y. Mashiko, *Bull. Chem. Soc. Japan*, 1969, **42**, 84.
[27] D. W. Scott, G. B. Guthrie, J. F. Messerly, S. S. Todd, W. T. Berg, I. A. Hossenlopp, and J. P. McCullough, *J. Phys. Chem.*, 1962, **66**, 911.
[28] H. M. Huffman, M. E. Gross, D. W. Scott, and J. P. McCullough, *J. Phys. Chem.*, 1961, **65**, 495.
[29] W. K. Wong and E. F. Westrum, jun., *J. Phys. Chem.*, 1970, **74**, 1303.
[30] J. F. Messerly, S. S. Todd, and H. L. Finke, *J. Phys. Chem.*, 1965, **69**, 353.
[31] J. F. Messerly, S. S. Todd, and H. L. Finke, *J. Phys. Chem.*, 1965, **69**, 4304.
[32] H. L. Finke, J. F. Messerly, and S. S. Todd, *J. Phys. Chem.*, 1965, **69**, 2094.
[33] M. Frankosky and J. G. Aston, *J. Phys. Chem.*, 1965, **69**, 3126.
[34] P. Goursot, H. L. Girdhar, and E. F. Westrum, jun., *J. Phys. Chem.*, 1970, **74**, 2538.
[35] R. A. Arndt and A. C. Damask, *J. Chem. Phys.*, 1966, **45**, 755.

[a] at 282.6 K.

Table 2 (*cont.*)

Formula	*Compound*	T(t) / K	ΔH(t) / J mol^{-1}	ΔS(t) / J K^{-1} mol^{-1}	T(m) / K	ΔH(m) / J mol^{-1}	ΔS(m) / J K^{-1} mol^{-1}	$S^{\ominus}$(298.15 K) / J K^{-1} mol^{-1} c or l	$S^{\ominus}$(298.15 K) / J K^{-1} mol^{-1} g	*Reference*
$C_{16}H_{16}$	[2.2]-Paracyclophane	55.5	213	4.18				265.7		36
$C_{16}H_{32}$	n-Decylcyclohexane				271.43	38597	142.2	540.2	694.4	31, 32
$C_{16}H_{34}$	n-Hexadecane				291.33	53359	183.2	619.7		24
$C_{17}H_{36}$	n-Heptadecane	284.22	10 942	38.50	295.14	40164	136.1	652.2		24
$C_{18}H_{38}$	n-Octadecane				301.33	61706	204.8	480.2		24
$C_{20}H_{14}$	Triptycene				527.18	30275	57.43	274.0		37
CH_4O	Methanol	157.34	636.0	4.04	175.59	3215	18.31	127.2		38
		103 (G)								39
C_3O_2	Carbon suboxide				160.96	5401	33.56		259.9[a]	40
C_3H_7O	Propylene oxide				161.22	6533	40.52	196.3	287.4	41
C_3H_8O	Propan-1-ol	118 (G)			148.75	5372	36.11	192.8	321.7	42
C_3H_8O	Propan-2-ol				185.20	5410	29.21	180.6	305.6	43, 44
$C_3H_8O_2$	Dimethoxymethane				168.01	8332	49.59	244.0	335.7	45
$C_4H_6O_4$	Succinic acid							167.3		46
C_4H_8O	Butan-2-one				186.48	8439	45.25	238.0	338.1	47
					186.47	8385	44.98	239.0	338.2	48
$C_4H_{10}O$	Butan-1-ol				184.51	9372	50.79	225.7	363.2	49
$C_4H_{10}O$	Butan-2-ol	114 (G)			184.70	5970	32.32	214.7	351.4	50
$C_4H_{10}O$	2-Methylpropan-1-ol	139 (G)			171.18	6322	36.93	214.5	350.0	42
$C_4H_{10}O$	2-Methylpropan-2-ol	286.1	828	2.89	298.97	6703	22.42	170.9		51
		281.5	649							
		294.5	490							
$C_4H_{10}O$	Diethyl ether				s 156.92	7190	45.82	253.5	342.2	52
					m 149.86	6820	45.51			
$C_5H_{10}O$	2-Methylfuran				181.90	8552	47.02	213.9		53
$C_5H_{10}O$	Pentan-2-one	110	137.7	2.18	196.31	10632	54.14	274.1	378.2	48
$C_5H_{10}O$	Pentan-3-one	118.5	110.9	0.96	234.16	11594	49.50	266.0	370.1	48
		180	9.6	0.04						
$C_5H_{10}O$	3-Methylbutan-2-one				180.01	9343	51.88	268.5	369.9	48
[illegible]	Valeric acid				[illegible]	[illegible]	[illegible]		[illegible]	[illegible]

C_6H_6O	Phenol				314.07	11514	36.66	144.0		55
$C_6H_8O_7,H_2O$	Citric acid monohydrate							283.4		56
C_6H_{12}	Cyclohexanol	150 (G)								
		244.8	8640							
		265.50	8827	33.24	299.09	1783	5.96	185.1[b]	331.1[b]	57
$C_6H_{14}O$	Hexan-2-one				217.69	14900	68.44	308.1		58
$C_6H_{14}O$	Hexan-3-one	145	682	4.73	217.72	13490	61.98	305.3	409.6	58
$C_6H_{14}O$	3,3-Dimethylbutan-2-one				221.74	11330	51.10	282.4		58
$C_7H_6O_2$	Benzoic acid									59
C_7H_8O	*o*-Cresol				304.20	15820	52.00	165.4	354.8	60

[36] J. T. S. Andrews and E. F. Westrum, jun., *J. Phys. Chem.*, 1970, **74**, 2170.
[37] J. T. S. Andrews and E. F. Westrum, jun., *J. Chem. Thermodyn.*, 1970, **2**, 245.
[38] H. G. Carlson and E. F. Westrum, jun., *J. Chem. Phys.*, 1971, **54**, 1464.
[39] M. Sugisaki, H. Suga, and S. Seki, *Bull. Chem. Soc. Japan*, 1968, **41**, 2586.
[40] L. A. McDougall and J. E. Kilpatrick, *J. Chem. Phys.*, 1965, **42**, 2311.
[41] F. L. Oetting, *J. Chem. Phys.*, 1964, **41**, 149.
[42] J. F. Counsell, E. B. Lees, and J. F. Martin, *J. Chem. Soc. (A)*, 1968, 1819.
[43] R. J. L. Andon, J. F. Counsell, and J. F. Martin, *Trans. Faraday Soc.*, 1963, **59**, 1555.
[44] J. H. S. Green, *Trans. Faraday Soc.*, 1963, **59**, 1559.
[45] D. M. McEachern jun. and J. E. Kilpatrick, *J. Chem. Phys.*, 1964, **41**, 3127.
[46] C. E. Vanderzee and E. F. Westrum, jun., *J. Chem. Thermodyn.*, 1970, **2**, 681.
[47] G. C. Sinke and F. L. Oetting, *J. Phys. Chem.*, 1964, **68**, 1354.
[48] R. J. L. Andon, J. F. Counsell, and J. F. Martin, *J. Chem. Soc. (A)*, 1968, 1894.
[49] J. F. Counsell, J. L. Hales, and J. F. Martin, *Trans. Faraday Soc.*, 1965, **62**, 1869.
[50] R. J. L. Andon, J. E. Connett, J. F. Counsell, E. B. Lees, and J. F. Martin, *J. Chem. Soc. (A)*, 1971, 661.
[51] F. L. Oetting, *J. Phys. Chem.*, 1963, **67**, 2757.
[52] J. F. Counsell, D. A. Lee, and J. F. Martin, *J. Chem. Soc. (A)*, 1971, 313.
[53] H. G. Carlson and E. F. Westrum, jun., *J. Chem. and Eng. Data*, 1965, **10**, 134.
[54] L. A. McDougall and J. E. Kilpatrick, *J. Chem. Phys.*, 1965, **42**, 2307.
[55] R. J. L. Andon, J. F. Counsell, E. F. G. Herington, and J. F. Martin, *Trans. Faraday Soc.*, 1963, **59**, 830.
[56] D. M. Evans, F. G. Hoare, and T. P. Melia, *Trans. Faraday Soc.*, 1962, **58**, 1511.
[57] K. Adachi, H. Suga, and S. Seki, *Bull. Chem. Soc. Japan*, 1968, **41**, 241.
[58] R. J. L. Andon, J. F. Counsell, E. B. Lees, and J. F. Martin, *J. Chem. Soc. (A)*, 1970, 833.
[59] H. Suga and S. Seki, *Bull. Chem. Soc. Japan*, 1965, **38**, 1000.
[60] R. J. L. Andon, J. F. Counsell, E. B. Lees, J. F. Martin, and C. J. Mash, *Trans. Faraday Soc.*, 1967, **63**, 1115.

[a] at 230 K [b] at 280 K

Table 2 (*cont.*)

Formula	*Compound*	$T(t)$/K	$\Delta H(t)$/J mol^{-1}	$\Delta S(t)$/J K^{-1} mol^{-1}	$T(m)$/K	$\Delta H(m)$/J mol^{-1}	$\Delta S(m)$/J K^{-1} mol^{-1}	$S^{\ominus}$(298.15 K)/J K^{-1} mol^{-1} c or l	$S^{\ominus}$(298.15 K)/J K^{-1} mol^{-1} g	*Reference*
C_7H_8O	*m*-Cresol				285.40	10707	37.52	212.6	348.1	60
C_7H_8O	*p*-Cresol				307.94	12707	41.26	167.3	342.3	60
$C_7H_{16}O$	2,4-Dimethylpentan-3-one				204.81	11180	54.58	318.0		58
$C_8H_{14}O$	3-Oxabicyclo[3.2.2]-nonane	208.50	7017	34.4	448.43	6753	15.1	236.3		61
$C_9H_{20}O$	Nonan-5-one	110	373	3.39	269.31	24930	92.56	401.4		58
$C_{12}H_{14}O_4$	Diethyl phthalate	180 (G)			269.92	17984	66.64	425.08		62
$C_{29}H_{41}O_2$	Galvinoxyl radical	81.5	1504.6	18.67				670.10		63
$C_{29}H_{42}O_2$	Phenol derivative							662.24		63
C_2H_3N	Acetonitrile	216.9	897.9		229.315	8167	35.61	149.6	245.5	64
$C_3H_2N_2$	Malononitrile	260.3	1264	4.81	304.9	10795	35.40	131.0	286.6	65
C_3H_5N	Propionitrile	176.964	1707	9.64	180.37	5046	27.98	189.3	287.7	66
$C_4H_4N_2$	Succinonitrile	233	6201	26.57	331.30	3703	11.21	191.6	330.7	67
C_4H_5N	Pyrrole	65.5	12.6		249.74	7908	31.66	156.4	270.5	68
C_5H_9N	Trimethylacetonitrile	{213	232	1.09	292.13	9288	31.80	232.0	333.2	69
		{232.74	1732	7.78						
$C_5H_6N_2$	Dimethylmalononitrile	302.60	9866	32.59	307.47	4054	13.18	187.9	349.1	70
$C_5H_6N_2$	Glutaronitrile				244.21	12585	51.54	239.5	368.6	71
C_6H_7N	2-Methylpyridine				206.446	9724	47.10	217.9	325.0	72
C_6H_7N	3-Methylpyridine				255.01	14180	55.61	216.3	325.0	73
$C_6H_{12}N_2$	1,4-Diazabicyclo-[2.2.2]octane	351.08	10548	30.08	432.99	7431	17.15	157.6		74
$C_7H_{13}N$	1-Azabicyclo-[2.2.2]octane	196	5226	26.52	430	5800	13.39	207.0		61
$C_8H_{15}N$	3-Azabicyclo-[3.2.2]nonane	297.78	14481	48.66	467.12	6916	14.85	234.9		75, 76
$C_2H_5NO_2$	Nitroethane				183.69	9853				77

6

CF_4	Carbon tetrafluoride	76.221	1462	19.19	89.515	705.4	7.880		262.0	80
		76.23	1709	22.54	89.56	712.1	7.95		226.1[c]	81
C_3F_8	Octafluoropropane	99.37	3556	35.78	125.46	477.4	3.805		371.3	82
C_4H_9Cl	t-Butyl chloride	182.91	1872	10.21	247.53	2071	8.37			82*a*
		219.25	5883	26.82						
$C_5H_8F_4$	Pentaerythrityl fluoride	249.40		52.97	367.43	5142	14.01	290.0		83
$C_5H_8Cl_4$	Pentaerythrityl chloride	235	257.3	1.08	368.23	22259	60.46	257.5		84, 85

61 E. F. Westrum, jun., W. K. Wong, and E. Morawetz, *J. Phys. Chem.*, 1970, **74**, 2542.
62 S. S. Chang, J. A. Horman, and A. B. Bestul, *J. Res. Nat. Bur. Stand.*, 1967, **71A**, 293.
63 A. Kosaki, H. Suga, S. Seki, K. Mukai, and Y. Deguchi, *Bull. Chem. Soc. Japan*, 1969, **42**, 1525.
64 W. E. Putnam, D. M. McEeachern, jun., and J. E. Kilpatrick, *J. Chem. Phys.*, 1965, **42**, 749.
65 H. L. Girdhar, E. F. Westrum, jun., and C. A. Wulff, *J. Chem. and Eng. Data*, 1968, **13**, 239.
66 L. A. Weber and J. E. Kilpatrick, *J. Chem. Phys.*, 1962, **36**, 829.
67 C. A. Wulff and E. F. Westrum jun., *J. Phys. Chem.*, 1963, **67**, 2376.
68 D. W. Scott, W. T. Berg, I. A. Hossenlopp, W. N. Hubbard, J. F. Messerly, S. S. Todd, D. R. Douslin, J. P. McCullough, and G. Waddington, *J. Phys. Chem.*, 1967, **71**, 2263
69 E. F. Westrum, jun. and A. Ribner, *J. Phys. Chem.*, 1967, **71**, 1216.
70 A. Ribner and E. F. Westrum, jun., *J. Phys. Chem.*, 1967, **71**, 1208.
71 H. L. Clever, C. A. Wulff, and E. F. Westrum, jun., *J. Phys. Chem.*, 1965, **69**, 1983.
72 D. W. Scott, W. N. Hubbard, J. F. Messerly, S. S. Todd, I. A. Hossenlopp, W. D. Good, D. R. Douslin, and J. P. McCullough, *J. Phys. Chem.*, 1963, **67**, 680.
73 D. W. Scott, W. D. Good, G. B. Guthrie, S. S. Todd, I. A. Hossenlopp, A. G. Osborn, and J. P. McCullough, *J. Phys. Chem.*, 1963, **67**, 685.
74 J. C. Trowbridge and E. F. Westrum, jun., *J. Phys. Chem.*, 1963, **67**, 2381.
75 C. M. Barber and E. F. Westrum, jun., *J. Phys. Chem.*, 1963, **67**, 2373.
76 C. A. Wulff and E. F. Westrum, jun., *J. Phys. Chem.*, 1964, **68**, 430.
77 K. F. Liu and W. T. Zeigler, *J. Chem. Eng. and Data*, 1966, **11**, 187.
78 E. P. Egan *and* jun., Z. T. Wakefield, and T. D. Farr, *J. Chem. and Eng. Data*, 1965, **10**, 138.
79 R. H. Valentine, G. E. Brodale, and W. F. Giaque, *J. Phys. Chem.*, 1962, **66**, 392.
80 H. Ehokido, T. Shinoda, and Y. Mashiko, *Bull. Chem. Soc. Japan*, 1969, **43**, 3415.
81 J. H. Smith and E. L. Pace, *J. Phys. Chem.*, 1969, **73**, 4232.
82 E. L. Pace and A. C. Plaush, *J. Phys. Chem.*, 1967, **47**, 38.
82a A. Dworkin and M. Guillamin, *J. Chim. phys.*, 1966, **63**, 53.
83 J. C. Trowbridge and E. F. Westrum, jun., *J. Phys. Chem.*, 1964, **68**, 255.
84 D. H. Payne and E. F. Westrum, jun., *J. Phys. Chem.*, 1962, **66**, 748.
85 H. L. Clever, W. K. Wong, and E. F. Westrum, jun., *J. Phys. Chem.*, 1965, **69**, 1209.

[c] At 145.12 K

Table 2 (*cont.*)

Formula	*Compound*	T(t) K	ΔH(t) J mol^{-1}	ΔS(t) J K^{-1} mol^{-1}	T(m) K	ΔH(m) J mol^{-1}	ΔS(m) J K^{-1} mol^{-1}	$S^\ominus$(298.15 K) J K^{-1} mol^{-1} c or l	$S^\ominus$(298.15 K) g	*Reference*
$C_5H_8Br_4$	Pentaerythrityl bromide	320	1372	4.35	433.45	27966	64.52	291.1		84, 85
$C_5H_8I_4$	Pentaerythrityl iodide	385	632	1.72				316.7		84, 85
C_6F_6	Hexafluorobenzene				278.14	11435	41.11	285.5		86
					278.25	11590	41.65	279.9	383.2	87
					278.30	11585	41.63	280.8		88
C_6F_5H	Pentafluorobenzene				225.83	10853	48.06	275.9	377.8	89
C_6F_5Cl	Pentafluorochloro-	191.1	1246		257.29	8397	32.64	306.5		86, 90
	benzene	{191	3636	17.91	257.49	8355	32.45	300.7	407.7	91
		245	983	4.01						
$C_6F_3Cl_3$	Trifluorotrichlorobenzene				333.76	19849	59.47	245.3		92
$C_6F_2H_4$	1,2-Difluorobenzene				226.01	11045	48.87	222.6	321.9	93
$C_6F_2H_4$	1,3-Difluorobenzene	186.77	827.1	4.427	204.03	8581	42.06	223.8	321.4	88
$C_7F_5H_3$	Pentafluorotoluene	70.3	210.3	2.94	243.35	12990	53.38	306.4	413.6	94
C_7FH_7	4-Fluorotoluene				216.49	9351	43.19	237.1	339.5	95
CF_2O	Carbonyl fluoride				161.89	6708	41.43		238.9	96
C_3F_6O	Hexafluoroacetone				147.70	8383	56.75		375.0	97
C_6F_5HO	Pentafluorophenol	287	1134	3.93	310.62	16410	52.83	227.1		91
		245.15	1485	6.07	305.96	12845	41.98	242.8		98
C_2F_3N	Trifluoroacetonitrile				128.73	4969	38.60		272.0	99
$C_5F_{11}N$	Perfluoropiperidine	{161.0	6627	41.16	274.12	2816	10.27	393.4	485.8	100
		171.9	1837	10.69						
$C_5H_9O_2N$	L-Proline							164.1		103
$C_5H_{11}O_2N_2$	L-Valine							178.9		104
$C_6H_{13}O_2N$	L-Leucine							211.8		104
$C_6H_{13}O_2N$	L-Isoleucine							208.0		104
$C_9H_{11}O_2N$	L-Phenylalanine							213.6		103

$C_4H_{12}NCl$	Tetramethylammonium chloride	75.76 184.85	116.3 108.4	1.548 0.586	190.7	101
$C_4H_{12}NBr$	Tetramethylammonium bromide				200.8	101
$C_4H_{13}NCl_2$	Tetramethylammonium hydrogen dichloride				253.7	102
$C_6H_9O_2N_3,HCl$	L-Histidine hydro-chloride				276.1	105
$C_6H_{14}O_2N_2,HCl$	L-Lysine hydrochloride				264.5	105
$C_6H_{14}O_2N_4,HCl$	L-Arginine hydrochloride				286.3	105

86 I. E. Paukov and L. K. Glukhikh, *Zhur. Vsesoyuz. Khim. obshch. im D. I. Mendeleeva*, 1967, **12**, 236.
87 J. F. Counsell, J. H. S. Green, J. L. Hales, and J. F. Martin, *Trans. Faraday Soc.*, 1965, **61**, 212.
88 J. F. Messerly and H. L. Finke, *J. Chem. Thermodyn.*, 1970, **2**, 867.
89 J. F. Counsell, J. L. Hales, and J. F. Martin, *J. Chem. Soc.* (*A*), 1968, 2042.
90 I. E. Paukov and L. K. Glukhikh, *Russian J. Phys. Chem.*, 1969, **43**, 754.
91 R. J. L. Andon, J. F. Counsell, J. L. Hales, E. B. Lees, and J. F. Martin, *J. Chem. Soc.* (*A*), 1968, 2357.
92 I. E. Paukov and L. K. Glukhikh, *Russian J. Phys. Chem.*, 1969, **43**, 120.
93 D. W. Scott, J. F. Messerly, S. S. Todd, I. A. Hossenlopp, A. Osborn and, J. P. McCullough, *J. Chem. Phys.*, 1963, **38**, 532.
94 J. F. Counsell, J. L. Hales, E. B. Lees, and J. F. Martin, *J. Chem. Soc.* (*B*), 1968, 2994.
95 D. W. Scott, J. F. Messerly, S. S. Todd, I. A. Hossenlopp, D. R. Douslin, and J. P. McCullough, *J. Chem. Phys.*, 1962, **37**, 867.
96 E. L. Pace and M. A. Reno, *J. Chem. Phys.*, 1968, **48**, 1231.
97 A. C. Paush and E. L. Pace, *J. Chem. Phys.*, 1967, **47**, 44.
98 I. E. Paukov, M. N. Lavent'ava, and M. P. Anisimov, *Russian J. Phys. Chem.*, 1969, **43**, 436.
99 E. L. Pace and R. J. Bobka, *J. Chem. Phys.*, 1961, **35**, 454.
100 W. D. Good, S. S. Todd, J. F. Messerly, J. L. Lacina, J. P. Dawson, D. W. Scott, and J. P. McCullough, *J. Phys. Chem.*, 1963, **67**, 1306.
101 S. S. Chang and E. F. Westrum, jun., *J. Chem. Phys.*, 1962, **36**, 2420.
102 S. S. Chang and E. F. Westrum, jun., *J. Chem. Phys.*, 1962, **36**, 2571.
103 A. G. Cole, J. O. Hutchens, and J. W. Stout, *J. Phys. Chem.*, 1963, **67**, 1852.
104 J. O. Hutchens, A. G. Cole, and J. W. Stout, *J. Phys. Chem.*, 1963, **67**, 1128.
105 A. G. Cole, J. O. Hutchens, and J. W. Stout, *J. Phys. Chem.*, 1963, **67**, 2245.

Table 2 (*cont.*)

Formula	*Compound*	$T(t)$ / K	$\Delta H(t)$ / J mol^{-1}	$\Delta S(t)$ / J K^{-1} mol^{-1}	$T(m)$ / K	$\Delta H(m)$ / J mol^{-1}	$\Delta S(m)$ / J K^{-1} mol^{-1}	$S^{\ominus}(298.15\ K)$ / J K^{-1} mol^{-1} c or l	g	*Reference*
$[C_{10}H_{16}N_2]ClO_4$	Wurster's Blue perchlorate	189.9	1710	9.16						106
C_2H_6S	Ethanethiol				125.25	4976	39.73	207.02	296.1	107
$C_5H_{10}S$	Cyclopentanethiol				m 151.6	7213				
					s 155.39	7831	50.39	256.9	361.4	108
$C_5H_{12}S$	1-Pentanethiol				197.45	17531	88.79	310.4	415.2	107
$C_5H_{12}S$	2-Thiahexane				175.30	12452	71.03	307.5	411.8	109
$C_5H_{12}S$	3-Thiahexane				156.10	10581	67.79	309.5	413.5	109
$C_5H_{12}S$	3,3-Dimethyl-2-thiabutane				190.84	8414	44.09	276.1	373.3	110
$C_5H_{12}S$	2-Methyl-2-butanethiol	145	33.5							
		159.1	7979	50.15	169.37	608·4	3.592	290.1	386.9	111
C_6H_8S	2,5-Dimethylthiophen				m 204.87	7401				
					s 210.55	8192	38.91	244.7		112
$C_6H_{12}S$	Cyclohexanethiol				189.64	10000	52.73	258.6	364.7	113, 114
$C_6H_{12}S$	2,4-Dimethyl-3-thiapentane				195.07	10414	53.39	313.0	415.5	113, 114
$C_6H_{14}S$	4-Thiaheptane				170.44	12142	71.24	338.3	448.4	109
$C_6H_{14}S$	1-Hexanethiol				192.62	18012	93.51	343.2	453.8	107
$C_7H_{16}S$	1-Heptanethiol				229.92	25384	110.4	375.3	492.2	107
$C_8H_{18}S$	5-Thianonane				198.13	19426	98.05	405.1	525.9	109
$C_{10}H_{22}S$	1-Decanethiol				247.86	33317	134.4	473.1	609.7	107
$C_8H_{12}S_6$	1,3,5,7-Tetramethyl-2,4,6,8,9,10-hexathia-adamantane							321.1		115
$C_{14}H_{16}S_4$	1,3,5,7-Tetramethyl-2,4,6,8-tetrathia-adamantane							300.8		115
C_2H_6SO	Dimethyl sulphoxide				[illegible]	[illegible]	[illegible]	[illegible]	[illegible]	[illegible]

$C_2H_6SO_2$	Dimethyl sulphone			[illegible]	[illegible]	[illegible]	[illegible]		[illegible]
		169.3	0.17						
CH_4N_2S	Thiourea	171.2	0.04						
		200	0.71				115.8		117
C_3H_3NS	Thiazole	145 to 175		239.53	9591	40.04	170.0	278.1	118
C_4H_5NS	2-Methylthiazole			m 246.50	11347	46.02			
				s 248.42	12163	48.95	211.9		119
C_7H_5NS	Benzothiazole			275.65	12782	46.37	209.8		120
CH_3POF_2	Methylphosphoryl-difluoride			236.34	11878	50.26	208.34	312.7	121
CH_3POCl_2	Methylphosphoryl-dichloride			306.14	18076	59.04	164.84	339.7	121
CH_3POClF	Methylphosphoryl-chlorofluoride			250.70	11853	47.28	216.40	335.0	121

[106] H. Chihara, M. Nakamura, and S. Seki, *Bull. Chem. Soc. Japan*, 1965, **38**, 1776.
[107] H. L. Finke, J. P. McCullough, J. F. Messerly, G. B. Guthrie, and D. R. Doulsin, *J. Chem. Thermodyn.*, 1970, **2**, 27.
[108] W. T. Berg, D. W. Scott, W. N. Hubbard, S. S. Todd, J. F. Messerly, I. A. Hossenlopp, A. Osborn, D. R. Douslin, and J. P. McCullough, *J. Phys. Chem.*, 1961, **65**, 1425.
[109] J. P. McCullough, H. L. Finke, W. N. Hubbard, S. S. Todd, J. F. Messerly, D. R. Douslin, and G. Waddington, *J. Phys. Chem.*, 1961, **65**, 784.
[110] D. W. Scott, W. D. Good, S. S. Todd, J. F. Messerly, W. T. Berg, I. A. Hossenlopp, J. L. Lacina, A. Osborn, and J. P. McCullough, *J. Chem. Phys.*, 1962, **36**, 406.
[111] D. W. Scott, D. R. Douslin, H. L. Finke, W. N. Hubbard, J. F. Messerly, I. A. Hossenlopp, and J. P. McCullough, *J. Phys. Chem.*, 1962, **66**, 1334.
[112] H. G. Carlson and E. F. Westrum, jun., *J. Phys. Chem.*, 1965, **69**, 1524.
[113] J. F. Messerly, S. S. Todd, and G. B. Guthrie, *J. Chem. Eng. Data*, 1967, **12**, 426.
[114] D. W. Scott and G. A. Crowder, *J. Chem. Phys.*, 1967, **46**, 1054.
[115] S. S. Chang and E. F. Westrum, jun., *J. Phys. Chem.*, 1962, **66**, 524.
[116] H. L. Clever and E. F. Westrum, jun., *J. Phys. Chem.*, 1970, **74**, 1309.
[117] E. F. Westrum, jun. and E. Chang, *Colloq. Internat. du C.N.R.S.*, 1967, No. 156, 163.
[118] P. Goursot and E. F. Westrum, jun., *J. Chem. and Eng. Data*, 1968, **13**, 471.
[119] P. Goursot and E. F. Westrum, jun., *J. Chem. and Eng. Data*, 1968, **13**, 468.
[120] P. Goursot and E. F. Westrum, jun., *J. Chem. and Eng. Data*, 1969, **14**, 1.
[121] G. T. Furukawa, M. L. Reilly, J. H. Piccirelli, and M. Tenenbaum, *J. Res. Nat. Bur. Stand.*, 1964, **68A**, 367.

Table 2 (*cont.*)

Formula	*Compound*	$T(t)$ / K	$\Delta H(t)$ / J mol^{-1}	$\Delta S(t)$ / J K^{-1} mol^{-1}	$T(m)$ / K	$\Delta H(m)$ / J mol^{-1}	$\Delta S(m)$ / J K^{-1} mol^{-1}	$S^\ominus(298.15\ K)$ / J K^{-1} mol^{-1} c or l	g	*Reference*
$C_6H_{18}OSi_2$	Hexamethyldisiloxane				204.93	11922	58.18	433.8	534.9	122
$BO_3C_6H_{12}N$	Triethanolamineborate	466.54	4774	10.23	511.86	24100	47.08	183.2		123, 124
C_3H_9Al	Trimethylaluminium				288.43	8791	30.48	209.4		125
$C_6H_{18}Al_2$	Trimethylaluminium dimer								524.9	125
$SrCa_2(C_2H_5CO_2)_6$	Strontium calcium propionate	104.2	1067	13.4				949.3		126
		282.6	668	2.55						
$PbCa_2(C_2H_5CO_2)_6$	Lead calcium propionate	191.5	4853	24.4				983.2		126
$BaCa_2(C_2H_5CO_2)_6$	Barium calcium propionate	266.9	7284	27.2						126
$C_3H_9N.BH_3$	Trimethylamine borane	219								
		350.1	2535	7.24						
		360.4	5939	16.48	368.70	4947	13.42	169.6	304.2	127
$C_6H_{15}N.BH_3$	Triethylamine borane				269.48	14906	55.32	301.7	433.5	127

[122] D. W. Scott, J. F. Messerly, S. S. Todd, G. B. Guthrie, I. A. Hossenlopp, R. T. Moore, A. Osborn, W. T. Berg, and J. P. McCullough, *J. Phys. Chem.*, 1961, **65**, 1320.

[123] H. L. Clever, W.-K. Wong, C. A. Wulff, and E. F. Westrum, jun., *J. Phys. Chem.*, 1964, **68**, 1967.

[124] P. Castle, R. Stoesser, and E. F. Westrum, jun., *J. Phys. Chem.*, 1964, **68**, 49.

[125] J. P. McCullough, J. F. Messerly, R. T. Moore, and S. S. Todd, *J. Phys. Chem.*, 1963, **67**, 677.

[126] N. Nakamura, H. Suga, H. Chihara, and S. Seki, *Bull. Chem. Soc. Japan*, 1965, **38**, 1779.

[127] H. L. Finke, S. S. Todd, and J. F. Messerly, *J. Chem. Thermodyn.*, 1970, **2**, 129.

transitions have been published.[128–134] Of the 168 compounds listed in Table 2, over 60 exhibit transitions in the solid state.

Order–Disorder Transitions.—The thermodynamic properties of globular molecules have been studied extensively by E. F. Westrum, jun., and his collaborators (Part XVII of the series [29]). Many of these substances have two crystalline forms with a high entropy of transition between the two phases. The high-temperature crystals are characterized by a low entropy of fusion ($<21\ \mathrm{J\,K^{-1}\,mol^{-1}}$), high vapour pressure, cubic or hexagonal symmetry, ease of deformation, and high self-diffusion rates, and have been called 'plastic crystals' by Timmermans.[135] Values of entropy of transition, $\Delta S(\mathrm{t})$ have been used in the interpretation of order–disorder transitions for which $\Delta S(\mathrm{t})$ is approximately equal to $R \ln N$, where N is the ratio of the number of states of disorder in the two phases. Values of N can sometimes be deduced from considerations of possible disordered states arising from positional disorder, over-all molecular orientation, and internal molecular conformation.[136] Examples of simple order–disorder transitions in which N is small have been given by Guthrie and McCullough,[136] who developed the method further to include more complex ($N \geqslant 10$) order–disorder transitions found in plastic crystals. For these compounds, the high value of the entropy of transition implies a large number of distinguishable orientations which can be interpreted by considerations of molecular and lattice symmetry, and steric interactions between neighbouring molecules.

A theoretical treatment of order–disorder phenomena in molecular crystals has been developed by Pople and Karasz.[137, 138] The theory considered disorder in both the positions and orientations of the molecules and it was assumed that each molecule could take up one or two orientations on the normal, α-, and the interstitial, β-, sites of the two-lattice model proposed by Lennard-Jones and Devonshire in their treatment of the melting of inert gas crystals. The theory introduced a single non-dimensional parameter, ν- related to the relative energy barriers for the

[128] Conference on 'Plastic Crystals and Rotation in the Solid State,' *J. Phys. and Chem. Solids*, 1961, **18**, 1–92.

[129] E. F. Westrum, jun., and J. P. McCullough, 'Thermodynamics of Crystals' in 'Physics and Chemistry of the Organic Solid State,' eds. D. Fox, M. M. Labes, and A. Weissberger, Interscience, New York, 1963, Vol. 1, Chap. 1.

[130] E. F. Westrum, jun., *Pure Appl. Chem.*, 1964, **8**, 167.

[131] Conference on 'Mouvements et Changements de Phase dans les Solides Moléculaires,' *J. Chim. phys.*, 1966, **63**, 9–205.

[132] E. F. Westrum, jun., *Ann. Rev. Phys. Chem.*, 1967, **18**, 135.

[133] J. G. Aston, 'Plastics Crystals' in 'Physics and Chemistry of the Organic Solid State,' eds. D. Fox, M. M. Labes, and A. Weissberger, Interscience, New York, 1963, Vol. 1, Chap. 9.

[134] C. N. R. Rao and K. J. Rao, 'Phase Transformation in Solids, in 'Progress in Solid-State Chemistry,' Pergamon, Oxford, 1967, Vol. 4, Chap. 4.

[135] J. Timmermans, *J. Phys. and Chem. Solids*, 1961, **18**, 1.

[136] G. B. Guthrie and J. P. McCullough, *J. Phys. and Chem. Solids*, 1961, **18**, 53.

[137] J. A. Pople and F. E. Karasz, *J. Phys. and Chem. Solids*, 1961, **18**, 28.

[138] F. E. Karasz and J. A. Pople, *J. Phys. and Chem. Solids*, 1961, **20**, 294.

rotation of a molecule and for its diffusion in a crystal lattice. For small values of this parameter, the theory makes quantitative predictions for a solid-state rotational transition and a solid–liquid transition. For large values, the two transitions coalesce into a single transition (fusion). The theory of Pople and Karasz was applied to carbon tetrafluoride,[81] which was found experimentally to have an orientationally disordering transition at 76.23 K and a positionally disordering fusion at 89.56 K. The carbon tetrafluoride molecule was considered to reside in two potential wells, having corresponding potential barriers hindering the vibration of the molecule on its lattice site and its translation to an interstitial site. Calculation of these two potential barriers in terms of the Pople and Karasz parameter ν gave theoretical entropies of transition and fusion consistent with the experimental values. Recently, the theory has been further developed[139] to include the possibility of more than two molecular orientations, which were defined by the method of Guthrie and McCullough.[136] The theory gave values for entropies of transitions in very good agreement with calorimetric values.

Glass Transition.—Many organic liquids can be supercooled to temperatures well below the melting temperature of the crystalline phase and at a certain temperature, the glass transformation temperature, there are marked changes in properties such as thermal expansivity, heat capacity, and viscosity. The glassy solid retains some of the disorder of the liquid state and the entropy does not approach zero as $T \to 0$.

Some organic liquids, *e.g.* propan-1-ol[42] (see Figure 3), 2-methylpropan-1-ol,[42] and butan-2-ol,[50] form glasses very readily and crystallize only with difficulty. The glass transformation is a relaxation phenomenon and the properties of the glass are dependent on the history of the sample, particularly on the duration of the experiment. Diethyl phthalate[62] has been carefully studied in liquid, crystalline, and glass (quenched and annealed) forms. The heat capacities of the annealed glass were lower than those of the quenched and the residual entropies were 20 J K^{-1} mol^{-1} and 23 J K^{-1} mol^{-1} respectively. The glass transformation temperatures were 176.5 K for the annealed and 180.8 K for the quenched glass.

Seki and collaborators have devised apparatus for heat capacity measurements of the glass form of substances which cannot normally be prepared as glasses. They described a calorimeter in which a glass was formed in the sample measurement cell by carefully controlled condensation of the sample vapour.[39] With this apparatus, measurements were made on amorphous samples of methanol[39] and water,[140] which exhibited glass transition phenomena. Methanol, which was deposited from methanol vapour in the measurement cell at 95 K, showed a glass transition phenomenon at 103 K, at which temperature the heat capacity rose by 26 J K^{-1}

[139] L. M. Amzel and L. N. Becka, *J. Phys. and Chem. Solids*, 1969, **30**, 521.
[140] M. Sugisaki, H. Suga, and S. Seki, *Bull. Chem. Soc. Japan*, 1968, **41**, 2591.

mol^{-1}. Rapid crystallization started at 105 K accompanied by an exothermic effect amounting to 1.54 kJ mol^{-1}.

Several crystalline substances have been found to supercool and transform into glass-like forms which exhibit glass transition temperatures with sharp changes of heat capacity. Examples of organic substances in this class are cyclohexanol[57] and 2-methylthiophen.[141]

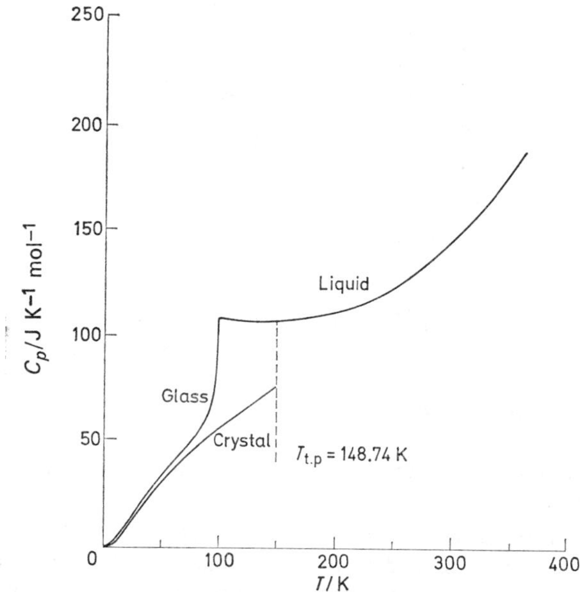

Figure 3 *Heat capacity of propan-*1*-ol; crystal, liquid, and glass*

Cyclohexanol crystallized in a high-temperature form (crystal I), a low-temperature form (crystal II), and a metastable form (crystal III). It was found that crystal I could be supercooled and the supercooled crystals showed a glass-like transition at 150 K and a residual entropy of 4.72 J K^{-1} mol^{-1} at $T = 0$. The term 'glassy crystal' was applied to the supercooled crystal I phase. The glass-like transition in the heat capacity curve of 2-methylthiophen was found in the normal crystal and not in a supercooled phase. The heat capacity of 2-methylthiophen plotted against temperature gave a continuous inflected curve with no maximum in the transition region. A slow energy release was observed during the heating of the sample near the transition region. A similar heat evolution was observed during heat capacity measurements of pentan-2-one and pentan-3-one[48] but, for these compounds, the effect was apparently caused by supercooling of a high-temperature form. If the crystals were annealed near

[141] H. G. Carlson and E. F. Westrum, jun., *J. Chem. and Eng. Data*, 1968, **13**, 273.

the temperature at which the heat evolution occurred, they were transformed into the low-temperature phase which exhibited a small lambda-type transition at a temperature higher than the annealing temperature. Pentan-2-one (see Figure 4) was found to have a non-isothermal transition

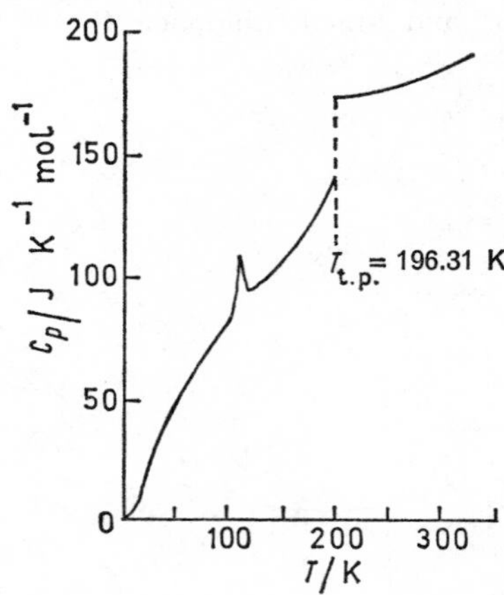

Figure 4 *Heat capacity of pentan-2-one*

with a maximum heat capacity at 118.5 K and an entropy of transition of 0.96 J K^{-1} mol^{-1}.

Investigation of Phase Transition by N.M.R. Spectroscopy.—Several benzene derivatives have been studed in the crystalline state by techniques other than calorimetry, particularly n.m.r. spectroscopy and the measurement of dielectric properties.[131] Pentafluorochlorobenzene, for example, has been investigated by heat capacity calorimetry from 10 to 350 K,[86, 91] and by n.m.r. spectroscopy between 77 K and its melting temperature.[142] The heat capacity curve (Figure 5) shows complex phase transformations starting below 190 K. The heat capacity of the crystals rises to very high maxima at 191 K and 245 K with a 'hump' at 206 K. The second moments of the n.m.r. absorption line shapes are also plotted on Figure 5. From the n.m.r. measurements it was concluded that the molecule of pentafluorochlorobenzene is held rigidly in the crystal lattice at 77 K. The slight increase in second moment between 77 K and 190 K was explained by thermal expansion of the lattice and/or by molecular vibrations. At 190 K, near the temperature of the largest C_p anomaly, the n.m.r. absorption linewidth begins to narrow and the second moment falls sharply. This change in second moment was accounted for by reduction of the nuclear dipole–dipole interactions by rotation of the pentafluorochlorobenzene molecule

[142] I. J. Lawrenson and C. Lewis, *Proc. Phys. Soc.*, 1966, **89**, 923.

about an axis perpendicular to the plane of the molecule. This interpretation was in agreement with the measured entropy of transition at 191 K, *viz.* 17.9 J K^{-1} mol^{-1}, which is close to $R \ln 6$ (14.9 J K^{-1} mol^{-1}). The curve for the variation of the second moment with temperature shows no effect corresponding to the heat capacity anomaly at 245 K and it was suggested that this transition may be associated with the onset of molecular motion too slow to narrow the absorption line or with a change in crystal

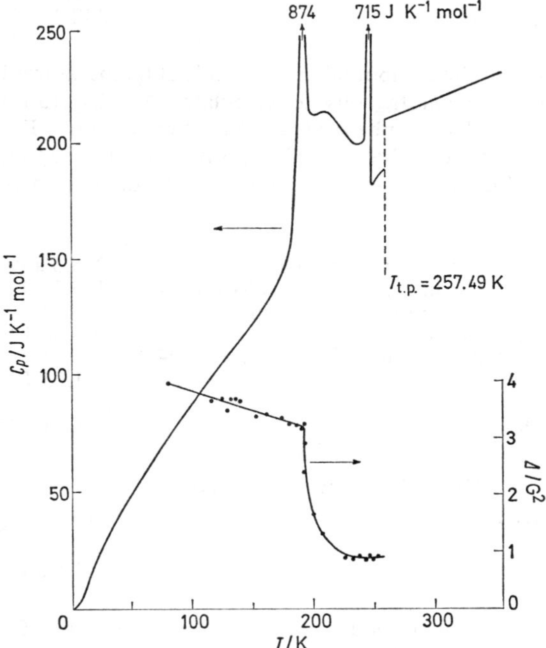

Figure 5 *Heat capacity C_p and n.m.r. second moment Δ of pentafluorochlorobenzene*

structure. This explanation is supported by unusually long equilibration times observed during heat capacity measurements in the temperature range 193 to 244 K.

Entropy in Ideal Gas State.—The computation of the entropy of an organic compound in the gas phase from its entropy in a condensed phase was referred to in the introduction. The additional experimental data required for the computation are sometimes available from vapour-pressure measurements and from vapour-flow calorimetry. For example, the entropy of diethyl ether in the ideal gas state (298.15 K, 101.325 kPa) was calculated as follows:[52]

$S^\ominus$(l, 298.15 K) from low-temperature heat capacity	253.5 J K^{-1} mol^{-1}
$\Delta H_v/T$ from measurements of ΔH_v by vapour-flow calorimetry	90.86 J K^{-1} mol^{-1}
R ln p from vapour pressure measurements	− 2.88 J K^{-1} mol^{-1}
$S_i - S_r = p\,dB/dT$, derived from vapour heat capacities measured by vapour-flow calorimetry	0.7 J K^{-1} mol^{-1}
$S^\ominus$(g, 298.15 K)	342.2 J K^{-1} mol^{-1}

If heat capacities of the compound are available at temperatures higher than 298 K, it may be advantageous to calculate $S^\ominus$(g, 298.15 K) from the entropy of the condensed phase at a higher temperature. Errors arising from extrapolation of enthalpies of vaporization and vapour pressures may thus be reduced. For example, the entropy of propan-1-ol gas was calculated as follows:[42]

$S^\ominus$(l, 350 K) from C_p measurements	218.5 J K^{-1} mol^{-1}
$\Delta H_v/T$ at 350 K from vapour-flow calorimetry	123.72 J K^{-1} mol^{-1}
R ln p at 350 K from vapour pressure measurements	− 6.86 J K^{-1} mol^{-1}
$S_i - S_r$ at 350 K from vapour-flow calorimetry	1.05 J K^{-1} mol^{-1}
$S^\ominus$(g, 350 K)	336.41 J K^{-1} mol^{-1}
$S^\ominus$(g, 350 K) − $S^\ominus$(g, 298.15 K) $= \int_{298.15\,\mathrm{K}}^{350\,\mathrm{K}} (C_p^\ominus/T)\,dT$ from vapour-flow calorimetry	14.85 J K^{-1} mol^{-1}
$S^\ominus$(g, 298.15 K)	321.6 J K^{-1} mol^{-1}

Data on gaseous entropies are required for calculations of equilibrium constants of chemical reactions. The equilibrium constants of the reaction, 2EtOH(g) = EtOEt(g) + H_2O(g) have been recently calculated[52] using the value above for $S^\ominus$(g, 298.15 K) of diethyl ether together with enthalpies of formation, entropies, and vapour heat capacities of the compounds taking part in the reaction. The values obtained agreed well within the limits of experimental error with direct measurements of the equilibrium constant of the reaction. Reasonably good agreement was also obtained for the values of $S^\ominus$(g, 298.15 K) for (±)-butan-2-ol calculated from low-temperature heat capacities[50] and obtained from measurements of equilibrium constants of the reaction:[143]

$$CH_3{\cdot}CH_2{\cdot}CHOH{\cdot}CH_3(g) = CH_3{\cdot}CH_2{\cdot}CO{\cdot}CH_3(g) + H_2O(g).$$

[143] E. Buckley and E. F. G. Herington, *Trans. Faraday Soc.*, 1965, **61**, 1620.

The 'third-law' value of $S^{\ominus}$(g, 298.15 K) was found to be 357.2 J K^{-1} mol^{-1}, which may be compared with 358.5 J K^{-1} mol^{-1} from equilibrium studies.

Calculation of Heat Capacities of Crystalline Compounds.—The method developed by Lord, Ahlberg, and Andrews [144] for the calculation of the heat capacity of crystalline benzene has been extended by Wulff [145] to the determination of barrier heights from low-temperature heat capacities. The heat capacity at constant pressure, C_p, of a crystalline phase is divided into contributions arising from (i) the intermolecular vibrations (lattice vibrations) corresponding to the translational and over-all rotational contributions for the gas, (ii) the intramolecular vibrations, (iii) the expansion of the lattice, and (iv) the internal rotations. Wulff derives an equation for the heat capacity and describes in detail the procedure for its use. He showed that the barrier heights derived from thermal data for toluene, *o*, *m*-, and *p*-xylenes, mesitylene, 1,1-dichloroethane, and 1,1,1-trifluoroethane were in agreement with those estimated from comparisons of 'third-law' and statistically determined entropies of the ideal gas state. Similar calculations have been published [146] for mesitylene, toluene, benzenethiol, methanethiol, and trifluoromethylthiol. For these compounds, the precision of the low barrier values was better than for those determined from gas-phase entropy calculations and that of the high barrier values was similar to the precision of gas-phase entropy values. Recently, the magnitudes of the potential energy barriers hindering internal rotation of methyl groups in dimethyl sulphoxide and dimethyl sulphone were estimated by Wulff's method and by comparison of experimental and statistically calculated values of the 'third-law' entropies in the ideal gas phase.[116] The values by Wulff's method were 17.2 and 17.6 J K^{-1} mol^{-1} for dimethyl sulphoxide and dimethyl sulphone respectively and 11.7 and 14.2 J K^{-1} mol^{-1} by the 'third-law' method. Values obtained by spectroscopic methods were in closer agreement with those obtained by the 'third-law' method than those based on heat capacities for the crystals.

[144] R. C. Lord, jun., J. E. Anlberg, and D. H. Andrews, *J. Chem. Phys.*, 1937, **5**, 649.
[145] C. A. Wulff, *J. Chem. Phys.*, 1963, **39**, 1227.
[146] K. O. Simpson and E. T. Beynon, *J. Phys. Chem.*, 1967, **71**, 2796.

5
The p, V, T Behaviour of Single Gases

BY J. D. COX AND I. J. LAWRENSON

1 Introduction

Chemical thermodynamics is concerned with the interrelation of variables of state for chemical substances, and in this chapter we deal with the interrelation of pressure p, volume V, and temperature T for single gases.

In other reviews of the p, V, T relations of gases, the emphasis has often been on chemical engineering aspects or, in some instances, on the molecular physics aspect. In this chapter, the standpoint is adopted that p, V, T studies of gases are an important component of the total determination of the thermodynamic properties of chemical substances in all physical states. In later sections we shall show how knowledge of the p, V, T relations of a gas permits calculation of the dependence of real-gas thermodynamic properties on pressure and temperature, and also how p, V, T studies give information on interactions at the molecular level. Our treatment will concern mostly, though not exclusively, the subcritical temperature range. Hence, in pressure terms, our emphasis will be on the 'low' (about 1 bar) and 'medium' (1 to 100 bar) ranges.

2 The Units of p, V, T

Pressure.—The SI unit of pressure [1] is the newton per square metre, N m^{-2}, which by recent international agreement may be called the pascal, Pa. For p, V, T studies the pascal is inconveniently small, and the most appropriate multiple of the unit would appear, according to SI rules, to be the megapascal, MPa = 10^6 Pa. However, the MPa is not yet in common use and we shall employ instead the bar, equal to 10^5 Pa. In SI, the bar has the status of a unit that may be used as an (inferior) alternative to an SI unit and its *recommended* multiples. The bar is related to some other, now obsolescent, units of pressure as follows:[2]

$$1 \text{ bar} = 0.986\,923 \text{ standard atmosphere}$$

$$1 \text{ bar} = 750.064 \text{ Torr ('mmHg')}$$

$$1 \text{ bar} = 1.019\,716 \text{ kgf cm}^{-2} \text{ ('at.')}$$

$$1 \text{ bar} = 14.5038 \text{ lbf in}^{-2} \text{ ('p.s.i.')}$$

[1] M. L. McGlashan, 'Physicochemical Quantities and Units,' Royal Institute of Chemistry, London, 2nd Edn., 1971.

[2] P. Anderton and P. H. Bigg, 'Changing to the Metric System,' Her Majesty's Stationery Office, London, 1969, 3rd Edn.

Volume and Density.—The SI unit of volume [1] is the cubic metre, m^3, and the SI unit of density is $kg\,m^{-3}$.

A 'unit' of volume (strictly a ratio of volumes) that is unique to p, V, T work is the Amagat volume, much used in the accurate studies by Michels and his colleagues at the van der Waals Laboratory, Amsterdam. The Amagat volume of a gas is the ratio between the volume of a fixed mass of the gas at any temperature and pressure and its volume at some reference temperature and pressure—273.15 K and 1 standard atmosphere in Amsterdam work. The usefulness of the Amagat volume in reporting p, V, T measurements was explained in a report [3] prepared by Professor Michels shortly before his death.

Amount of Substance.—The SI unit of amount of substance [1] is the mole, symbolized as mol. Hence the SI unit of molar volume is $m^3\,mol^{-1}$. The inverse of this unit, $mol\,m^{-3}$, is the SI unit of amount density or concentration.

The second virial coefficient of a gas (defined later) has the dimension of molar volume and hence its SI unit is $m^3\,mol^{-1}$. However, the submultiple $dm^3\,mol^{-1}$ will often be more convenient to use.

The symbol to be used here for molar volume in general is V_m.

Temperature.—The SI unit of thermodynamic temperature [1] is the kelvin, K, which may be used for specifying both temperature (in the absolute sense) and temperature interval. The degree Celsius, °C, identical with K, is often used to express values of Celsius temperature defined as thermodynamic temperature less 273.15 K.

In experimental p, V, T work, as in other branches of experimental science, it is not thermodynamic temperatures but values on a practical temperature scale that are actually employed. To keep corresponding values closely in step, as the accuracy of thermometrology increases, it is necessary to adjust and redefine the practical scale from time to time. Such redefinitions were promulgated in 1948 and 1968 by the General Conference on Weights and Measures. As a result of the redefinitions, a given temperature measured on the 1948 International Practical Temperature Scale (IPTS 48) may well differ from that measured on the 1968 International Practical Temperature Scale (IPTS 68), the one now in force.[4] The difference varies over the range for which the scale is valid (currently, above 14 K). In consequence, any publication containing accurate p, V, T data (or constants derived therefrom) should be scrutinized to determine the particular practical temperature scale used by the investigator(s). Such scrutiny is especially important in relation to work published in the years immediately succeeding a redefinition of the practical temperature scale,

[3] A. Michels (translated by H. J. Michels and edited by S. Angus), 'A note on the Amagat unit of volume,' Report PC/D30, IUPAC Thermodynamic Tables Project Centre, Imperial College, London, 1970.

[4] *Metrologia*, 1969, **5** (2), 35.

since investigators tend to be reluctant to change to a new definition—*caveat emptor*!

The conversion of p, V, T data published earlier on to the 1968 International Practical Temperature Scale will require a painstaking review of the whole literature; a method of proceeding has been discussed by Angus.[5] In this article we mainly deal with p, V, T data that relate to differences between real- and ideal-gas properties and as these differences are generally not established to an accuracy such that changes in definition of temperatures are relevant, we avoid the problem of revision here. The larger problem of converting measured thermodynamic properties that depend on temperature from IPTS 48 to IPTS 68 has been discussed by Rossini.[6]

The Gas Constant, *R*.—The gas constant is among the less well established of the fundamental constants, and efforts to determine it more precisely are known to be in progress. The value used here is $R = 8.3143\ \mathrm{J\,K^{-1}\,mol^{-1}}$, for which three times the standard deviation is $0.0012\ \mathrm{J\,K^{-1}\,mol^{-1}}$. In thermodynamic calculations using SI it is often necessary to employ the equivalences $1\ \mathrm{J} = 1\ \mathrm{N\,m} = 1\ \mathrm{Pa\,m^3}$. Hence $R = 8.3143\ \mathrm{N\,m\,K^{-1}\,mol^{-1}} = 8.3143\ \mathrm{Pa\,m^3\,K^{-1}\,mol^{-1}}$.

3 Equations of State

The expression 'equation of state', though applicable to solids, liquids, or gases (pure or mixed), is here used to mean the equilibrium relation between pressure, volume, and temperature of a single gas, in the absence of a special force field. The simplest equation of state:

$$pV_{\mathrm{m}} = RT, \tag{1}$$

describes the behaviour of an ideal gas at any pressure, or the limiting behaviour of a real gas as pressure tends to zero. Much of this chapter will be concerned with discussing the departure of real-gas behaviour from equation (1). An extreme case of such departure is the condensation of a gas at subcritical temperatures to form a liquid, a phenomenon that is not implicit in equation (1). The modification of equation (1) so that the possibility of condensation is allowed for has been the preoccupation of numerous workers during the century since van der Waals published his classic work. The search for a useful equation of state has been described by Martin[7] as 'mathematically fascinating and particularly tantalizing because it seems so simple, at least at the start'.

Numerous equations of state have been proposed.[8, 9] We do not attempt

[5] S. Angus, 'A note on the 1968 International Practical Temperature Scale,' Report PC/D26, IUPAC Thermodynamic Tables Project Centre, Imperial College, London, 1970.

[6] F. D. Rossini, *J. Chem. Thermodyn.*, 1970, **2**, 447.

[7] J. J. Martin, *Ind. Eng. Chem.*, 1967, **59** (12), 34.

[8] J. R. Partington, 'An Advanced Treatise on Physical Chemistry,' Longmans Green, London, 1948, Vol. I, p. 660.

[9] J. A. Beattie and W. H. Stockmayer, 'A Treatise on Physical Chemistry', eds. H. S. Taylor and S. Glasstone, Van Nostrand, New York, 1951, 3rd Edn., Vol. II, p. 187.

to list them, but refer only to some which are frequently cited in the modern literature. Recent precise p, V, T measurements [10, 11] on gases up to very high densities provide a cautionary comment on attempts to represent such data analytically: Robertson and Babb conclude that *no known* equation of state adequately represents their results.

The van der Waals Equation.—Though rarely used nowadays for fitting to experimental p, V, T data, the van der Waals equation (2) is frequently cited in current discussions of gas + liquid phenomena. Indeed, allusion has been made [12] to a 'renaissance' in the use of the equation:

$$p = RT/(V_m - b) - a/V_m^2. \tag{2}$$

A fundamental requirement of any equation of state for use around the critical point is that it should be capable of showing a point of inflexion in the p, V isotherm for the critical temperature, T_c, since

$$(\partial p/\partial V)_T = 0 \qquad (T = T_c), \tag{3}$$

and

$$(\partial^2 p/\partial V^2)_T = 0 \qquad (T = T_c). \tag{4}$$

The van der Waals equation can show such an inflexion, and by application of (3) and (4) to (2) a set of three equations containing a and b is obtained. Alternative solutions for a and b are obtainable as follows:

(i) in terms of T_c and V_c (the critical molar volume),

$$a = 9RT_cV_c/8 \quad \text{and} \quad b = V_c/3,$$

(ii) in terms of T_c and p_c (the critical pressure),

$$a = 27R^2T_c^2/64p_c \quad \text{and} \quad b = RT_c/8p_c,$$

(iii) in terms of p_c and V_c,

$$a = 3p_cV_c^2 \quad \text{and} \quad b = V_c/3.$$

By substitution of these pairs of values of a and b into equation (2), three alternative formulations of the van der Waals equation in terms of the critical parameters are obtained:

(i) $$p = RT/(V_m - V_c/3) - 9RT_cV_c/8V_m^2, \tag{5}$$

(ii) $$p = RT/(V_m - RT_c/8p_c) - 27R^2T_c^2/64p_cV_m^2, \tag{6}$$

(iii) $$p = RT/(V_m - V_c/3) - 3p_cV_c^2/V_m^2. \tag{7}$$

Yet further formulations of the van der Waals equation are possible by introduction of reduced pressures p_r, reduced molar volumes V_r, and reduced temperatures T_r, defined by $p_r = p/p_c$, $V_r = V_m/V_c$, and $T_r = T/T_c$. Thus from equation (5)

$$p_r = (RT_c/p_cV_c)\,[3T_r/(3V_r - 1) - 9/8V_r^2]. \tag{8}$$

[10] S. L. Robertson, S. E. Babb, and G. J. Scott, *J. Chem. Phys.*, 1969, **50**, 2160.
[11] S. L. Robertson and S. E. Babb, *J. Chem. Phys.*, 1969, **50**, 4560.
[12] M. Rigby, *Quart. Rev.*, 1970, **24**, 416.

It is evident that at a critical point $p_r = V_r = T_r = 1$, so by substitution of these values into (8) we have

$$p_c V_c / RT_c = Z_c = 3/8, \tag{9}$$

where Z_c is the compression factor at the critical point. Combination of (8) and (9) then leads to

$$p_r = 8T_r/(3V_r - 1) - 3/V_r^2. \tag{10}$$

This so-called reduced equation of state implies that all gases when held at equal values of p_r and T_r should have equal values of V_r. This theorem, known as the Principle of Corresponding States, has provided the basis for many theories of the fluid state. In the next section we refer to some attempts to modify the van der Waals equation to obtain an improved fit to real-gas p, V, T data, while retaining the simplicity of the Principle of Corresponding States.

Generalized Equations of State.—Redlich and Kwong [13] proposed the following modified form of equation (2):

$$p = RT/(V_m - b) - a/T^{\frac{1}{2}}V_m(V_m + b). \tag{11}$$

By application of arguments similar to those that led from equation (2) to equation (8), the Redlich–Kwong equation leads to

$$p_r = \frac{T_r}{Z_c(V_r - 0.08664/Z_c)} - \frac{0.42748}{T_r^{\frac{1}{2}}V_r Z_c^2(V_r + 0.08664/Z_c)^2}. \tag{12}$$

Equation (12) is another example of a reduced, or generalized, equation of state and, just as substitution of $p_r = V_r = T_r = 1$ into equation (8) showed that a van der Waals gas should have $Z_c = 0.375$ [equation (9)], so similar substitution into equation (12) shows that a Redlich–Kwong gas should have $Z_c = 0.333$. Now the experimental value of Z_c for most simple gases is $\leqslant 0.29$, so to improve the overall fit it is often recommended that experimental values of Z_c be used in equation (12).

Whilst the Redlich–Kwong equation provides a better fit to real-gas behaviour than does the van der Waals equation, attempts to produce even better generalized equations of state have been made by many authors. An example, equation (13), is due to Martin,[7] and like equation (12) is best used with experimental values of Z_c, rather than the calculated value of 0.335:

$$p_r = \frac{T_r}{Z_c(V_r - 0.085/Z_c)} - \frac{9(4 - T_r)}{64Z_c^2(V_r + 0.04/Z_c)^2}. \tag{13}$$

Other examples build on the foundations of the Redlich–Kwong equation by addition of extra parameters. A recent instance is the work of Redlich and Ngo.[14] Another instance is the work of Sugie and Lu,[15] who proposed

[13] O. Redlich and J. N. S. Kwong, *Chem. Rev.*, 1949, **44**, 233.
[14] O. Redlich and V. B. T. Ngo, *Ind. Eng. Chem. Fundamentals*, 1970, **9** (2), 287.
[15] H. Sugie and B. C.-Y. Lu, *Ind. Eng. Chem. Fundamentals*, 1970, **9** (3), 428.

the following generalized equation of state:

$$p = \frac{RT}{V_m - b + c} - \frac{a}{T^{\frac{1}{2}}(V_m + c)(V_m + b + c)} + \frac{dT^2}{V_m^2} + \sum_{j=1}^{5} \frac{e_j + f_j T + g_j T^{-2}}{V_m^{j+1}}, \tag{14}$$

where

$$a = a^* R^2 T_c^{2.5}/p_c, \tag{14a}$$
$$b = b^* R T_c/p_c, \tag{14b}$$
$$c = c^* R T_c/p_c, \tag{14c}$$
$$d = d^* R^2/p_c, \tag{14d}$$
$$e_j = e_j^*(R^{j+1} T_c^{j+1}/p_c^j), \tag{14e}$$
$$f_j = f_j^*(R^{j+1} T_c^j/p_c^j), \tag{14f}$$
$$g_j = g_j^*(R^{j+1} T_c^{j+3}/p_c^j), \tag{14g}$$

and

$$a^* = 0.42748,$$
$$b^* = 0.08664,$$
$$c^* = (1 - 3Z_c)/3,$$
$$d^* = -\,4.9882 \times 10^{-3} + 7.4904 \times 10^{-2}\,\omega,$$
$$e_1^* = -\,6.7421 \times 10^{-2} - 5.5509 \times 10^{-2}\,\omega,$$
$$e_2^* = -\,4.2765 \times 10^{-2} - 1.0221 \times 10^{-2}\,\omega,$$
$$e_3^* = -\,2.2213 \times 10^{-4} - 5.8986 \times 10^{-3}\,\omega,$$
$$e_4^* = 4.1495 \times 10^{-4} + 4.1457 \times 10^{-4}\,\omega,$$
$$e_5^* = 4.7332 \times 10^{-5} + 3.6809 \times 10^{-5}\,\omega,$$
$$f_1^* = 8.6454 \times 10^{-2} + 1.9838 \times 10^{-1}\,\omega,$$
$$f_2^* = 1.0935 \times 10^{-2} - 6.6320 \times 10^{-2}\,\omega,$$
$$f_3^* = 3.7028 \times 10^{-3} + 1.8764 \times 10^{-2}\,\omega,$$
$$f_4^* = -\,5.3805 \times 10^{-4} - 1.7176 \times 10^{-3}\,\omega,$$
$$f_5^* = 2.5478 \times 10^{-5} - 1.1957 \times 10^{-5}\,\omega,$$
$$g_1^* = 3.1751 \times 10^{-2} - 1.9409 \times 10^{-1}\,\omega,$$
$$g_2^* = -\,6.8871 \times 10^{-4} + 5.4512 \times 10^{-2}\,\omega,$$
$$g_3^* = 7.0579 \times 10^{-3} - 5.5635 \times 10^{-3}\,\omega,$$
$$g_4^* = -\,1.9279 \times 10^{-3} + 7.9439 \times 10^{-4}\,\omega,$$
$$g_5^* = 1.0455 \times 10^{-4} - 9.1929 \times 10^{-5}\,\omega,$$

where ω is Pitzer's acentric factor.[16]

[16] K. S. Pitzer, D. Z. Lippman, R. F. Curl, jun., C. M. Huggins, and D. E. Petersen, *J. Amer. Chem. Soc.*, 1955, **77**, 3433.

The Sugie–Lu equation of state is obviously a product of the digital-computing era. It affords good fits to experimental p, V, T data for simple gases over wide ranges of the variables, but achieves this fit by departing far from the attractive simplicity of equation (8). We follow this path no further, but turn in the next section to consider a different approach to the accurate representation of p, V, T data.

Virial Equations of State.—The closed equations of state described so far (*i.e.* those with a prescribed number of constants) lack the flexibility required to describe the behaviour of a gas from relatively low temperature and pressure to relatively high temperature and pressure. Except for the critical region it is preferable to use an open form of equation of which the outstanding example is the so-called virial equation of state:

$$p = (RT/V_m)(1 + B/V_m + C/V_m^2 + D/V_m^3 + \ldots). \qquad (15)$$

The quantities B, C, D, ..., termed the second, third, fourth, ..., virial coefficients, are for a given gas functions of temperature only. Equation (15) can be described as the virial equation of state in the volume-explicit (or Leiden) form. A related equation of state, that in the pressure-explicit (or Berlin) form, is also in common use:

$$p = (1/V_m)(RT + B'p + C'p^2 + D'p^3 + \ldots), \qquad (16)$$

where the coefficients are distinguished from those in (15) by use of primes. The coefficients in *infinite* series of the two types (15) and (16) can be shown to be inter-related as follows:

$$B = B', \qquad (17)$$

$$C = B^2 + RTC', \qquad (18a)$$

or

$$C' = (C - B^2)/RT, \qquad (18b)$$

$$D = B^3 + RTBC' + R^2T^2D', \qquad (19a)$$

or

$$D' = (D - 3BC + 2B^3)/R^2T^2. \qquad (19b)$$

Thus it is feasible to write a Berlin-type virial equation of state with Leiden-type virial coefficients:

$$p = (RT/V_m)\{1 + B(p/RT) + (C - B^2)(p/RT)^2 + (D - 3BC + 2B^3)(p/RT)^3 + \ldots\}. \qquad (20)$$

There are theoretical grounds for preferring equation (15) to equation (16), and furthermore equation (15) can be used to represent high-pressure p, V, T data with fewer terms than equation (16) can. However, algebraic manipulation is often easier with equation (16) than with equation (15), and so it is preferred by some workers. The fact that both forms of the virial equation of state are in practical use has occasionally led to confusion.

Clearly there is no point in trying to fit more virial coefficients to experimental data than the uncertainty of the data warrants, so in practice truncated forms of the virial equations of state, often containing no terms higher than B, are employed in the analysis of data of modest accuracy covering fairly short ranges of pressure. However, gross truncation of the equations causes (17) to be inapplicable and it is not then justifiable to speak of 'the second virial coefficient' without specifying the series to which the coefficient applies. Thus equation (15) can be expressed in the form

$$Z = pV_m/RT = 1 + B/V_m + C/V_m^2 + D/V_m^3 + \dots \quad (21)$$

In general, a plot of the compression factor Z against $1/V_m$ will be a curve with an intercept of unity and a *limiting* slope of B as $(1/V_m) \to 0$. In many instances to be found in the literature, Z is plotted against $1/V_m$ and a linear least-squares procedure is applied, the slope (around the mean experimental value of $1/V_m$, not at zero) being taken as equal to B. Other instances occur where Z is plotted against p and the linear least-squares slope (around the mean experimental pressure, not at zero) is taken as equal to B'/RT [*cf.* equation (16)]. Scott and Dunlap [17] have shown that a linear least-squares analysis of certain experimental data by (i) a Z versus $1/V_m$ plot, and (ii) a Z versus p plot, gave values of B and B' respectively which differed by more than the combined errors of fit. Only when quadratic fitting was employed (*i.e.* when third virial coefficients were included) did $B = B'$, though the values of C and C' concomitantly produced were not especially meaningful in relation to the standard errors.

Cox has shown that [18, 19] if truncated forms of equations (15) and (16) are applied to measurements of p_1, V_1 and p_2, V_2 of a fixed amount of a gas at constant temperature T, then

$$B = RTr(1 - r)(p_1 r - p_2)/(r^2 p_1 - p_2)^2, \quad (22)$$

$$B' = RT(1/p_1 - r/p_2)/(r - 1), \quad (23)$$

where

$$r = V_1/V_2.$$

From (22) and (23) it can be shown that

$$B - B' = BB'(r^3 p_1 - p_2)/RTr(1 - r). \quad (24)$$

If $r = 2$, then $p_2 \approx 2p_1$. Substitution of these values into (24) indicates that $B - B' \approx 3p_1 BB'/RT$. Typical values of p_1 and B in Cox's work were 0.3 bar and $-$ 1.5 dm^3 mol^{-1} respectively; hence $B \approx B' + 0.06$ dm^3 mol^{-1} for these conditions. The quantity 0.06 dm^3 mol^{-1} exceeded the standard error of a measurement, so just as found by Scott and Dunlap the

[17] R. L. Scott and R. D. Dunlap, *J. Phys. Chem.*, 1962, **66**, 639.
[18] J. D. Cox, *Trans. Faraday Soc.*, 1960, **56**, 959.
[19] R. J. L. Andon, J. D. Cox, E. F. G. Herington, and J. F. Martin, *Trans. Faraday Soc.*, 1957, **53**, 1074.

difference between apparent values of B and B' is significant.* Henceforward in this chapter we shall write B_v for the *apparent* second virial coefficient in a volume-explicit virial equation truncated at the B term, and B_p for the *apparent* second virial coefficient in a pressure-explicit virial equation truncated at the B' term, whenever distinction between apparent and true second virial coefficients is a matter of significance.

Scott and Dunlap's findings,[17] and the analysis in the previous paragraph, expose a difficulty that an experimentalist must face when seeking to determine B, C, ..., by compression measurement: to measure departures from ideal-gas behaviour with greatest accuracy he must use as high a pressure as is convenient, yet to derive B with greatest accuracy he must carry his measurements down to as low a pressure as is possible. These stringent, and somewhat contradictory, requirements have not always been appreciated.

We round out this introduction to the virial equation of state by reference to its theoretical foundation. Thus statistical mechanics permits deduction of an expression for pV in terms of either the grand partition function or the radial distribution function.[20, 21] The leading term in the expansion of the latter function corresponds to pairwise interaction between molecules, and indicates the following relation between the second virial coefficient and the potential energy $u(r)$ of the interacting pair, when this depends only on the distance r between molecular centres:

$$B(T) = -2\pi N_A \int_0^\infty [1 - \exp\{-u(r)/kT\}]\, r^2\, dr, \qquad (25)$$

where N_A is Avogadro's constant and k is Boltzmann's constant. Similarly, interaction between three molecules at a time gives rise to a third virial coefficient, and so on.

We postpone discussion of equation (25) until a later section, and proceed to describe yet one more approach to the problem of describing non-ideal-gas behaviour.

Equilibria in a Polymerizing Gas.—The gross departures from ideal behaviour exhibited by some gases are ascribable to reversible polymerization of the gas, since the presence of more than one chemical species can be demonstrated spectroscopically, *e.g.* $2NO_2 = N_2O_4$. It is obviously sensible to treat the p, V, T behaviour of such systems according to the laws of chemical equilibrium. Other instances occur where the existence of polymers or clusters in a gas is likely, but unambiguous identification of the species present is difficult—examples are hydrogen fluoride and the lower alkanols. Even in such instances, treatment of non-ideal-gas behaviour in terms of

[20] J. O. Hirschfelder, C. F. Curtiss, and R. B. Bird, 'Molecular Theory of Gases and Liquids,' Wiley, New York, 1954.

[21] E. A. Mason and T. H. Spurling, 'The Virial Equation of State,' Pergamon, Oxford, 1969.

* Note that for an experiment to which (24) is applicable, $B > B'$, if B' is negative.

polymerization is often illuminating; it is known as the quasi-chemical approach. A general treatment of the problem was given by Woolley,[22] who assumed that the formation of clusters X_r from single gaseous molecules X is governed by the laws of chemical equilibrium and that species X and X_r behave ideally, *i.e.*

$$K_r = p(X_r)/\{p(X)\}^r, \quad (26)$$

$$p = \sum_r p(X_r), \quad (27)$$

$$1/V_m = (1/RT)\sum_r rp(X_r), \quad (28)$$

where K_r is an equilibrium constant, and $p(X_r)$ and $p(X)$ are partial pressures of the various cluster species and of the single molecules respectively. Woolley showed[22] that

$$Z = pV_m/RT = 1 - K_2p + (3K_2^2 - 2K_3)p^2 + (-10K_2^3 + 12K_2K_3 - 3K_4)p^3 + (35K_2^4 + 10K_3^2 - 60K_3K_2^2 + 20K_2K_4 - 4K_5)p^4 + \ldots \quad (29)$$

When this equation is compared with equation (20), it is apparent that

$$B = -K_2RT, \quad (30)$$

$$C = (4K_2^2 - 2K_3)(RT)^2, \quad (31)$$

$$D = (-20K_2^3 + 18K_2K_3 - 3K_4)(RT)^3. \quad (32)$$

In principle, therefore, knowledge of B, C, D, ..., permits derivation of K_2, K_3, K_4, ... It should be noted that Woolley's treatment relates to *clusters*, *i.e.* groups of molecules in close proximity to each other, which may result from either random motions or loose binding of monomer molecules to form dimers and higher polymers. Hirschfelder, McClure, and Weeks[23] gave a treatment that is specific for a gas dimerizing to a small extent:

$$B = b - RTK_2, \quad (33)$$

where K_2 is defined* as $p_{\text{dimer}}/(p_{\text{monomer}})^2$ and b is a volume not available for occupancy by other molecules (the 'hard spheres' assumption). An equation of the same general form as (33) emerges from the work of several authors who considered observed second virial coefficients to be compounded of a physical part (resulting from repulsive forces and attractive dispersion forces) and a chemical part giving rise to dimerization (resulting from dipole–dipole forces, or other specific interactions):

$$B_{\text{observed}} = B_{\text{physical}} - RTK_2. \quad (34)$$

[22] H. W. Woolley, *J. Chem. Phys.*, 1953, **21**, 236.
[23] J. O. Hirschfelder, F. T. McClure, and I. F. Weeks, *J. Chem. Phys.*, 1942, **10**, 201.
* In ref. 23, K_2 was defined for dissociation, in terms of concentrations.

Lambert *et al.*[24] assumed that B_{physical} is equal to the value of B predicted by the Berthelot equation (see Figure 9 and the appertaining discussion), but Prausnitz and Carter[25] assumed it equal to the value of B for a non-polar analogue of similar molecular size and shape (a 'homomorph'). Uncertainties in the estimated value of B_{physical} will not undermine the application of equation (34) to data for polar gases so long as $|B_{\text{physical}}| \ll RTK_2$, but in the event that $|B_{\text{physical}}| \gg RTK_2$, equation (34) is unlikely to be useful.

In the case of a gas that dimerizes and forms one higher polymer (again to a small extent only), the following equation may be derived:

$$pV_m/RT = 1 + (b/RT - K_2)p + (1 - r)K_r p^{r-1}, \qquad (35)$$

where an excluded volume b is allowed for, and the value of r (> 2) is assumed to be known. Equation (35), with $r = 4$, has found considerable application to the p, V, T relations of the lower alkanols. Many of the real-gas properties of methanol and its homologues are indeed well correlated[26] by a model which assumes the presence of monomers, dimers, and tetramers (the 1-2-4 model) around the normal boiling temperatures, *i.e.* $r = 4$, equivalent to a pressure-explicit virial equation with B and D terms but no C term. However, Kell and McLaurin[27] found that medium-pressure p, V, T data for methanol around 500 K are fitted better by an equation in B and C (a 1-2-3 model) than one in B and D. There are conflicting views[28] concerning the best model for low-pressure methanol vapour around 300 K. Since the contributions of the various terms in equation (35) are sensitive to both p and T, the varying interpretations of the real-gas behaviour of methanol may not be as incompatible as would appear at first sight.

4 Some Experimental p, V, T Methods

In the ensuing account of modern p, V, T methods we shall emphasize the use of the virial equation of state in interpreting measured quantities. For this purpose we shall utilize a version of equation (15) that explicitly cites the volume V occupied by an amount n of gas at pressure p and temperature T, namely

$$pV/nRT = 1 + nB/V + n^2C/V^2 + n^3D/V^3 + \ldots. \qquad (36)$$

The equivalence of (15) and (36) is evident, since $V_m = V/n$. It should be remembered that $B, C, D, \ldots$, are functions of temperature.

[24] J. D. Lambert, G. A. H. Roberts, J. S. Rowlinson, and V. J. Wilkinson, *Proc. Roy. Soc.*, 1949, **A196**, 113.

[25] J. M. Prausnitz and W. B. Carter, *Amer. Inst. Chem. Eng. J.*, 1960, **6**, 611.

[26] N. S. Berman, *Amer. Inst. Chem. Eng. J.*, 1968, **14**, 497.

[27] G. S. Kell and G. E. McLaurin, *J. Chem. Phys.*, 1969, **51**, 4345.

[28] E. E. Tucker, S. B. Farnham, and S. D. Christian, *J. Phys. Chem.*, 1969, **73**, 3820; V. Cheam, S. B. Farnham, and S. D. Christian, *J. Phys. Chem.*, 1970, **74**, 4157; A. N. Fletcher, *J. Phys. Chem.*, 1971, **75**, 1808.

An obvious experimental approach to the determination of B, C, D, ..., at a particular temperature T_1 would involve measurement of sufficient values of p, V, and n that the virial coefficients were uniquely determined.* Though this method has occasionally been employed, the difficulty of measuring the quantities, especially the volume, with the required accuracy makes this frontal approach to p, V, T measurement less attractive than it might initially seem. Instead, it is more customary to keep two of the quantities n, p, V, T constant and to determine the way the remaining two depend on each other. There are 6 ways of doing this but two are little used: keeping p and V constant is experimentally inconvenient and, for a pure gas, the subject of this review, no useful information is gained by keeping p and T constant as V/n is then invariant. The remaining binary combinations of n, p, V, and T are experimentally useful, and examples are listed below. We classify them according to the variables studied, mindful of the arbitrariness of such a classification, since a given apparatus can be used in more than one way. Our examples are mostly drawn from the post-1967 literature, because the pre-1967 literature is largely covered by Mason and Spurling's monograph.[21]

Experimental p, V, T measurements may be absolute, or may be relative to the properties of another gas, usually one that is more nearly ideal than the gas under investigation. Helium, argon, and nitrogen, with known values for $B(T)$ and $C(T)$, are the most commonly used reference gases. Sometimes the reference gas is chemically similar to the gas under investigation, but with better known p, V, T properties.

Methods that Determine p, V, T Relations Directly.—*Variation of n, p, and T.* Barton and Hsu [29] filled a glass bulb with the gas under study, and measured the pressure by means of a manometer, separated from the sample by a metal diaphragm. The mass of gas in the bulb was found by carefully weighing the bulb when evacuated and when filled with the gas, while the volume of the bulb was found by weighing it empty and when filled with distilled water. Measurements were made at temperatures between 50 and 125 °C and at pressures between 0.04 and 0.9 bar. Adsorption of the gas on the wall of the bulb is always a potential source of error with experiments of this type, but Barton and Hsu considered the effect of adsorption of their samples (formic and acetic acids) to be negligible at the temperatures studied.

A somewhat similar arrangement was used by Chevalier [30] in the temperature range 25 to 65 °C. A glass ampoule containing a known amount of sample was broken inside a stainless-steel vessel. The volume of the vessel was not measured directly, but by means of measurements on benzene,

[29] J. R. Barton and C. C. Hsu, *J. Chem. and Eng. Data*, 1969, **14**, 184.
[30] J.-L. Chevalier, *Compt. rend.*, 1969, **268C**, 747.

* In practice, measurement of n would involve measurement of the mass of a sample of gas of known molar mass, *i.e.* a sample of high purity.

whose second virial coefficient is known as a function of temperature. Pressures were measured to *ca.* 1 Pa using a quartz Bourdon-tube gauge.

Variation of V, T or p, T. When the amount of gas and either the pressure or the volume is kept constant, the experimental arrangement is that of a constant-pressure or a constant-volume gas thermometer. Virial coefficients have been measured by both methods in the past, though less commonly nowadays. Khodeeva, Lebedeva, and Belousova[31] have constructed an apparatus similar to a constant-volume thermometer. A stainless-steel vessel A (Figure 1) contains a weighed amount of gas. Mercury fills the bottom of the vessel and is contained in a tube B leading to a mercury reservoir C; here there is an interface between mercury and oil, and the latter transmits the pressure to a piston gauge. An electrical contact D monitors the mercury level in the vessel, and another contact E measures the position of the mercury–oil interface in C. The apparatus was used at temperatures up to 150 °C.

Schäfer and co-workers[32] have described a variant of the gas-thermometer principle, in which one glass bulb A (Figure 2) is filled with a known amount of nitrogen, the other B with a known amount of the gas under investigation. A differential pressure gauge C measures the difference in pressure between the two bulbs. The volumes of bulbs A and B, which are almost equal, are accurately measured. If the pressures in the two bulbs are equal at temperature T_1, then at temperature T_2 the pressures will, in general, differ. The pressures can be equalized by alteration of the mercury level in the calibrated tube D of the manometer, and from this change of volume can be found a value for the second virial coefficient of the gas at temperature T_2.

Variation of n, p. Kell, McLaurin, and Whalley[33] have developed a technique in which the volume of the container need not be measured independently. Sample fluid is injected from a calibrated screw injector, until a certain pressure is reached in a steel vessel held at constant measured temperature, when the amount of fluid remaining in the injector is found. More fluid is added until the pressure reaches another value, when the amount in the injector is again measured, and so on. The temperature is then changed and the experiments continued. The observations, after correction for expansion and dilatation of the 'fixed' volume, are analysed to give an equation of state. This technique was applied to steam,[33] D_2O,[34] and methanol,[27] at temperatures between 150 and 450 °C.

Variation of Density and p. Some workers have studied gas densities by measurement of the buoyancy of an evacuated bulb surrounded by the gas

[31] S. M. Khodeeva, E. S. Lebedeva, and Z. S. Belousova, *Russ. J. Phys. Chem.*, 1966, **40**, 1669.
[32] R. N. Lichtenthaler, B. Schramm, and K. Schäfer, *Ber. Bunsengesellschaft Phys. Chem.*, 1969, **73**, 36.
[33] G. S. Kell, G. E. McLaurin, and E. Whalley, *J. Chem. Phys.*, 1968, **48**, 3805.
[34] G. S. Kell, G. E. McLaurin, and E. Whalley, *J. Chem. Phys.*, 1968, **49**, 2839.

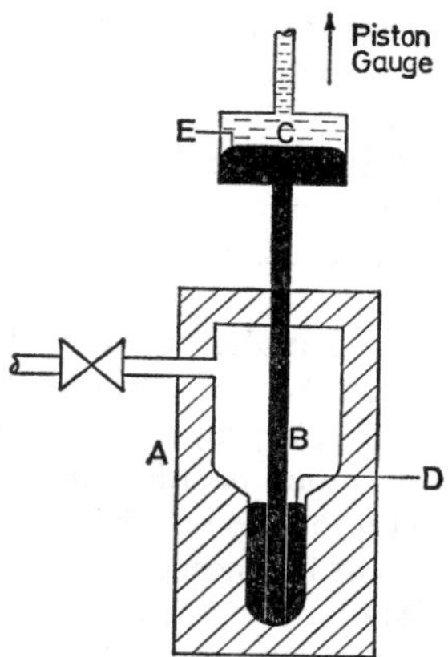

Figure 1 *p, V, T equipment based on a constant-volume gas thermometer*

under investigation, with application of Archimedes' principle. Di Zio *et al.*[35] used a precision analytical balance in a vacuum-tight chamber in a thermostatted bath to weigh an evacuated float. Weights required to null the balance were measured for various gas pressures, determined with a quartz Bourdon gauge. Results on n-hexane and argon were good, but results were less satisfactory with polar gases because of adsorption on the variety of materials used in the balance. Haworth and Sutton[36] have constructed a balance in which float, balance-beam, and suspension were

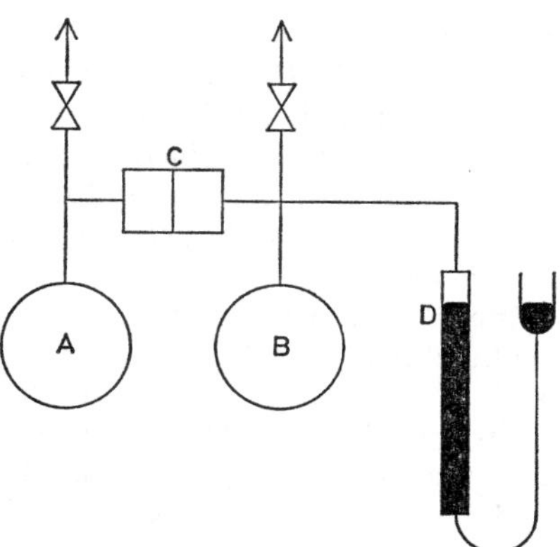

Figure 2 *A comparative method for p, V, T measurement based on constant-volume gas thermometry*

[35] S. F. Di Zio, M. M. Abbott, D. Zibello, and H. C. Van Ness, *Ind. Eng. Chem. Fundamentals*, 1966, **5**, 569.

[36] W. S. Haworth and L. E. Sutton, *J. Phys. (E)*, 1970, **3**, 271.

all made of quartz, as was a plate, fused to the beam on the side opposite to the float. The plate was designed to equalize the area moments on the two sides of beam, and thus to nullify the effects of any adsorption which might occur. Haworth and Sutton's balance was adjusted magnetically: a permanent magnet, sheathed in quartz, hung below the float, and the current through a solenoid surrounding this magnet was used to bring the balance to a null.

Variation of Substance and p. An alternative method of using a buoyancy balance is as follows. The chamber surrounding the balance is filled with a reference gas (with known p, V, T properties) to a measured pressure and the balance beam brought to a null position mechanically. The chamber is then evacuated and filled with the gas under investigation until the pressure is such that the beam is again balanced. Thus both gases have the same density and temperature. Hajjar and McWood [37] have used this technique for the study of halogenated methanes in the temperature range 40 to 130 °C. They used as reference gas tetrafluoromethane, whose virial coefficients in this temperature range are well established. Since its molar mass is similar to that of the gases studied, the pressure ratio will be neither very large nor very small.

Variation of p, V. When the amount of gas and the temperature are kept constant, experiments may be of either the compression or the expansion type. A typical apparatus for a compression experiment consists of two or more glass bulbs connected by narrow-bore glass tubing (Figure 3). A fixed amount of gas at constant temperature is made to occupy various volumes corresponding to combinations of the volumes of the bulbs, the pressure being measured. When the gas is compressed by mercury, the latter's vapour pressure puts an upper limit to the temperature at which such an apparatus may be used. (Mercury may go into solution in a compressed gas to an extent that is markedly in excess of the saturated vapour pressure,[38] and this unwanted effect increases rapidly as the temperature is raised.) Often a differential pressure detector is placed between the sample gas and the pressure gauge, and the pressure at which nitrogen nulls the differential detector is measured.

Knoebel and Edmister [39] used a six-bulb apparatus of this type. The second virial coefficient and the amount of gas were extracted from a least-squares analysis of a plot of pV against p; the volumes of the bulbs had to be measured absolutely. Hajjar *et al.*[40] used a similar technique, but measured the amount of sample directly. Attached to the series of glass bulbs was a calibrated capillary tube. The condensed liquid sample was contained over mercury in this capillary, and the liquid volume measured.

[37] R. F. Hajjar and G. E. McWood, *J. Chem. and Eng. Data*, 1970, **15**, 3.
[38] W. B. Jepson, M. J. Richardson, and J. S. Rowlinson, *Trans. Faraday Soc.*, 1957, **53**, 1586.
[39] D. H. Knoebel and W. C. Edmister, *J. Chem. and Eng. Data*, 1968, **13**, 312.
[40] R. F. Hajjar, W. B. Kay, and G. F. Leverett, *J. Chem. and Eng. Data*, 1969, **14**, 377.

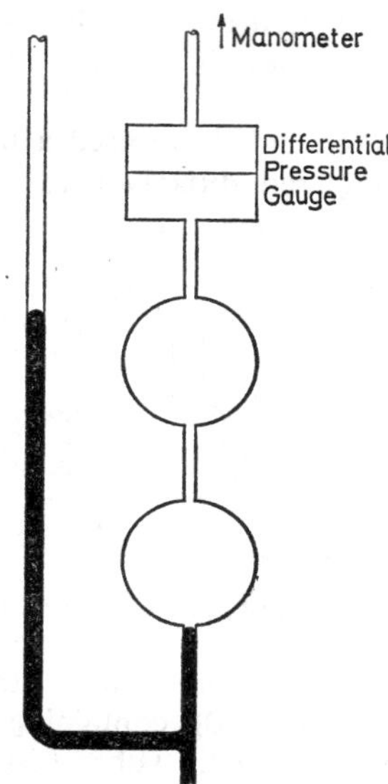

Figure 3 *p, V, T equipment using a compression method*

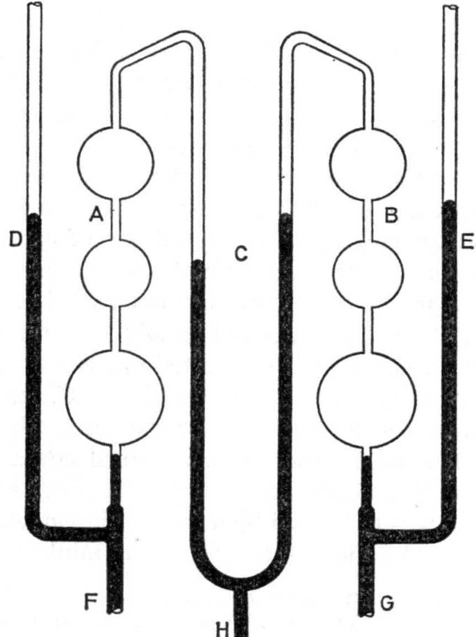

Figure 4 *A comparative compression method of p, V, T measurement*

Some earlier workers [19] have used only two bulbs; the absolute values of the two volumes need not be known, but their ratio may be determined with the aid of a reference gas.

McGlashan and co-workers [41, 42] have used a method whereby the gas under investigation is compared directly with an approximately equal amount of a reference gas (nitrogen). The apparatus (Figure 4) consisted

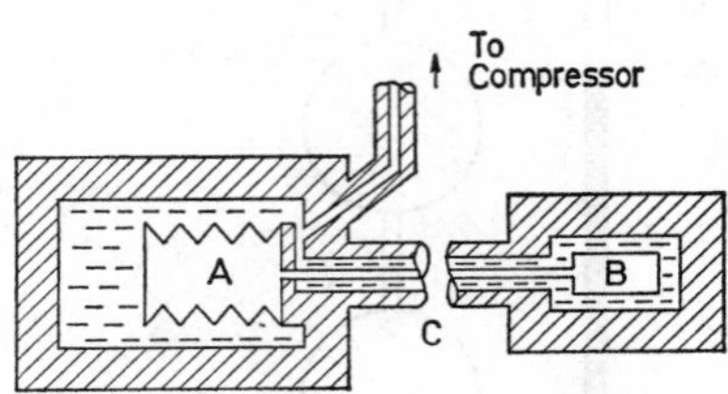

Figure 5 *Equipment for high-pressure p, V, T measurement*

of a set of three glass bulbs (A), containing the gas to be measured, separated from a similar set of three bulbs (B), containing the reference gas, by a differential mercury manometer (C). Tubes (D) and (E) were mercury manometers. The gases were compressed by mercury forced upwards through tubes F and G, until they occupied first three bulbs, then two bulbs, then one bulb. The absolute volumes of the bulbs did not have to be known but the volume ratios were required. After compression, the unknown and reference gases were at different pressures. The pressure difference was measured on the differential manometer C. Small changes of volume caused by alteration of the mercury levels in C could be allowed for, as the volume of the manometer had been previously calibrated. A similar method, using two bulbs for each gas, has been described by Schäfer *et al.*[32] However, these workers allowed one of the gases after compression to displace mercury from a calibrated capillary tube until the pressures of the two gases were equal. This small volume change, equal to the volume of capillary occupied by gas, was used to calculate the second virial coefficient.

Measurements at pressures up to 10 000 bar have been made on equipment designed by Babb *et al.*[43] A cell of variable volume A (Figure 5), held at a fixed temperature (35 °C) was connected to a cell of fixed volume B, at the temperature of measurement, by a capillary C. Both cells and the capillary were surrounded by liquid under pressure. The variable-volume cell A consisted of metal bellows, with electrical contacts for measuring change of length and thus change of volume. The pressure in the surrounding hydrostatic fluid was measured by a manganin gauge. The system was filled with the gas to be measured through a stainless steel U-tube (not

[41] M. L. McGlashan and D. J. B. Potter, *Proc. Roy. Soc.*, 1962, **A267**, 478.
[42] P. G. Francis, M. L. McGlashan, and C. J. Wormald, *J. Chem. Thermodyn.*, 1969, **1**, 441.
[43] S. E. Babb, G. J. Scott, C. D. Epp, and S. L. Robertson, *Rev. Sci. Instr.*, 1969, **40**, 670.

illustrated), which was subsequently immersed in liquid nitrogen so that a plug of solid material sealed off the system. This method of filling precluded an accurate measurement of the volume of the gas, so density values were obtained relative to values at 35 °C.

Weber[44] has reported measurements in the ranges 54 to 300 K, 0.1 to 33 bar, using an apparatus that depends on a combination of constant *n*, *V* and constant *n*, *T* techniques.[45, 46] A sequence of pressure and temperature

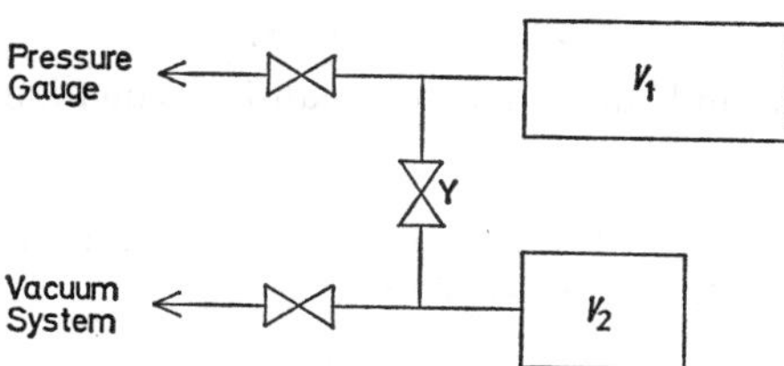

Figure 6 *Schematic diagram of Burnett's apparatus for p, V, T measurement*

readings was made on a fixed amount of sample in an almost constant volume. The total amount of gas was then determined by expanding it into vessels of accurately known volume; *p* (around 1 bar) and *T* were measured. If the equation of state of the gas at low pressures is known, the amount of gas can be deduced. Straty and Prydz[47] made an adaptation of this apparatus, where all the materials were suitable for use with fluorine gas.

Douslin and co-workers[48] have employed a compression method originated by Beattie.[49] A weighed sample contained in a cell is compressed by mercury, forced into the cell from a calibrated screw injector. Thus the change of volume of the gas is equal to the amount of mercury added by the injector. The mercury transmits the gas pressure to oil and thus to a piston gauge. Apparatus of this type can be used at pressures up to about 500 bar.

An expansion method with several attractive features is that due to Burnett.[50] This method does not require measurements of volumes, or of the amount of gas. As it does not use mercury as confining fluid the method can be used at relatively high temperatures. On the other hand, the method cannot be used in regions where pV is a rapidly varying function of p, *i.e.* in the critical region. Initially, a quantity of gas is contained in a cell of volume V_1 (Figure 6), and the temperature and pressure are measured. The gas is then allowed to expand through the valve Y into the evacuated

[44] L. A. Weber, *J. Res. Nat. Bur. Stand.*, 1970, **74A**, 93.
[45] R. D. Goodwin, *J. Res. Nat. Bur. Stand.*, 1961, **65C**, 231.
[46] L. Holborn and H. Schultze, *Ann. Physik.*, 1915, **47**, 1089.
[47] G. C. Straty and R. Prydz, *Rev. Sci. Instr.*, 1970, **41**, 1223.
[48] D. R. Douslin, R. H. Harrison, and R. T. Moore, *J. Chem. Thermodyn.*, 1969, **1**, 305.
[49] J. A. Beattie, *Proc. Amer. Acad. Arts Sci.*, 1934, **69**, 389.
[50] E. S. Burnett, *J. Appl. Mech.*, 1936, **A3**, 136.

volume V_2 and, after equilibrium is achieved, the pressure is measured. Valve Y is closed, V_2 is evacuated, and the gas in V_1 is again allowed to expand into V_2. This cycle is repeated until the pressure is as low as can be measured with good accuracy. The compression factor Z [equation (21)] may be written as

$$Z = pV/nRT, \tag{37}$$

and an 'apparatus constant' N may be defined by

$$N = (V_1 + V_2)/V_1. \tag{38}$$

If p_q is the pressure in V_1 after the qth expansion, equation (37) gives

$$p_q(V_1 + V_2) = Z_q n_q RT, \tag{39}$$

Z_q being the compression factor at p_q, T, and n_q being the amount of gas in $(V_1 + V_2)$. Before the qth expansion

$$p_{q-1}V_1 = Z_{q-1}n_{q-1}RT, \tag{40}$$

applies. Since $n_q = n_{q-1}$, division of equation (40) by equation (39) yields

$$p_{q-1}/p_q = NZ_{q-1}/Z_q. \tag{41}$$

Since the Z's become unity in the limit of sufficiently low pressures, N may be found by extrapolating a plot of p_{q-1}/p_q *versus* p_q to $p_q = 0$.

If equation (41) is applied repetitively to the first q expansions, the resulting equations can be combined to give

$$p_q N^q = Z_q(p_0/Z_0), \tag{42}$$

where the subscript 0 defines the initial condition, *i.e.* $p_0 = p_{q-1}$ for $q = 1$. In the limit as p_q tends to zero, the right-hand side of equation (42) tends to the value p_0/Z_0, as Z_q tends to one. A plot of $p_q N^q$ *versus* p_q will thus extrapolate to (p_0/Z_0). The compression factor Z_q at any experimental pressure p_q can then be determined from equation (42). Since the apparatus constant N is raised to the qth power, it is apparent that an accurate determination of N is required for good results with the Burnett apparatus. Burnett originally suggested measuring N in a separate experiment using a gas such as helium, and the suggestion has been followed by some workers. This procedure takes advantage of the relatively small values of the virial coefficients of helium; the plot of p_{q-1}/p_q *versus* p_q is then nearly linear, and an accurate value of N may be found by extrapolation. Thus Eubank and co-workers,[51, 52] in studies of polar gases, preferred to evaluate the apparatus constant from a run with helium. However, this procedure assumes that N will remain a constant over a period of time, and it is in general preferable to let each series of pressure measurements with the gas being studied determine the apparatus constant.

[51] L. N. Anderson, A. P. Kudchadker, and P. T. Eubank, *J. Chem. and Eng. Data*, 1968, **13**, 321.
[52] A. P. Kudchadker and P. T. Eubank, *J. Chem. and Eng. Data*, 1970, **15**, 7.

McKetta and co-workers[53, 54] have discussed various graphical and analytical methods of treating experimental data. Butcher and Dadson[55] and Canfield, Leland, and Kobayashi[56] have described graphical methods that yield more accurate values of N. The latter method has been used recently by Lange and Stein.[57] A computational method which extracts the virial coefficients from the experimental data by iteration has been described by Hoover *et al.*[58]

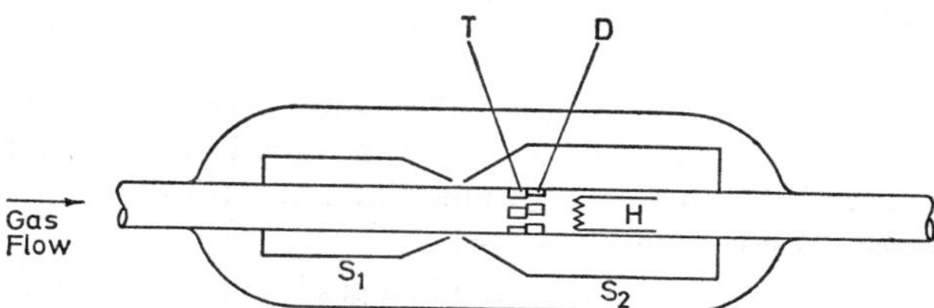

Figure 7 *Schematic diagram of equipment for measuring isothermal Joule–Thomson coefficients*

Methods that Determine p, V, T Relations Indirectly.—There are, besides the direct methods of measuring p, V, T properties described above, several indirect methods available.[21] Measurement of any property which is dependent on the non-ideality of a gas can in principle provide information about the equation of state.

Joule–Thomson Effect. Francis, McGlashan, and Wormald[42] have measured the isothermal Joule–Thomson coefficient

$$\phi = (\partial H/\partial p)_T = \mu C_p, \tag{43}$$

where C_p is the heat capacity at constant pressure, and

$$\mu = (\partial T/\partial p)_H, \tag{44}$$

is the isenthalpic Joule–Thomson coefficient.

In the apparatus of McGlashan and co-workers, gas at pressure p_1 passes through an adjustable throttle T (Figure 7) and assumes pressure p_2. The throttle T consists of a PTFE disc containing two holes. These holes can be brought, by external control, into coincidence with a corresponding pair of holes in the face D of the heater compartment H; the temperatures of the shields S_1 and S_2 are separately controlled. The Joule heat required to keep the temperature of the gas the same on both sides of the throttle is measured. The experiment provides a value of the quantity $\{H(p_1) - H(p_2)\}/(p_1 - p_2)$,

[53] I. H. Silberberg, K. A. Kobe, and J. J. McKetta, *J. Chem. and Eng. Data*, 1959, **4**, 314.
[54] I. H. Silberberg, D. C. K. Lin, and J. J. McKetta, *J. Chem. and Eng. Data*, 1967, **12**, 226.
[55] E. G. Butcher and R. S. Dadson, *Proc. Roy. Soc.*, 1964, **A277**, 448.
[56] F. B. Canfield, T. W. Leland, jun., and R. Kobayashi, *Adv. Cryog. Eng.*, 1963, **8**, 146.
[57] H. B. Lange, jun. and F. P. Stein, *J. Chem. and Eng. Data*, 1970, **15**, 56.
[58] A. E. Hoover, F. B. Canfield, R. Kobayashi, and T. W. Leland, jun., *J. Chem. and Eng. Data*, 1964, **9**, 568.

where $H(p_i)$ is the enthalpy at fixed temperature and the relevant pressure. In the limit of a very small pressure drop the measured quantity becomes the isothermal Joule–Thomson coefficient. It can be expressed as [42]

$$\phi = (B' - T\mathrm{d}B'/\mathrm{d}T) + (C' - T\mathrm{d}C'/\mathrm{d}T)\,\tfrac{1}{2}(p_1 + p_2) + \ldots, \tag{45}$$

where B', C', ..., are the virial coefficients in the Berlin form of the virial equation [equation (16)]. A plot of ϕ against $\frac{1}{2}(p_1 + p_2)$ thus gives the quantities $(B' - T\mathrm{d}B'/\mathrm{d}T)$ and $(C' - T\mathrm{d}C'/\mathrm{d}T)$, and indicates if higher terms are significant.

The advantages of this indirect p, V, T method over direct methods are (*i*) that the extent of the gas imperfection is measured directly, and not as a small difference between two large quantities, and (*ii*) that the results are uncomplicated by adsorption effects.

Sound Velocity. The velocity of sound in a gas is a property which depends on real-gas behaviour. The velocity of sound w is given by

$$w^2 = -(V_\mathrm{m}^2/M)(\partial p/\partial V_\mathrm{m})_S, \tag{46}$$

where M is the molar mass. The quantity w^2 can be further expressed as a power series in the pressure p:

$$w^2 = a_0 + a_1 p + a_2 p^2 + \ldots. \tag{47}$$

The coefficients in this equation are temperature dependent and are related to the virial coefficients. In particular $a_0 = \gamma RT/M$, where γ is the ratio of the heat capacity at constant pressure to that at constant volume, and

$$a_1 = \gamma\{2B + 2(\gamma - 1)\,T(\mathrm{d}B/\mathrm{d}T) + (\gamma - 1)^2 T^2(\mathrm{d}^2B/\mathrm{d}T^2)/\gamma\}/M. \tag{48}$$

Measurement of the velocity of sound in helium has been described by Gammon and Douslin.[59] They studied the gas in the ranges − 175 to 150 °C and 10 to 150 bar using an ultrasonic interferometer operating at 2.5 MHz.

Pressure Dependence of Vapour Heat Capacity. Measurement of vapour heat capacities as a function of pressure can give information on virial coefficients. For example, the relation

$$\lim_{p\to 0}(\partial C_p/\partial p)_T = -T\mathrm{d}^2B/\mathrm{d}T^2, \tag{49}$$

provides information about the second derivative of the second virial coefficient with respect to temperature. Measurements of this type are described in more detail by Counsell, in Chapter 6.

Enthalpy of Vaporization. The Clapeyron equation:

$$\mathrm{d}p_\mathrm{s}/\mathrm{d}T = \Delta H_\mathrm{v}/T(V_\mathrm{m} - V_\mathrm{l}), \tag{50}$$

[59] B. E. Gammon and D. R. Douslin, Proceedings of the 5th Symposium on Thermophysical Properties, sponsored by the Standing Committee on Thermophysical Properties, Heat Transfer Divn., Amer. Soc. Mech. Eng., Newton, Mass., 1970.

can provide values of total gas imperfections. ΔH_v is the enthalpy of vaporization, p_s is the saturated vapour pressure, and V_l is the molar volume of the liquid and V_m that of the *saturated* vapour. Rearrangement gives

$$RT(1 + B/V_m + C/V_m^2 + \ldots)/p_s = V_l + \Delta H_v/(T\,dp_s/dT). \quad (51)$$

Equation (51) permits evaluation of $(B + C/V_m \ldots)$ if ΔH_v, $p_s(T)$, and V_l are measured or are known with requisite accuracy. For example, Douslin and co-workers [48] derived values of B for hexafluorobenzene round the normal boiling temperature, on the assumption that C and higher virial coefficients have a negligible effect when $p_s \leqslant 1$ bar.

5 Some Recent *p, V, T* Measurements

In this section we give brief notes on some recent measurements of the *p, V, T* relations of real gases, to illustrate the intense activity in this subject. Only work that was not included in Dymond and Smith's [60] or Mason and Spurling's monographs [21] is mentioned. Hence the present notes are intended to complement and supplement these two earlier reviews.

Monatomic Gases.—Kudchadker and Eubank [52] calibrated their Burnett apparatus by means of helium; they reported values of B for helium at 100, 150, and 200 °C. Kalfoglou and Miller [61] also studied the compressibility of helium; they reported values of B for temperatures between 30 and 500 °C.

Streett [61a] measured the density of neon at pressures up to 2000 bar from 80 to 130 K.

Robertson *et al.*[10] measured the density of argon at temperatures between 35 and 400 °C, at pressures between 1500 and 10 000 bar. Kalfoglou and Miller [61] studied the compressibility of argon at temperatures between 30 and 500 °C, at pressures between 3 and 80 bar, and derived values of B and C.

Diatomic Gases.—Weber [44] measured the density of oxygen at close intervals of temperature in the range 54 to 300 K, and at pressures in the range 1 to 330 bar. His derived values of B and C (85 to 300 K) seem extremely accurate.

Robertson and Babb [11] measured the density of nitrogen at temperatures between 35 and 400 °C, at pressures between 1500 and 10 000 bar. Tsiklis and Polyakov [62] measured the density of nitrogen over the same range as Robertson and Babb. The densities found by the Russian workers are higher (at a given pressure) at 100 and 200 °C and lower at 300 and 400 °C, than those found by the American workers. Weber [63] measured the density

[60] J. H. Dymond and E. B. Smith, 'The Virial Coefficients of Gases. A Critical Compilation,' Clarendon, Oxford, 1969.
[61] N. K. Kalfoglou and J. G. Miller, *J. Phys. Chem.*, 1967, **71**, 1256.
[61a] W. B. Streett, *J. Chem. and Eng. Data*, 1971, **16**, 289.
[62] D. S. Tsiklis and E. V. Polyakov, *Doklady Akad. Nauk S.S.S.R.*, 1967, **176**, 308.
[63] L. A. Weber, *J. Chem. Thermodynamics*, 1970, **2**, 839.

of nitrogen at various temperatures around the critical point at pressures between 36 and 110 bar.

Robertson and Babb[64] measured the density of carbon monoxide at 35, 100, 200, and 300 °C, at pressures between 1500 and 10 000 bar. Their paper includes the mournful statement that corrosive attack of the gas on the steel pressure vessel 'led to the eventual retirement of the *pVT* apparatus'.

Prydz *et al.*[65] measured the *p*, *V*, *T* behaviour of fluorine (53 to 300 K). The use of special materials of construction was mandatory.

Triatomic Gases.—Kell *et al.*[33] measured mass, *p*, and *T* of water vapour in a nearly constant volume for various temperatures between 150 and 450 °C, and pressures in the range 0.6 to 94 bar. They calculated *B*, *C*, and *D*, and expressed *B* and *C* as functions of *T*. The same workers[34] carried out a similar study of deuterium oxide vapour, the range of variables being 150 to 500 °C and 1.0 to 101 bar. They calculated *B* and *C*, and showed that *B* for D_2O is very slightly more negative than *B* for H_2O.

Lewis and Fredericks[66] measured mass, *p*, and *T* of hydrogen sulphide in a fixed volume for various temperatures between 100 and 220 °C (the supercritical region), and pressures in the range 90 to 1700 bar. They calculated *B*, *C*, *D*, *E*, and *F*. Clarke and Glew[67] correlated published data (but not ref. 66) on hydrogen sulphide in terms of the Redlich–Kwong equation, and predicted the constants for deuterium sulphide.

Hajjar *et al.*[40] studied the low-pressure compressibility of carbon disulphide at various temperatures between 40 and 200 °C. Their calculated values of B_v vary more rapidly with temperature than do values selected from earlier work.[60]

Polyatomic Gases.—*Alkanes and Alkenes*. Robertson and Babb[68] measured the density of methane at 35, 100, and 200 °C at pressures between 1000 and 10 000 bar. McMath and Edmister[69] reported *B* and *C* for methane at 267, 278, and 289 K. Vennix and Kobayashi[70] analysed *p*, *V*, *T* data for methane and developed a six-constant equation that fits the data well over wide ranges of *p* and *T*.

Babb and Robertson[71] measured the density of propane at 35, 100, and 200 °C at pressures between 750 and 10 500 bar; the values for 35 °C refer to the liquid.

Hajjar *et al.*[40] studied the low-pressure compressibilities of n-pentane, n-hexane, and n-heptane at various temperatures between 40 and 200 °C. The results for B_v are broadly in agreement with selected values.[60] Belou-

[64] S. L. Robertson and S. E. Babb, *J. Chem. Phys.*, 1970, **53**, 1094.
[65] R. Prydz, G. C. Straty, and K. D. Timmerhaus, *J. Chem. Phys.*, 1970, **53**, 2359; R. Prydz and G. C. Straty, *J. Res. Nat. Bur. Stand.*, 1970, **74A**, 747.
[66] L. C. Lewis and W. J. Fredericks, *J. Chem. Eng. Data*, 1968, **13**, 482.
[67] E. C. W. Clarke and D. N. Glew, *Canad. J. Chem.*, 1970, **48**, 764.
[68] S. L. Robertson and S. E. Babb, *J. Chem. Phys.*, 1969, **51**, 1357.
[69] H. G. McMath, jun. and W. C. Edmister, *Amer. Inst. Chem. Eng. J.*, 1969, **15**, 370.
[70] A. J. Vennix and R. Kobayashi, *Amer. Inst. Chem. Eng. J.*, 1969, **15**, 926.
[71] S. E. Babb and S. L. Robertson, *J. Chem. Phys.*, 1970, **53**, 1097.

sova and Zaalishvili [72] studied the compressibility (in the pressure range 5 to 10 bar) of n-heptane at 200, 210, 220, and 230 °C, and calculated values of B_v.

Babb and Robertson [71] measured the density of ethylene at 35 °C at pressures between 1500 and 8000 bar. McMath and Edmister [69] reported B and C for ethylene at 267, 278, and 289 K.

Robertson and Babb [68] measured the density of propene at 35, 100, and 200 °C at pressures between 1000 and 10 000 bar.

Cyclic Hydrocarbons. Hajjar *et al.*[40] studied the low-pressure compressibility of cyclohexane at various temperatures between 43 and 200 °C. The derived values of B_v are appreciably more negative at lower temperatures than selected values.[60]

Several new p, V, T investigations of benzene have been reported. Thus Belousova and Zaalishvili [72] measured the compressibility at various temperatures between 160 and 230 °C, at 3.5 to 9 bar pressure. The compressibility at lower pressures was measured by Hajjar *et al.*[40] (40 to 200 °C) and Knoebel and Edmister [39] (40 to 100 °C). Francis, McGlashan, and Wormald [42] reported values of both B (from low-pressure compressibility measurements, 54 to 150 °C) and $(B - T\,\mathrm{d}B/\mathrm{d}T)$ (from Joule–Thomson effect measurements, 60 to 130 °C). In an analysis of published second virial coefficients, published values of $T^2\,\mathrm{d}^2B/\mathrm{d}T^2$ (from measurements of the pressure coefficient of heat capacity), and their own values of $(B - T\,\mathrm{d}B/\mathrm{d}T)$, Francis *et al.*[42] concluded that at least some of the measurements must be wrong. Re-examination of the situation, with inclusion of recently published values of B, led us to the same conclusion.

Organic Hydroxy-compounds. Kell and McLaurin [27] measured mass, p, and T of methanol in a nearly constant volume for temperatures between 150 and 300 °C, and pressures between 0.4 and 31 bar. They carefully analysed their results to see if use of second and fourth virial coefficients gave a better fit than use of second and third, as had been claimed earlier. They concluded that use of second and third virial coefficients was slightly to be preferred, and gave values of association constants for dimerization and trimerization. Tucker *et al.*[28] gave values of association constants for various combinations of r-mers, based on their measurements of low-pressure compressibility at 15, 25, and 35 °C; they found that a postulated mixture of monomers, trimers, and octamers fitted their data best. Kudchadker and Eubank,[52] and Knoebel and Edmister [39] measured the compressibility of methanol (in the ranges 25 to 200 °C, and 40 to 100 °C respectively) and presented values of the second virial coefficient only.

Knoebel and Edmister [39] studied the low-pressure compressibility of ethanol at 60, 80, and 100 °C, and reported values of B_p. Counsell *et al.*[73] determined the molar volume of ethanol at the saturation vapour pressure,

[72] Z. S. Belousova and Sh. D. Zaalishvili, *Russ. J. Phys. Chem.*, 1967, **41**, 1290.
[73] J. F. Counsell, J. O. Fenwick, and E. B. Lees, *J. Chem. Thermodyn.*, 1970, **2**, 367.

and also the pressure coefficient of heat capacity, from which they derived B and D as functions of T.

Stiel *et al.*[73a–c] measured the density of methanol (200 to 300 K), ethanol (200 to 350 K), and propan-2-ol (200 to 300 K) at pressures up to 650 bar. Alkanols whose molar volumes at saturation and pressure coefficients of heat capacity have been determined include the following: n-propanol studied by Matthews and McKetta,[74] propan-2-ol, studied by Hales, Cox, and Lees[75] and by Berman, Larkham, and McKetta,[76] n-butanol, studied by Counsell, Hales, and Martin,[77] 2-methylpropan-1-ol and pentan-1-ol studied by Counsell, Fenwick, and Lees,[73] butan-2-ol studied by Berman and McKetta,[78] and 2-methylpropan-2-ol studied by Beynon and McKetta.[79] In most of these publications, the data were interpreted on the assumption that monomers, dimers, and tetramers were present in the vapours, though for pentan-1-ol a monomer–dimer treatment was adequate.

Andon *et al.*[80] studied the low-pressure compressibility of phenol (at 155, 165, and 181 °C) and *o*-methylphenol (at 165 and 181 °C), and derived values of B_v.

Ethers. Haworth and Sutton[80a] determined B_p for dimethyl ether (25 to 55 °C). Knoebel and Edmister[39] measured the low-pressure compressibility of diethyl ether at 60, 80, and 100 °C, and derived values of B_p. Chevalier[30] calculated B_p from direct measurements of mass, pressure, volume, and temperature (25 to 55 °C). Counsell *et al.*[81] determined the molar volume of diethyl ether at saturation (8 to 34 °C), and also the pressure coefficient of heat capacity, from which they deduced B_p. They also presented an equation affording a least-squares fit of all published values of $B(T)$ for diethyl ether.

Chevalier[30] calculated values of B_p from direct measurements of mass, p, V, and T for the following ethers: ethyl propyl ether (25 to 55 °C), dipropylether (35 to 55 °C), di-isopropyl ether (35 to 65 °C), ethyl butyl ether (35 to 55 °C), dibutyl ether (65 °C). Hales *et al.*[82] determined the molar volume of methyl phenyl ether at saturation (366 to 426 K), and also the pressure coefficient of heat capacity, from which they deduced B_p.

[73a] R. S. Finkelstein and L. I. Stiel, *Chem. Eng. Prog., Symp. Series No. 98*, 1970, **66**, 11.
[73b] H. Y. Ho and L. I. Stiel, *Ind. Eng. Chem. Fundamentals*, 1969, **8**, 713.
[73c] J. K. Tseng and L. I. Stiel, *Amer. Inst. Chem. Eng. J.*, 1971, **17**, 1283.
[74] J. F. Matthews and J. J. McKetta, *J. Phys. Chem.*, 1961, **65**, 758.
[75] J. L. Hales, J. D. Cox, and E. B. Lees, *Trans. Faraday Soc.*, 1963, **59**, 1544.
[76] N. S. Berman, C. W. Larkham, and J. J. McKetta, *J. Chem. and Eng. Data*, 1964, **9**, 218.
[77] J. F. Counsell, J. L. Hales, and J. F. Martin, *Trans. Faraday Soc.*, 1965, **61**, 1869.
[78] N. S. Berman and J. J. McKetta, *J. Phys. Chem.*, 1962, **66**, 1444.
[79] E. T. Beynon and J. J. McKetta, *J. Phys. Chem.*, 1963, **67**, 2761.
[80] R. J. L. Andon, D. P. Biddiscombe, J. D. Cox, R. Handley, D. Harrop, E. F. G. Herington, and J. F. Martin, *J. Chem. Soc.*, 1960, 5246.
[80a] W. S. Haworth and L. E. Sutton, *Trans. Faraday Soc.*, 1971, **67**, 2907.
[81] J. F. Counsell, D. A. Lee, and J. F. Martin, *J. Chem. Soc. (A)*, 1971, 313.
[82] J. L. Hales, E. B. Lees, and D. J. Ruxton, *Trans. Faraday Soc.*, 1967, **63**, 1876.

Ketones. The p, V, T relations of acetone have been studied recently by three groups. Thus Knoebel and Edmister [39] measured the low-pressure compressibility at 40, 60, 80, and 100 °C, and derived B_p. A similar study (40 to 200 °C) by Hajjar *et al.*[40] was interpreted in terms of B_v. Anderson *et al.*[51] measured the compressibility of acetone, 25 to 150 °C, at pressures between 0.07 and 7 bar; from their results they deduced B', C' and D', and constants for dimerization.

Hales *et al.*[82] determined the molar volumes at saturation of diethyl ketone (335 to 375 K), ethyl propyl ketone (354 to 396 K), and methyl isopropyl ketone (327 to 367 K), and also the pressure coefficients of heat capacity, from which they deduced values of B_p.

Carboxylic Acids. Barton and Hsu [29] measured the molar volumes of both formic and acetic acids at 50, 65, 95, 110, and 125 °C, at various pressures. They noted that, contrary to the simplest theory, the association constants for dimer formation were not independent of total pressure, so they derived true dimerization constants by extrapolation of apparent values to zero pressure. The pressure dependence of apparent values of K_2 for acetic acid has also been remarked upon by Miksch *et al.*[83] who criticized other workers for attempting to express the real-gas behaviour in the form of a virial equation with no term higher than B [*cf.* equation (29) and the discussion following].

Organohalogen Compounds. The compression of tetrafluoromethane has been recently studied by two groups. Kalfoglou and Miller [61] covered the temperature range 300 to 770 K (pressures 3 to 50 bar), whilst Lange and Stein [57] covered 203 to 368 K (pressures 6 to 80 bar); values of B and C were calculated from both sets of results.

Low-pressure densities of trifluoromethane were measured by Hajjar and MacWood,[37] and values of B for 40, 70, 100, and 130 °C were derived. The compressibility of trifluoromethane (243 to 368 K; 6 to 45 bar) was measured by Lange and Stein,[57] who derived values of B and C.

The density of difluoromethane (298 to 473 K; 8 to 20 bar) was measured by Mears *et al.*,[84] who gave constants for a Martin–Hou equation of state.

The densities of five chlorofluoromethanes, *viz.* CHF_2Cl, $CHFCl_2$, $CFCl_3$, CF_3Cl, and CF_2Cl_2, were studied by Hajjar and MacWood [37] in the temperature range 40 to 130 °C. From the results, values of B_p were calculated. Kunz and Kapner [85] derived values of B for three of these compounds from an analysis of technical data.

Haworth and Sutton [80a] measured B_p (25 to 50 °C) for CHF_3, CHF_2Cl, CF_3Cl, CF_2Cl_2, $CF_2 : CH_2$, CH_3Br, and C_2H_5Cl.

Shank [86] measured the compressibility of 1,1,1,2,2-pentafluoropropane

[83] G. Miksch, F. Ratkovics, and F. Kohler, *J. Chem. Thermodyn.*, 1969, **1**, 257.
[84] P. F. Malbrunot, P. A. Meunier, G. M. Scatena, W. H. Mears, K. P. Murphy, and J. V. Sinka, *J. Chem. and Eng. Data*, 1968, **13**, 16.
[85] R. G. Kunz and R. S. Kapner, *J. Chem. and Eng. Data*, 1969, **14**, 190.
[86] R. L. Shank, *J. Chem. and Eng. Data*, 1967, **12**, 474.

between 0 and 240 °C, at pressures in the range 2 to 70 bar. The data were fitted to a Benedict–Webb–Rubin equation of state, and a graph of compression factors against temperature was also given.

Brown [87] measured the density of perfluoropropane (347 to 439 K; 17 to 57 bar) and derived constants for a Martin–Hou equation of state. Dantzler and Knobler [88] deduced values of B(233 to 373 K) from this equation.

Taylor and Reed [89] studied the compressibilities (1 to 18 bar, 350 to 450 K) of four isomeric perfluorohexanes and of perfluorocyclohexane, and reported values of B and C; they also reproduced $B(T)$ data for the isomeric perfluorocyclohexanes, taken from a doctoral dissertation.

Scott *et al.*[90] determined the molar volume at saturation of 1,2-difluorobenzene (326 to 367 K) and also the pressure coefficient of heat capacity; they deduced values of B_v.

Douslin *et al.*[48] measured the compressibility of hexafluorobenzene between 225 and 350 °C, at pressures in the range 20 to 350 bar; they derived values of B, C, and D. Counsell *et al.*[91] derived values of B_p for hexafluorobenzene from determinations of the molar volume at saturation (316 to 353 K) and of the pressure coefficient of heat capacity. Counsell *et al.*[92–94] obtained values of B by a similar procedure for the following derivatives of hexafluorobenzene: methylpentafluorobenzene (349 to 390 K), chloropentafluorobenzene (349 to 391 K), and pentafluorobenzene (320 to 356 K).

Hajjar *et al.*[40] studied the low-pressure compressibility of perfluoromethylcyclohexane at various temperatures between 40 and 200 °C, and derived values of B_v.

Miscellaneous Compounds. Hajjar and MacWood [37] deduced B_p of sulphur hexafluoride at 70 and 100 °C from measurements of gas density at low pressure. Mears *et al.*[95] measured the density of sulphur hexafluoride (25 to 130 °C, 10 to 75 bar) and expressed the results in terms of a Martin–Hou equation of state;[7] they also compared derived values of B and C with literature values. Haworth and Sutton [80a] measured B_p of trimethylamine at 25 °C.

Miscellaneous compounds for which values of B_v have been derived from the molar volumes at saturation and the pressure coefficients of heat

[87] J. A. Brown, *J. Chem. and Eng. Data*, 1963, **8**, 106.
[88] E. M. Dantzler and C. M. Knobler, *J. Phys. Chem.*, 1969, **73**, 1335.
[89] Z. L. Taylor, jun., and T. M. Reed, *Amer. Inst. Chem. Eng. J.*, 1970, **16**, 738.
[90] D. W. Scott, J. F. Messerly, S. S. Todd, I. A. Hossenlopp, A. Osborn, and J. P. McCullough, *J. Chem. Phys.*, 1963, **38**, 532.
[91] J. F. Counsell, J. H. S. Green, J. L. Hales, and J. F. Martin, *Trans. Faraday Soc.*, 1965, **61**, 212.
[92] J. F. Counsell, J. L. Hales, E. B. Lees, and J. F. Martin, *J. Chem. Soc.* (*A*), 1968, 2994.
[93] R. J. L. Andon, J. F. Counsell, J. L. Hales, E. B. Lees, and J. F. Martin, *J. Chem. Soc.* (*A*), 1968, 2357.
[94] J. F. Counsell, J. L. Hales, and J. F. Martin, *J. Chem. Soc.* (*A*), 1968, 2042.
[95] W. H. Mears, E. Rosenthal, and J. V. Sinka, *J. Phys. Chem.*, 1969, **73**, 2254.

capacity include thiacyclopentane[96] (349 to 394 K), cyclopentane thiol[97] (360 to 405 K), pyrrole[98] (362 to 402 K), and hexamethyldisiloxane[99] (332 to 374 K).

The compressibility of trimethyl borate between 225 and 300 °C (30 to 210 bar) was measured by Griskey *et al.*[100] Values of B were derived from these measurements by Kunz and Kapner.[85]

6 Correlation and Prediction of Real-gas Behaviour

In Section 3 we emphasized three different approaches to the description of real-gas behaviour: (*i*) generalized equations of state, (*ii*) the virial equation of state, (*iii*) the quasi-chemical equation of state. Each approach has some utility in relation to the correlation and interpretation of experimental measurements, leading to the prediction of real-gas behaviour for substances (or for ranges of p, V, and T) for which measured values are lacking. Approach (*i*) is best for considering p, V, T behaviour along and near the saturation line and in the critical region. Approach (*ii*) is best for regions of low and medium density (the so-called dilute gas), away from the critical region. Approach (*iii*) is best reserved for gases showing definite evidence of association.

Correlation and Prediction *via* the Principle of Corresponding States.— Equations (10), (12), (13), and (14) are examples of generalized equations based on the Principle of Corresponding States. Such equations, if reliable, permit the calculation of the p, V, T relations of any gas for which values of p_c, V_c and T_c are available.* Intercomparisons of p, V, T values (often in terms of Z) from the several generalized equations, and comparisons with experimental values, are published from time to time, *e.g.* refs. 7 and 102. It is evident that for gases of simple molecular structure (monatomic, homopolar diatomic, and pseudo-spherical non-polar polyatomic) this approach to the prediction of real-gas behaviour is useful, but for gases of more complex molecular structure it is less reliable. A distinct advantage is gained by abandoning attempts at analytical representation of Corresponding States behaviour, in favour of tabular or graphical representa-

[96] W. N. Hubbard, H. L. Finke, D. W. Scott, J. P. McCullough, C. Katz, M. E. Gross, J. F. Messerly, R. E. Pennington, and G. Waddington, *J. Amer. Chem. Soc.*, 1952, **74**, 6025.

[97] W. T. Berg, D. W. Scott, W. N. Hubbard, S. S. Todd, J. F. Messerly, I. A. Hossenlopp, A. Osborn, D. R. Douslin, and J. P. McCullough, *J. Phys. Chem.*, 1961, **65**, 1425.

[98] D. W. Scott, W. T. Berg, I. A. Hossenlopp, W. N. Hubbard, J. F. Messerly, S. S. Todd, D. R. Douslin, J. P. McCullough, and G. Waddington, *J. Phys. Chem.*, 1967, **71**, 2263.

[99] D. W. Scott, J. F. Messerly, S. S. Todd, G. B. Guthrie, I. A. Hossenlopp, R. T. Moore, A. Osborn, W. T. Berg, and J. P. McCullough, *J. Phys. Chem.*, 1961, **65**, 1320.

[100] R. G. Griskey, W. E. Gorgas, and L. N. Canjar, *Amer. Inst. Chem. Eng. J.*, 1960, **6**, 128.

[101] R. C. Reid and T. K. Sherwood, 'The Properties of Gases and Liquids. Their Estimation and Correlation,' McGraw-Hill, New York, 1966, 2nd Edn.

[102] K. K. Shah and G. Thodos, *Ind. Eng. Chem.*, 1965, **57**, (3), 30.

* If experimental values are unavailable, p_c, V_c, and T_c can be estimated by various empirical procedures.[101]

tions. Indeed, charts showing one reduced quantity plotted against another are much used in chemical engineering, especially for gases (*e.g.* refrigerants) whose properties along the saturation line are of cardinal importance. An example of a plot of this type — adapted from Kennedy and Thodos's [103] diagram for carbon dioxide — is shown in Figure 8. The curves in this plot of reduced density against reduced temperature are for constant reduced pressures between 0.1 ($p \approx 7$ bar) and 50 ($p \approx 3600$ bar).

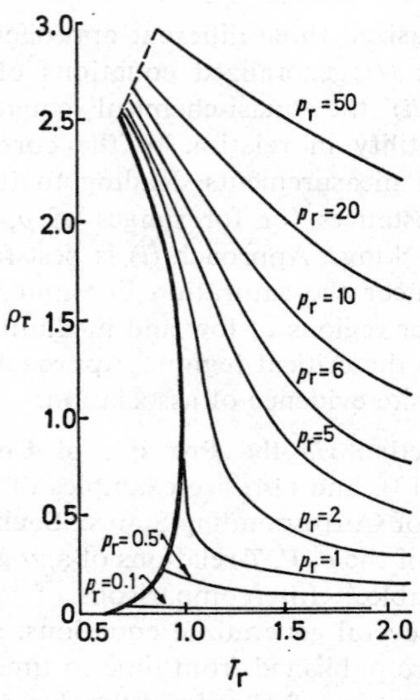

Figure 8 *Reduced density* versus *reduced temperature curves for carbon dioxide*

The curves for $p_r \leqslant 1$ terminate at a curve representing orthobaric densities. The latter curve can itself be predicted using the Principle of Corresponding States; equations for this purpose have been proposed by Guggenheim.[104]

So far in this review we have referred to the Principle of Corresponding States solely as an outcome of the van der Waals equation of state. However, Pitzer showed in 1939 that the Principle is a consequence of the forces acting between molecules, if certain assumptions on the nature of these forces hold good.[105] One assumption is that the potential energy, u, of a fluid is expressible as the product of an energy parameter, ε, and a function of distance r (in dimensionless form) between molecular centres, *i.e.* $u(r) = \varepsilon \times f(r/\sigma)$. Many expressions for $u(r)$ in terms of the two parameters ε and σ have been proposed and some are discussed more fully later. A given

[103] J. T. Kennedy and G. Thodos, *J. Chem. and Eng. Data*, 1960, **5**, 293.
[104] E. A. Guggenheim, *J. Chem. Phys.*, 1945, **13**, 253.
[105] T. W. Leland, jun., and P. S. Chappelear, *Ind. Eng. Chem.*, 1968, **60**(7), 15.

expression leads to a relation between reduced second virial coefficient (B/V_c) and reduced temperature, *via* equation (25). For example, the dotted curve in Figure 9 is thecurve* predicted by the Lennard-Jones 12 – 6 potential (described in more detail in the next section) for a gas conforming to the Principle of Corresponding States, but this method of estimating $B(T)$ is of limited utility because only gases of simple structure (see above) conform well to the Principle. However, non-conforming gases can be

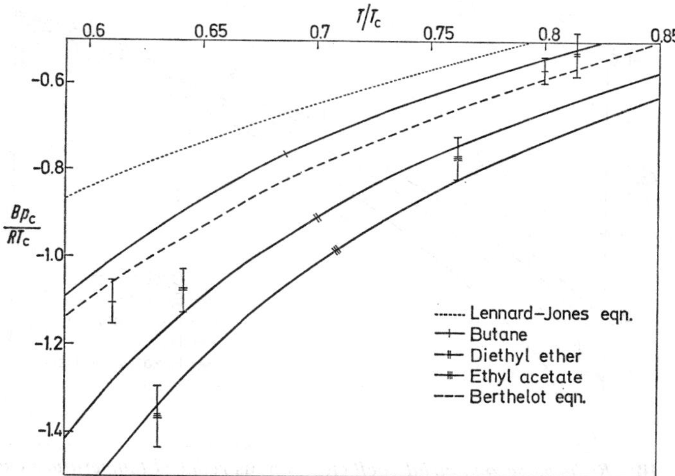

Figure 9 *Reduced second virial coefficient* versus *reduced temperature curves*

treated by an abridged method which considers their potential energies to depend on a parameter additional to ε and σ. The extra parameter, called by Pitzer the 'acentric factor' and symbolized as ω, takes account of non-centrally acting forces between molecules; ω for a given fluid is defined[16] in terms of the reduced vapour pressure, $p_{s,r}$, at $T_r = 0.7$ by the relation

$$\omega = -\{1 + \log_{10} p_{s,r}\}_{T_r=0.7}. \qquad (52)$$

A fluid that conforms to the Principle of Corresponding States has $p_{s,r} = 0.1$ at $T_r = 0.7$, *i.e.* $\omega = 0$ for such a fluid. Non-conforming fluids have $\omega > 0$, and their second virial coefficients can conveniently be derived[106] from Pitzer and Curl's equation:*

$$Bp_c/RT_c = (0.1445 + 0.073\omega) - (0.330 - 0.46\omega)T_r^{-1} - (0.1385 + 0.5\omega)T_r^{-2} - (0.0121 + 0.097\omega)T_r^{-3} - 0.0073\omega T_r^{-8}. \qquad (53)$$

[106] K. S. Pitzer and R. F. Curl, jun., *J. Amer. Chem. Soc.*, 1957, **79**, 2369.

* It will be noted that the volume parameter used to reduce B is here RT_c/p_c instead of V_c; this is advantageous, since reliable values of T_c and p_c are more plentiful than reliable values of V_c.

Plots of equation (53) for representative C_4 and C_6 compounds are given in Figures 9 and 10; molecules of pseudo-spherical, non-spherical, non-polar, dipolar, and multipolar types are represented. The vertical bars correspond to selected experimental values, the bar heights indicating experimental uncertainties. Comparison of the bars with the curves to which they relate reveals that Pitzer and Curl's equation provides a reasonable prediction of

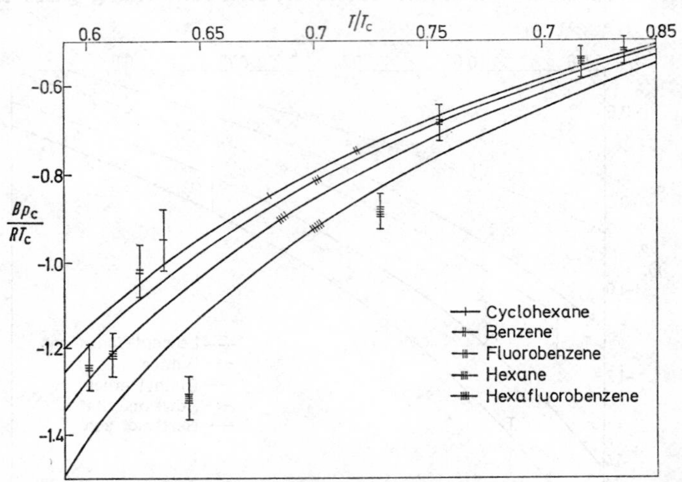

Figure 10 *Reduced second virial coefficient* versus *reduced temperature curves*

Bp_c/RT_c (and hence of B itself) for all the compounds, though the slopes of the curves are generally too small. The dashed curve in Figure 9 is that given by the Berthelot equation of state

$$p = RT/V_m + (9/128)(RT_c/p_c)(1 - 6T_r^{-2})p. \qquad (54)$$

The Berthelot curve is found to correspond approximately to a Pitzer and Curl curve with $\omega = 0.21$: this fact shows why the Berthelot equation has had comparatively wide applicability, since many polyatomic gases have ω values around 0.2.

McGlashan and co-workers [107, 108] have shown that by use of the carbon number n of organic compounds forming an homologous series, B/V_c may be expressed as a function of T_r by a single equation, valid for all members of the series. For n-alkanes the equation is

$$B/V_c = 0.430 - 0.886T_r^{-1} - 0.694T_r^{-2} - 0.0373(n - 1)T_r^{-4.5}, \qquad (55)$$

[107] M. L. McGlashan and D. J. B. Potter, 'Proceedings of the Joint Conference on Thermodynamic and Transport Properties of Fluids,' Institution of Mechanical Engineers, London, 1958.
[108] M. L. McGlashan and C. J. Wormald, *Trans. Faraday Soc.*, 1964, **60**, 646.

and an equation of similar form would appear to be applicable to the alk-1-ene and perfluoro-alkane series also.[88] The parameter n in equation (55) plays a rôle similar to that of the parameter ω in equation (53); in some applications n need not be integral.[109] Equation (55) with $n = 1$ provides an approximate representation of B/V_c for a Corresponding States gas, just as equation (53) with $\omega = 0$ provides a representation of $B/(RT_c/p_c)$ for such a gas.

To sum up, several good methods of estimating $B(T)$ by modified versions of the Principle of Corresponding States are available. The situation with regard to $C(T)$ is much less satisfactory. Chueh and Prausnitz[110] have proposed an equation relating C/V_c^2 to T_r, which involves a parameter dependent on molecular size, shape, and polarizability. However, the reliability of their equation as applied to polyatomic gases cannot be tested until additional, and more reliable, data accrue.

Prediction of Second Virial Coefficients from Knowledge of Intermolecular Forces.—We have already indicated that $B(T)$ data may be predicted from equations such as (53) and (55), which depend on common patterns in the intermolecular potentials of molecules of certain types. In this section we consider the feasibility of accurately predicting $B(T)$ data from intermolecular potentials that are tailored to fit the properties of particular molecular structures. Much of our discussion will relate perforce to the converse problem: 'What can be learned about intermolecular forces from measured $B(T)$ and other data?' We shall start with the assumption that intermolecular forces act as if from molecular centres and that the potential energy u depends only on the distance r between the centres of pairs of interacting molecules. Later we shall consider the effect of non-centrally acting intermolecular forces. The reader will find detailed expositions of the subject in Hirschfelder, Curtiss, and Bird's book[20] and in Mason and Spurling's more recent monograph.[21] Our treatment is intended to focus on the problem of predicting $B(T)$ for substances for which data are lacking (inevitably these are mostly compounds of complex molecular structure); we shall also refer to some recently published developments of general importance.

Molecules with Central Force Fields. For these molecules equation (25) is directly applicable; it is necessary to assume a functional form for $u(r)$, then to perform the required integration. From general observations of the properties of matter it is obvious that u is large positive when r is small. It is also obvious that $u = 0$ when $r = \infty$. Many equations to describe $u(r)$ between $r = \infty$ and $r =$ 'small' have been proposed and plots of some of them are shown in Figure 11. The 'hard sphere' and 'soft sphere' representations of $u(r)$ (curves a and b) allow for repulsive forces only. The other representations illustrated in Figure 11 allow additionally for

[109] E. A. Guggenheim and C. J. Wormald, *J. Chem. Phys.*, 1965, **42**, 3775.
[110] P. L. Chueh and J. M. Prausnitz, *Amer. Inst. Chem. Eng. J.*, 1967, **13**, 896.

attractive forces, which give rise to the minima at negative $u(r)$ seen in curves *c, d, e*, and *f*. These minima may be single-valued or may occur over ranges of r; a recent model of intermolecular forces proposed by Kreglewski [111] is unusual in leading to *two* flat minima (curve *e*). By common consent, attractive intermolecular forces must exist in molecular fluids, to account for the existence of the liquid state, and from a large body of

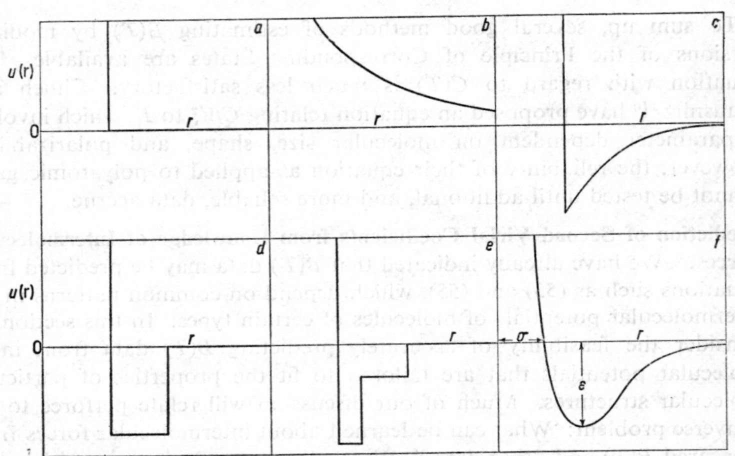

Figure 11 *Some potential energy versus distance curves;* (a) *hard-spheres potential,* (b) *soft-spheres potential,* (c) *Sutherland potential,* (d) *square-well potential* (e) *Kreglewski potential,* (f) *bireciprocal potential*

theoretical and experimental work it emerged long ago that a continuous curve similar to *f* in Figure 11 is the most realistic. However, discussion of the other types of curve illustrated has played, and continues to play, an important simplifying rôle in the overall theory.

Curves of type *f*, known as Mie or bireciprocal potentials, have the general form:

$$u(r) = \varepsilon\{1 - (m/n)\}^{-1}(n/m)^{m/(n-m)}\{(\sigma/r)^n - (\sigma/r)^m\}, \qquad (56)$$

where $n > m$. Lennard-Jones took $m = 6$, and since there is theoretical backing for this choice most other workers have taken $m = 6$, though a few have taken $m = 7$. Concerning the best value for n there is far less unanimity. Lennard-Jones considered n to lie between 8 and 14, and opted for $n = 12$. Some other workers have taken $n > 12$; for example, recent work [112] based on experimental data for methane (120 to 600 K) indicated 12 and 48 as outer limits for n, with $n = 21$ as the optimum value. The version of equation (56) with $m = 6$ is conveniently spoken of

[111] A. Kreglewski, *J. Phys. Chem.*, 1969, **73**, 608.
[112] R. C. Ahlert, G. Biguria, and J. W. Gaston, jun., *J. Phys. Chem.*, 1970, **74**, 1639.

as the Lennard-Jones $n - 6$ potential; when $n = 12$, the case most frequently cited in the literature, the potential takes the form

$$u(r) = 4\varepsilon\{(\sigma/r)^{12} - (\sigma/r)^{6}\}, \tag{57}$$

where ε and σ have the significance shown in Figure 11 curve *f*.

The integration on the right-hand side of equation (25), using $u(r)$ from equation (57), was first carried out by Lennard-Jones.[20, 21] The integration yielded values of the reduced quantity B^*, equal to $B/(2\pi N_A\sigma^3/3)$, as a function of the reduced quantity T^*, equal to $T/(\varepsilon/k)$. The function $B^*(T^*)$, which formed the basis of the dotted curve in Figure 9, is available in tabular form [20] for a wide range of T^* values. A quick test to see whether $B(T)$ data for a given gas fit a Lennard-Jones $12 - 6$ potential consists in placing a tracing of $\log_{10} B$ *versus* $\log_{10} T$ over a plot of $\log_{10} B^*$ *versus* $\log_{10} T^*$. If the two curves superimpose (to be sure of which data covering a wide temperature range are needed), then the scale factors $2\pi N_A\sigma^3/3$ and ε/k can be deduced from the relative positions of the axes of the two plots. If the curves do not superimpose at all, it is likely that the Lennard-Jones potential is inapplicable to $B(T)$ for that gas. Let us suppose that the potential *is* applicable to $B(T)$ for a particular gas. Once ε and σ have been deduced (*e.g.* by the log–log plot or by other methods [113]), the availability of $B^*(T^*)$ for a wide range of T^* obviously permits the accurate prediction of $B(T)$ at temperatures other than the experimental ones. Can the procedure be used for *a priori* prediction of $B(T)$? Clearly the answer to this question depends on whether (*i*) the applicability of the Lennard-Jones potential can itself be predicted and (*ii*) values of ε and σ can be predicted or 'borrowed' from the results of other measurements, *e.g.* of gas viscosity, η. In regard to (*i*), it is now known that for only relatively few gases (monatomic, homopolar diatomic, and a few polyatomic) does the potential apply with good accuracy. In regard to (*ii*), the situation is disappointing, since borrowed values of ε and σ often lead to predicted $B(T)$ values that are at variance with experimental values. This arises because $B(T)$ values are more (or less) sensitive to a particular part of the potential than are, say, $\eta(T)$ values. Shortcomings of a given potential function are thereby exposed, and it must be concluded that the bireciprocal potential (either in the $12 - 6$ form or in other forms) is insufficiently applicable to gases in general to provide a useful means of predicting $B(T)$. We turn now to consider whether some more complicated potential functions fare any better.

Molecules with Pseudo-central Force Fields. Up until this point in the discussion, the potential energy of a pair of molecules has been taken to depend on the distance between their centres, and not on their mutual orientation. Interaction models that offer a semblance of orientational dependence yet retain the basic feature of central potentials (for reasons of

[113] Ref. 21, pp. 250–7.

mathematical tractability) have been called 'pseudo-central', by Mason and Spurling.[21] We shall refer to two such models.

Kihara[114] supposed that molecules have impenetrable convex cores of volume v_0, surface area s_0, and mean curvature M. Interaction between two cores is held to follow a $12 - 6$ law involving ρ, the shortest distance between cores. Then

$$B(T)/N_A = (v_0 + s_0M/4\pi) + (s_0 + M^2/4\pi)\,\rho_m F_1(T^*) + M\rho_m^2 F_2(T^*) + 2\pi\rho_m^3 F_3(T^*)/3, \qquad (58)$$

where ρ_m is the value of ρ when the potential energy is a minimum, and $F_j(T^*)$ values are given in tabular form by Connolly and Kandalic.[115] Kihara[114] has given values of v_0, s_0, and M for various solid geometries. Prausnitz and Keeler[116] have given values appropriate for a dumb-bell shaped core, and have shown that second virial coefficients of 2,2,4-trimethylpentane can be successfully correlated by the Kihara potential applied to a core of that shape. On the other hand, Hoover *et al.*[117] found that $B(T)$ data for ethane, using any of four core geometries, are correlated less well by the Kihara potential than by the Lennard-Jones potential. Hoover's $B(T)$ and $C(T)$ values for methane are correlated well by a Kihara potential based on spherical geometry and, indeed, many users of the Kihara potential have assumed spherical cores, even for molecules that are decidedly non-spherical. For the special case of a spherical core, diameter a, Douslin, Moore, and Waddington[118] showed that

$$u(r) = 4\varepsilon[\{\lambda\sigma/(r - a)\}^{12} - \{\lambda\sigma/(r - a)\}^6], \qquad (59)$$

where λ is a parameter defined as ρ_0/r_0 at $u(r) = 0$. They derived best values of ε/k, b $(= 2\pi N_A\sigma^3/3)$, and λ from $B(T)$ for perfluorocyclobutane over a wide temperature range, using an adaptation of the method of superimposed log–log plots described above in connection with the Lennard-Jones potential. With the latter potential, the experimentally based curve has to be fitted to a single theoretical curve, but for a three-parameter potential, like Kihara's, the fit has to be made to a family of curves for differing values of λ. The spherical-core Kihara potential was found to fit the experimental data for perfluorocyclobutane to within the experimental accuracy. The same potential has been used to correlate $B(T)$ data for effectively spherical molecules[57, 119] like CF_4 and CHF_3, and for much less spherical molecules[39] like CH_3OH, C_2H_5OH, $(CH_3)_2CO$, and $(C_2H_5)_2O$. Storvick and co-

114 Ref. 21, p. 245.

115 J. F. Connolly and G. A. Kandalic, *Phys. Fluids*, 1960, **3**, 463. (Also: Documentation Institute, U.S. Library of Congress, Document No. 6307.)

116 J. M. Prausnitz and R. N. Keeler, *Amer. Inst. Chem. Eng. J.*, 1961, **7**, 399.

117 A. E. Hoover, I. Nagata, T. W. Leland, jun., and R. Kobayashi, *J. Chem. Phys.*, 1968, **48**, 2633.

118 D. R. Douslin, R. T. Moore, and G. Waddington, *J. Phys. Chem.*, 1959, **63**, 1959.

119 D. R. Douslin, R. H. Harrison, R. T. Moore, and J. P. McCullough, *J. Chem. Phys.*, 1961, **35**, 1357.

workers [120, 121] have discussed the addition of terms for dipole–dipole interaction to the Kihara potential.

The second pseudo-central potential to be discussed is that due to de Rocco and Hoover,[122] who supposed that molecules can be considered as spherical shells, diameter d, with force centres distributed over the surfaces of the shells. Three parameters characterize the potential energy of the interaction between two shells, namely ε, d, and R_m, which is the value of the centre-to-centre distance, R, for the minimum in u. A table of $B^*(T^*, R_m/d)$ is available.[122] De Rocco and Hoover showed that their spherical-shell potential fitted $B(T)$ data for thirteen compounds with quasi-spherical molecules better than did the Lennard-Jones potential; the only molecule of this class for which the fit was worse was methane. The spherical-shell potential was also shown to provide a good fit to $B(T)$ data for compounds with linear or plate-shaped molecules. McKinley and Reed [123] utilized the spherical-shell potential to calculate $B(T)$ for gases having molecules of the type XY_4, X_2Y_6, *etc.* Molecules like these would be expected to conform closely to the model, since their main centres of interaction are likely to be on the periphery. McKinley and Reed deduced X–X, X–Y, and the predominant Y–Y interaction parameters from $B(T)$ data for model compounds and predicted $B(T)$ for the compounds of interest. Agreement with experimental $B(T)$ data was fair. A somewhat related calculation of X–X, X–Y, and Y–Y interactions, for CH_4, has been reported by Snook and Spurling.[124] They claim to have made the first calculation of $B(T)$ for a polyatomic molecule without reference to bulk properties.

In summary, pseudo-central potentials with three disposable parameters offer powerful means of correlating and extrapolating measured $B(T)$ data for polyatomic gases. Accurate prediction of $B(T)$ data *via* these potentials seems a more remote prospect, though McKinley and Reed's approach is worthy of further investigation.

Molecules with Orientation-dependent Force Fields. In 1941, Stockmayer showed how to modify the Lennard-Jones 12 – 6 potential to allow for orientation-dependent forces that result from the presence of permanent dipoles of moment μ embedded at molecular centres:

$$u(r, \theta, \phi) = 4\varepsilon\{(\sigma/r)^{12} - (\sigma/r)^6\} - g\mu^2/r^3, \tag{60}$$

where $g = 2\cos\theta_1\cos\theta_2 - \sin\theta_1\sin\theta_2\cos\phi$, θ_1 and θ_2 being the angles between the respective dipole axes and the line joining centres, and ϕ the azimuthal angle between the dipole axes.[20, 21] A subtle distinction is to be noted between ε and σ in equation (60) and their counterparts in equation

[120] K. W. Suh and T. S. Storvick, *J. Phys. Chem.*, 1967, **71**, 1450.
[121] T. S. Storvick and T. H. Spurling, *J. Phys. Chem.*, 1968, **72**, 1821.
[122] A. G. de Rocco and W. G. Hoover, *J. Chem. Phys.*, 1962, **36**, 916.
[123] M. D. McKinley and T. M. Reed, *J. Chem. Phys.*, 1965, **42**, 3891.
[124] I. K. Snook and T. H. Spurling, *Austral. J. Chem.*, 1970, **23**, 819.

(57). Thus, as indicated in Figure 11 curve *f*, ε in equation (57) is the maximum depth of the potential well and σ is the value of r for $u = 0$. But ε in equation (60) is the value of u when $\mathrm{d}\{u(r) + \mu^2 g r^{-3}\}/\mathrm{d}r = 0$ and σ is the value of r when $u(r) + \mu^2 g r^{-3} = 0$, *i.e.* in equation (60) ε is the maximum depth that the potential well would have, and σ is the value that the collision diameter would have, if the dipole–dipole energy were zero.

From statistical mechanics, the version of equation (25) that allows for orientation-dependent forces is

$$B(T) = (N_A/4)\int_0^\infty r^2\,\mathrm{d}r\int_0^\pi \sin\theta_1\,\mathrm{d}\theta_1\int_0^\pi \sin\theta_2\,\mathrm{d}\theta_2\int_0^{2\pi}\{1-\exp(-u/kT)\}\,\mathrm{d}\phi. \tag{61}$$

Tables of B^* [$= B/(2\pi N_A\sigma^3/3)$] as a function of T^* [$= T/(\varepsilon/k)$] and a parameter t^* [$= \mu^2/(\varepsilon\sqrt{8}\sigma^3)$] are available.[20, 125] Some workers have followed the Stockmayer model as literally as possible by regarding μ as the observed dipole moment of the gas, in which circumstance t^* is related to ε and σ; other workers have regarded t^* as a disposable parameter that takes account of all types of orientational force, so the Stockmayer equation then becomes applicable to non-dipolar compounds! Examples of successful use of the Stockmayer potential for correlating $B(T)$ data abound: propyne,[126] chloroalkanes,[127] chlorofluoroalkanes,[128] alkanols,[27, 129] esters, ketones, ethers, and aliphatic nitrogen compounds [130] are amongst dipolar compounds treated by Stockmayer's method. However, a note of caution is in order, because in no instance did the experimental $B(T)$ data cover a sufficient temperature range to test the adequacy of Stockmayer's potential. Amongst non-dipolar compounds treated by Stockmayer's method, using an empirical value of t^*, are CF_4,[119] cyclo-C_4F_8,[118] and C_6F_6.[48] For CF_4 and cyclo-C_4F_8 Douslin and co-workers found the Kihara and Stockmayer potentials fitted their $B(T)$ data equally well; for C_6F_6 the Stockmayer potential was the better.

Buckingham and Pople [131] gave a more complete treatment than Stockmayer's of the interaction between two molecules having non-central force fields. Their treatment allows not only for dipole–dipole interaction but also for dipole–quadrupole, quadrupole–quadrupole, dipole–induced dipole, steric, anisotropic, and other effects. An important feature of the Buckingham–Pople procedure is that the various integrals needed can be calculated from a master table that they give. Buckingham and Pople applied their method to four polar compounds for which values of the dipole moment and the polarizability were known, and used $B(T)$ to

125 J. A. Barker and F. Smith, *Austral. J. Chem.*, 1960, **13**, 171.
126 J. Brewer, *J. Chem. and Eng. Data*, 1965, **10**, 113.
127 A. Pérez Masiá and M. Díaz Peña, *Anales de Quím.*, 1959, **55B**, 229.
128 R. F. Hajjar and G. E. McWood, *J. Chem. Phys.*, 1968, **49**, 4567.
129 J. D. Cox, *Trans. Faraday Soc.*, 1961, **57**, 1674.
130 R. F. Blanks and J. M. Prausnitz, *Amer. Inst. Chem. Eng. J.*, 1962, **8**, 86.
131 A. D. Buckingham and J. A. Pople, *Trans. Faraday Soc.*, 1955, **51**, 1173.

determine ε and σ. They were unable to test their method in respect of quadrupole effects because quadrupole moments Θ were not established at that time. More recently, values of Θ have become available, some *via* a route that involves $B(T)$ data. Thus for a molecule like CO_2 it is feasible to take ε and σ values from a Lennard-Jones fit to $\eta(T)$ data,* then to seek a value of Θ which correlates $B(T)$ data according to the Kielich[133] or Buckingham–Pople[131] treatments of interactions involving quadrupoles. This procedure has been used[134–137] to determine Θ for N_2, O_2, NO, CO, F_2, CO_2, N_2O, C_2H_2, C_2H_4, CS_2, H_2O, NH_3, SO_2, and MeOH, and a related procedure has been used to determine[138, 139] the octupole moments of CH_4, CF_4, and CMe_4. A value of Θ for benzene has been derived with the aid of a spherical-shell potential.[140]

Hence, potential models that allow for various kinds of orientation-dependent force have been used successfully to correlate $B(T)$ data. The calculation of $B(T)$ data from measurements of other properties would seem to be feasible for at least some polyatomic polar gases.

Correlation and Prediction *via* the Quasi-chemical Approach.—Equations (29) and (35), whose derivations involved differing definitions and assumptions, express p and V in terms of the equilibrium constants for the formation of groups of r molecules from single molecules at a particular temperature T, but for p, V, T correlation over a temperature range these equations are of limited utility, since the K's are likely to be temperature dependent. In relation to equation (35), the variation of the K's with temperature can be expressed by the thermodynamic relation

$$K_r = \exp(\Delta S_r^\circ/R - \Delta H_r^\circ/RT), \qquad (62)$$

where ΔS_r° and ΔH_r° (respectively the standard entropy and enthalpy of formation of r-mers from monomers) can be assumed constant over, say, a 100 K temperature range. Thus equation (35) may then be written in the form

$$pV_\mathrm{m}/RT = 1 + [b/RT - \exp(\Delta S_2^\circ/R - \Delta H_2^\circ/RT)]\,p + (1 - r)\,[\exp(\Delta S_\mathrm{r}^\circ/R - \Delta H_\mathrm{r}^\circ/RT)]\,p^{r-1}, \qquad (63)$$

i.e.

$$B(T) = b - RT\exp(\Delta S_2^\circ/R - \Delta H_2^\circ/RT), \qquad (64)$$

[132] L. Monchick and E. A. Mason, *J. Chem. Phys.*, 1961, **35**, 1676.
[133] S. Kielich, *Physica*, 1965, **31**, 444.
[134] T. H. Spurling and E. A. Mason, *J. Chem. Phys.*, 1967, **46**, 322.
[135] G. A. Bottomley and T. H. Spurling, *Austral. J. Chem.*, 1967, **20**, 1789.
[136] A. L. Tsykalo, *Russ. J. Phys. Chem.*, 1968, **42**, 261.
[137] Y. Singh and A. Das Gupta, *J. Chem. Phys.*, 1970, **52**, 3064.
[138] T. H. Spurling, A. G de Rocco, and T. S. Storvick, *J. Chem. Phys.*, 1968, **48**, 1006.
[139] H. Enokido, T. Shinoda, and Y. Mashiko, *Bull. Chem. Soc. Japan*, 1969, **42**, 3415.
[140] T. H. Spurling and A. G. de Rocco, *J. Chem. Phys.*, 1968, **49**, 2867.

* It has been well demonstrated that gas transport properties are much less sensitive to that part of the potential involving electric moments than are second virial coefficients.[132]

and

$$R(T) = (1 - r)RT\exp(\Delta S_r^\circ/R - \Delta H_{ri}^\circ/RT), \tag{65}$$

where R is any virial coefficient, higher than the second, that corresponds to r. Equation (64) has found wide use for correlating $B(T)$ values for substances for which dimerization is the predominant polymerization mode,[24, 25, 141, 142] *e.g.* amines, esters, carbonyl compounds, nitriles. Since the physical significance of b depends on the model assumed, it is permissible to regard b as an empirically adjustable constant. Use of (64) for a dimerizing gas would therefore involve a search for a value of b that caused a plot of $\log_{10}\{[b - B(T)]/RT\}$ to be linear with $1/T$. Alternatively, if b is regarded as the 'physical' part of B, calculated values of $b(T)$ should give a linear plot of $\log_{10}\{[b(T) - B(T)]/RT\}$ *versus* $1/T$. In either event, ΔH_2° would be evaluated as $-2.303R \times$ the slope, and ΔS_2° would be evaluated as $2.303R \times$ the ordinate value at $1/T = 0$. Thus Prausnitz and Carter[25] evaluated $b(T)$ for acetonitrile from $B(T)$ of its homomorph propane;* a plot of $\log_{10}\{[b(T) - B(T)/RT]\}$ *versus* $1/T$ gave $\Delta H_2^\circ = -21.4$ kJ mol^{-1} and $\Delta S_2^\circ = -80.5$ J K^{-1} mol^{-1}. These values are not untypical of those found generally for the formation of gas-state dimers: the formation of dimer from monomer involves an appreciable loss of standard entropy and a decrease of enthalpy which is small compared with that accompanying the formation of a covalent bond. It is evident from equation (64) that as T increases, the exponential term will become dominated by the (negative) ΔS_2°, so that $B \rightarrow b$, *i.e.* dimerization will become less and less significant.

Consideration of hydroxy-compounds has been left until last, because these compounds necessitate consideration of more than one gaseous polymeric species. We referred above to the differing conclusions to be found in the literature on the species in methanol vapour. Probably the real situation is that the vapour of a given hydroxy-compound contains a series of polymer species with $r = 2$, upwards. Some species may be favoured over others for energetic (ΔH°) reasons and some for configurational (ΔS°) reasons. Therefore the overall composition of the vapour will depend in a subtle way on temperature, and because of the p^{r-1} term will be sensitive to pressure also. Quite conceivably one species may predominate in certain ranges of p and T to the extent that other species may be ignored. Thus it is well established that dimers, and to a lesser extent tetramers, are the significant species in the vapours of the lower alkanols at temperatures around the normal boiling temperature and $p = 0.2$ to 1 bar. More details may be found in Chapter 6.

141 J. D. Lambert and E. D. T. Strong, *Proc. Roy. Soc.*, 1950, **A200**, 566.
142 J. D. Lambert, J. S. Clarke, J. F. Duke, C. L. Hicks, S. D. Lawrence, D. M. Morris, and M. G. T. Shone, *Proc. Roy. Soc.*, 1959, **A249**, 414.

* In this instance a C_nH_{2n+2} homomorph with an integral value of n was selected. However, it is quite feasible to employ a non-integral value of n, using equation (55), if a more accurate match in the molar volumes of substance and homomorph is obtained thereby.

It may be concluded that the quasi-chemical approach offers a means of correlating p, V, T for gases which show marked departures from ideality. Thus Prausnitz *et al.*[143, 144] have calculated $B(T)$ for both steam and ammonia, *via* equation (34). They utilized the Kihara approach,[114] coupled with a Buckingham–Pople[131] treatment of orientation-dependent forces, to derive the first term on the right-hand side. The calculated values for steam cover an extremely wide temperature range.

7 Thermodynamic Properties of a Real Gas

In this section, we show how a knowledge of p, V, T relations permits calculation of the thermodynamic properties of a real gas. The gas chosen as an example is perfluorocyclobutane, the thermodynamic properties of which have been calculated by Harrison and Douslin[145] from p, V, T measurements by Douslin, Moore, and Waddington.[118]

Real-gas thermodynamic properties may be expressed as functions of the variables of state, according to relations which may be developed from first principles. A comprehensive list of such relations has been given by Beattie and Stockmayer.[9] For example, the molar enthalpy H, the molar entropy S, and the molar Gibbs energy G of a real gas can be written in terms of p, T, and ρ (the amount density) as follows:

$$H - H^\circ = \int_0^\rho [\{p - T(\partial p/\partial T)_\rho\}/\rho^2]\,\mathrm{d}\rho + p/\rho - RT, \qquad (66)$$

$$S - S^\circ = \int_0^\rho [\{R\rho - (\partial p/\partial T)_\rho\}/\rho^2]\,\mathrm{d}\rho - R\ln(RT\rho/p^\circ), \qquad (67)$$

$$G - G^\circ = \int_0^\rho \{(p - RT\rho)/\rho^2\}\,\mathrm{d}\rho + p/\rho - RT + RT\ln(RT\rho/p^\circ). \qquad (68)$$

In these equations, the real-gas properties are expressed relative to the standard state values, H°, S°, and G°, which are, respectively, the molar enthalpy, the molar entropy, and the molar Gibbs energy values which the gas would have at a standard pressure p° (1.013 25 bar) if it were ideal.

Douslin's experimental data[118] consisted of values of the pressure, measured at fixed values of amount density between 0.75 and 7.0 kmol m^{-3} and at rounded values of Celsius temperature (IPTS 48) between the critical temperature (115.22 °C) and 350 °C. For application of equations (66) to (68) the pressure at each value of density was expressed[145] in terms of temperature on the thermodynamic scale by a suitable function, which was then differentiated to obtain values for $(\partial p/\partial T)_\rho$. After the values of the

143 J. P. O'Connell and J. M. Prausnitz, *Ind. Eng. Chem. Fundamentals*, 1970, **9**, 579.

144 C. S. Lee, J. P. O'Connell, C. D. Myrat, and J. M. Prausnitz, *Canad. J. Chem.*, 1970, **48**, 2993.

145 R. H. Harrison and D. R. Douslin, 'Perfluorocyclobutane: the Thermodynamic Properties of the Real Gas,' U.S. Dept. of the Interior, Bureau of Mines, Washington, 1964.

derivatives had been smoothed, the integrals in equations (66), (67), and (68) were graphically evaluated. In order to establish the values of the integrands at the lower limits of integration the values were calculated from the second virial coefficient and/or its temperature derivative.

Values of real-gas enthalpies and entropies are needed for the construction of Mollier and other diagrams for use in chemical engineering calculations. Values of fugacity coefficients are needed for calculations relating to chemical equilibria, phase relationships, and chemical kinetics in systems involving high-pressure gases. Illustrative values [145] of $H - H°$, $S - S°$, and the fugacity coefficient γ for real-gas perfluorocyclobutane are given in the Table. The fugacity coefficient, which is the ratio between the fugacity and the pressure, is given by

$$\gamma = \exp\{(G - G°)/RT\}/(1.013\,25p/\text{bar}). \tag{69}$$

Table *Some thermodynamic properties of real-gas perfluorocyclobutane*

t/°C	115.22	150.00	200.00	250.00	300.00	350.00
$\dfrac{\rho}{\text{kmol m}^{-3}}$			$-(H - H°)$/kJ mol^{-1}			
1	4.30	3.73	3.28	2.94	2.66	2.42
2	7.93	6.87	6.04	5.43	4.91	4.46
3	10.73	9.41	8.35	7.53	6.79	6.12
4	12.76	11.52	10.29	9.24	8.23	7.30
5	14.56	13.29	11.76	10.35	8.94	7.67
			$-(S - S°)$/J K^{-1} mol^{-1}			
1	339.7	33.65	33.98	34.45	34.78	35.50
2	445.0	43.25	43.07	43.32	43.68	44.11
3	517.6	50.03	49.69	49.86	50.16	50.55
4	570.1	55.75	55.54	55.76	56.06	56.47
5	621.2	61.48	61.49	61.79	62.09	62.59
			γ			
1	0.748	0.782	0.827	0.867	0.903	0.933
2	0.672	0.678	0.724	0.780	0.837	0.891
3	0.665	0.615	0.647	0.714	0.792	0.874
4	0.659	0.544	0.572	0.662	0.777	0.905
5	0.534	0.431	0.505	0.654	0.846	1.073

In previous sections we have emphasized the importance of the virial equation of state. However, for accurate calculation of properties of real gases at high density, the virial equation of state is useful only if reliable values of the virial coefficients above the third (and their temperature derivatives) are available, which is rarely the case — that is why Douslin's calculations, referred to above, employed graphical, rather than analytical, integration. Naturally, calculation of the properties of a *dilute* gas can be performed to good accuracy in terms of just $B(T)$ and $C(T)$ or just $B'(T)$ and $C'(T)$. For example, the real-gas constant-pressure heat capacity, C_P, and the isenthalpic Joule–Thomson coefficient, μ, can be evaluated from the following equations:[9]

$$C_p - C_p^\circ = -T\int_0^p (\partial^2 V/\partial T^2)_p\,\mathrm{d}p, \tag{70}$$

$$= -pT\,\mathrm{d}^2B'/\mathrm{d}T^2 - \tfrac{1}{2}p^2T\,\mathrm{d}^2C'/\mathrm{d}T^2, \tag{71}$$

$$\mu = T^2\{\partial(V/T)/\partial T\}_p/C_p, \tag{72}$$

$$= \frac{T\,\mathrm{d}B'/\mathrm{d}T - B' + p(T\,\mathrm{d}C'/\mathrm{d}T - C')}{C_p^\circ - pT\,\mathrm{d}^2B'/\mathrm{d}T^2 - \tfrac{1}{2}p^2T\,\mathrm{d}^2C'/\mathrm{d}T^2}. \tag{73}$$

For application of (71) and (73) it is convenient to fit B' and C' to polynomials in T, and to evaluate the differential coefficients therefrom.

6
Modern Vapour-flow Calorimetry

BY J. F. COUNSELL

1 Introduction

Flow calorimetry is now the commonly used method for the determination of the heat capacities of gases and vapours at essentially constant pressure, C_p. Many methods have been described for the determination of C_p and pieces of apparatus and techniques have been developed so that, particularly in the last twenty years, accuracies of 0.1 per cent or better have been claimed for the results. Enthalpies of vaporization to a slightly higher level of accuracy are often determined during the course of the heat capacity measurements.

Heat capacity measurements made by flow calorimetry make an important contribution to the complete thermodynamic description of a vapour or gas. Values of $(\partial C_p/\partial p)_T$ derived from the measurements are related to the non-ideal gas behaviour of the substance. The heat capacity of the ideal gas, C_p°, which equals the heat capacity of the real gas at zero pressure, is also of considerable use in the study of fundamental molecular quantities, such as in the investigation of vibrational assignments and in the study of barrier heights to internal rotations in molecules.

2 Review of Experimental Methods

Partington and Shilling's 'Specific Heat of Gases' [1] reviews the methods and results of heat capacity measurements made in the first quarter of this century. Masi,[2] in an article published in 1954, covered the period from 1925 to 1952 and compared experimental and calculated values of heat capacities for many simple gases. He also briefly described ten possible methods of measuring heat capacities. In the last ten years Rowlinson [3] devoted a chapter of his book, 'The Perfect Gas', to the measurement of heat capacities, and a detailed account of measurements and treatment of results in vapour-flow calorimetry was written by McCullough and Waddington in 'Experimental Thermodynamics', Volume I.[4]

Three main types of calorimeter have been described for the measurement of heat capacities by the vapour-flow method. In these methods the

[1] J. R. Partington and W. G. Shilling, 'The Specific Heat of Gases', Benn, London, 1924.
[2] J. F. Masi, *Trans. A.S.M.E.*, 1954, 1067.
[3] J. S. Rowlinson, 'The Fluid State', Pergamon, London, 1963.
[4] J. P. McCullough and G. Waddington, in 'Experimental Thermodynamics', Butterworth, London, 1968, Vol. I.

temperature rise in a vapour caused by passing it over a heater is measured. In the first type of apparatus the heat lost by the vapour to its surroundings is greatly reduced by the use of controlled shields, as in the design due to Scott and Mellors.[5] The method due to Scheel and Heuse[6] and Bennewitz and Rossner[7] allows partial heat exchange between vapour entering and leaving the calorimeter so that the heat loss is proportional to F^{-2}, where F is the flow rate. This design has not been further developed. The third type of apparatus was initiated by Pitzer[8] and developed by Waddington, Todd, and Huffman[9] of the U.S. Bureau of Mines, Bartlesville. Further development has been carried out at Bartlesville[10] and has resulted in a great number of reported experimental measurements. An essentially similar apparatus was set up at the National Physical Laboratory by Hales, Cox, and Lees,[11] and it is this type of calorimeter, the non-adiabatic type in which the heat loss depends on F^{-1}, which will be discussed here.

3 The Theory of Flow Calorimetry in Non-adiabatic Calorimeters

As with all vapour-flow calorimeters one of the main features of the Pitzer type is the provision of a constant flow of vapour and a means of determining the rate of flow. A constant flow is achieved by boiling a liquid to produce vapour under steady-state conditions and the vapour flow rate is calculated from the power supplied to the boiler and the enthalpy of vaporization; the latter is determined in a separate experiment in the same apparatus. To maintain constant conditions in the boiler the vapour is condensed and returned as liquid to the boiler.

The calorimeter contains a platinum resistance thermometer (T1), a heater, and two further platinum resistance thermometers (T2, T3) which are all situated in the flowing gas (see Figure 1). The aim is to achieve a steady-state condition in which the rate of loss of heat by the vapour is small and constant. This steady state has been reached when the temperatures T_2 and T_3 indicated by T2 and T3 are constant. Even though corrections are made for heat losses it is advisable to design the apparatus so that these are as small as possible. This is achieved by making the walls of the calorimeter out of poor heat conductors of low emissivity. Under these circumstances the residual heat loss occurs mainly by radiation.

When an experiment is carried out a correction has to be made for the heat lost by the vapour, at temperature T, to its surroundings, at temperature T_1. It is assumed that the heat lost per unit time by a unit length of the

[5] R. B. Scott and J. W. Mellors, *J. Res. Nat. Bur. Stand.*, 1945, **34**, 243.
[6] K. Scheel and W. Heuse, *Ann. Physik*, 1912, **37**, 79.
[7] K. Bennewitz and W. Rossner, *Z. Phys. Chem.*, 1938, **B39**, 126.
[8] K. S. Pitzer, *J. Amer. Chem. Soc.*, 1941, **63**, 2413.
[9] G. Waddington, S. S. Todd, and H. M. Huffman, *J. Amer. Chem. Soc.*, 1947, **69**, 22.
[10] J. P. McCullough, D. W. Scott, H. L. Finke, W. N. Hubbard, M. E. Gross, C. Katz, R. E. Pennington, J. F. Messerly, and G. Waddington, *J. Amer. Chem. Soc.*, 1953, **75**, 1818.
[11] J. L. Hales, J. D. Cox, and E. B. Lees, *Trans. Faraday Soc.*, 1963, **59**, 1544.

vapour in a calorimeter of constant bore is $Q(T - T_1)$. This is a good assumption for heat lost by conduction and also for heat lost by radiation so long as $(T - T_1) \ll T_1$.

Figure 2 is a diagrammatic representation of the calorimeter. The vapour enters from the left and passes over a uniform heater of length L and power

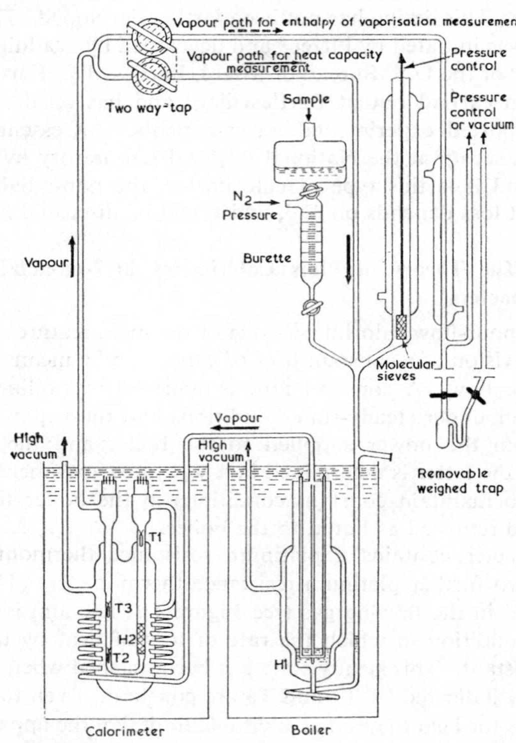

Figure 1 *The non-adiabatic flow calorimeter*

W. The variable l represents the distance of any point from the origin (see Figure 2), F is the molar flow rate, and C_p is the molar heat capacity of the vapour.

Equation (1) applies to the conditions when the vapour is passing over the heater:

$$C_p F \, dT = \{(W/L) - Q(T - T_1)\} \, dl, \qquad (1)$$

where a portion of heater of length dl causes a temperature rise dT in the vapour. Equation (1) may be rearranged and integrated to give:

$$\ln\{(W/QL) - (T - T_1)\} = -(Ql)/(C_p F) + \text{constant}(1). \qquad (2)$$

When $l = 0$, $T = T_1$ and hence

$$\ln\{(W/QL) - (T - T_1)\} = -(Ql)/(C_pF) + \ln(W/QL) \qquad (3)$$

After the vapour has passed the heater similar considerations lead to:

$$\ln(T - T_1) = -(Ql)/(C_pF) + \text{constant}(2). \qquad (4)$$

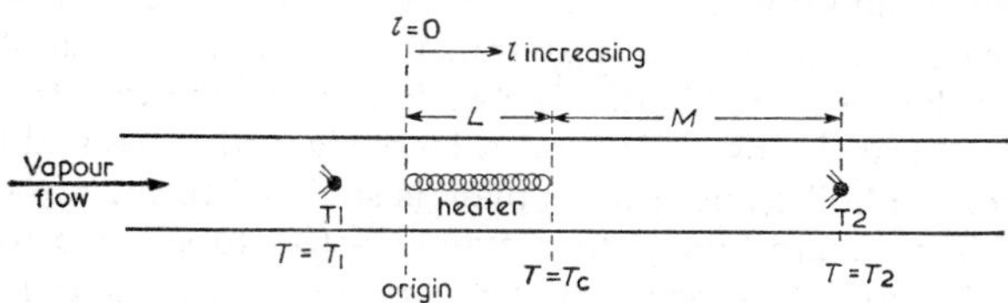

Figure 2 *Diagrammatic representation of flow calorimeter. At distance l to right of origin vapour has temperature T*

At the end of heater further from the origin the temperatures given by equations (3) and (4) have the same value T_c. Thus

$$(T_c - T_1) = (W/QL)\,[1 - \exp\{-(QL)/(C_p\,F)\}]$$
$$= \exp\{-(QL)/(C_pF) + \text{constant}(2)\}. \qquad (5)$$

The substitution of the integration constant(2) into equation (4) gives:

$$(T - T_1) = (W/QL)\,[1 - \exp\{-(QL)/(C_pF)\}]\exp\{-Q(l - L)/(C_pF)\}. \qquad (6)$$

The apparent value of the heat capacity C_p(app) measured at the point $l\,(= L + M)$ in the calorimeter is given by:

$$C_p(\text{app}) = W/\{F(T_2 - T_1)\}$$
$$= \exp\{(QM)/(C_pF)\}\,QL/[F - F\exp\{-(QL)/(C_pF)\}]. \qquad (7)$$

Expansion of the exponential terms leads to the approximate relation

$$C_p(\text{app}) = C_p + k/F + k^2/(2C_pF^2) + \ldots, \qquad (8)$$

where $k = Q(M + L/2)$.

If k is kept small and the flow rates are reasonably great then quadratic terms in equation (8) may be ignored and C_p can be obtained from the plot of C_p(app) against F^{-1} at the point where $F^{-1} = 0$. For example, for ethanol[12] at 475 K the largest value of the k/F term was $0.07C_p$, and ignoring the $k^2/(2C_p\,F^2)$ term in the calculation would have caused an error of about 0.04 per cent in C_p. For substances of low heat capacity the quadratic term in equation (8) plays a more important rôle.

4 The Construction of a Non-adiabatic Flow Calorimeter

The construction of the vapour-flow calorimeter used by Hales, Cox, and Lees[11] at N.P.L. will be described. The apparatus (see Figure 1) consists

[12] J. F. Counsell, J. O. Fenwick, and E. B. Lees, *J. Chem. Thermodyn.*, 1970, **2**, 367.

of a boiler for producing a constant stream of vapour, a calorimeter in which a steady flow of vapour is set up and heat capacities are measured, thermostats, a pressure controller, and various taps and heated lines.

Boiler.—The boiler is made of glass and is designed to produce a steady stream of vapour with no entrainment of liquid droplets. The inner vessel contains the liquid sample and the space between this vessel and the outer vessel is evacuated to reduce heat transfer between the sample and the surrounding thermostatted bath. The walls between the vessels are silvered to reduce heat transfer by radiation. A large glass-to-metal seal is attached to the top of the boiler and a brass plate is soldered to the top of the seal. Through this plate are brazed four stainless-steel tubes, two for carrying thermocouples used to measure the temperatures of the vapour and liquid and two which contain the electrical leads to the heater.

The heater is designed to produce a steady stream of small bubbles of vapour and for this purpose several arrangements have been described.[4, 9, 11, 13] The heater employed at N.P.L. is made by attaching to each end of a piece of Karma resistance wire (30 Ω, 250 mm) a piece of platinum wire (0.2 mm, 25 mm). To each free end of the platinum is attached a piece of 0.38 mm diameter copper wire, and 30 mm along each copper wire is attached a piece of 0.71 mm diameter copper wire. The construction of the heater is so arranged that the copper junctions are at the surface of the liquid in the boiler. This assembly is inserted into woven glass sleeving and then into a stainless-steel tube. The centre portion of the tube is bent to make a flat coil of about 330 mm length. The ends of the tube are brought through and brazed to the brass plate at the top of the boiler in such a position that the coil is about 15 mm below the surface of the liquid in the boiler. The thick copper wires serve as the 'current' leads to the heater and the thin copper wires as the 'potential' leads. The coiled stainless steel heater is covered with a glass necklace to promote nucleation in the boiling liquid. This necklace is constructed by threading a fine glass capillary tube (length 400 mm) on to a piece of Nichrome wire and cutting the glass into 4 mm segments. Each segment is collapsed on to the wire by heating this assembly in a small flame and the necklace is wound around the heater coil and secured with a ceramic cement. The heater covered with the necklace is allowed to dry and is then baked in an oven.

Calorimeter.—The calorimeter consists of a thin-walled glass U-tube inside an outer vessel. The space between the U-tube and the vessel is evacuated and the walls are silvered. In the inlet arm of the U-tube there is a small sensitive platinum resistance thermometer (T1), a heater (H2), and a mixing device to promote thermal equilibrium in the vapour. In the outlet arm there are two platinum resistance thermometers (T2, T3). The electrical leads of the heater and thermometers are taken out through the tops of the U-tube.

[13] G. Waddington, J. C. Smith, K. D. Williamson, and D. W. Scott, *J. Phys. Chem.*, 1962, **66**, 1074.

The heater consists of a coil of Nichrome wire (150 Ω) wound upon a mica former. Two coils of platinum wire are attached to the ends of the heater and leads from these go to the electrical outlets. The coils serve to prevent heat leaks away from the heater as the incoming vapour, at temperature T_1, passes over them before it reaches the heater. During experiments the calorimeter heater is supplied from the same stabilized potential source as the boiler heater so that the temperature rise of the vapour at the heater is almost independent of flow rate. A baffle situated between T1 and H2 serves to prevent heat transfer by radiation from the heater to the thermometer.

System.—Vapour is generated in the boiler and passes through heated glass lines, the temperatures of which are controlled at close to that of the calorimeter and above that of the boiler. The vapour then passes into a silver coil which surrounds the calorimeter in its thermostatted oil bath. The temperatures of both the oil baths are controlled to about 0.001 K. On leaving the coil the vapour enters the calorimeter and passes over platinum resistance thermometer T1, heater H2, and platinum resistance thermometers T2 and T3 and then, on leaving the calorimeter, it passes through heated lines to the two-way tap. If a measurement of heat capacity is being made the vapour follows the lower path in Figure 2, is condensed in the water-cooled condenser, and returns to the boiler through a long glass coil of small bore, which allows the condensed liquid to warm to the temperature of the bath. When an enthalpy of vaporization is being measured the vapour, after passing through the two-way tap, is diverted for a measured period of time and condensed in the removable trap.

Control Equipment.—The two thermostatted baths are similar in design and both contain an inner metal tank surrounded by heated mats and insulating material. The silicone oil in a bath is circulated by forcing it up a tube and over a heater and a temperature-sensitive resistance wire. The resistance wire acts as one arm of a Hallikainen temperature controller. The pulsed output of this instrument is fed *via* a transformer to the heater in the bath. The resulting temperature is steady to about ± 0.001 K.

Nitrogen pressure in the system is controlled by a Texas Instruments Precision Pressure Controller 156-01-050 with a stability of at worst 3 Pa. In the course of a measurement an interface between the nitrogen and the vapour of the substance under test occurs in the condenser. The nitrogen pressure is then communicated to the liquid in the boiler by the vapour.

Measuring Equipment.—The main measurements required are the temperatures of T1, T2, T3, and of two thermometers situated in the thermostatted baths (T4 in the boiler bath and T5 in the calorimeter bath), and the powers supplied to the boiler and calorimeter heaters. During enthalpy of vaporization measurements the time during which the vapour is diverted to the weighed trap is measured.

The resistances of thermometers T1, T2, T3, T4, T5 are measured by a CROPICO Resistance Bridge Smith No. 3. The potential applied to the boiler heater (up to 50 V) is divided by a 50/1 potential divider and the current through the heater is passed through a standard 0.5 Ω resistance, which, like the potential divider, is situated in an air thermostat. A similar arrangement is made for the calorimeter heater. The potentials across the potential dividers and the standard resistors, together with the amplified potential difference from a differential thermocouple (DT) placed with one end in the thermostatted bath and the other end in the vapour a little above the surface of the liquid in the boiler, are scanned by a Solartron data-transfer unit, read by a Solartron digital voltmeter (LM1867), and printed on a teletype. Automatic methods will soon be introduced to record the resistances of the five thermometers so that measurements of the values of the resistances and potentials can be printed on a teletype roll. The data will be conveniently arranged in columns so that they can be inspected to determine when a steady state has been reached.

The pressure in the system is measured automatically by the pressure controller but a small correction is made to the reading to allow for the pressure drop across the boiler. This pressure drop which is usually small and never exceeds 600 Pa is determined by measurement of the height of liquid in the boiler return arm, using a periscope.

5 Enthalpies of Vaporization

Measurements of enthalpy of vaporization with the apparatus illustrated are made at temperatures corresponding to the pressures at which subsequent heat capacity measurements will be made (usually about 25, 50, 100 kPa). The removable weighed trap is pumped, weighed, and attached to the system. With the two-way tap in its normal position which allows vapour circulation, the collecting system is pumped to test that the two-way tap and the trap connections do not leak. The thermostatted bath containing the boiler is set at a temperature about 0.1 K above the boiling point of the liquid at the operating pressure, so as to prevent condensation of vapour in the neck of the boiler. The calorimeter and heated lines are controlled at a temperature about 20 K above that of the boiler. The pressure is controlled on both sides of the two-way tap, the boiler heater energized, and the system brought to steady-state operating conditions. The two-way tap is operated for 10 s to flush away any remaining nitrogen from the two-way tap. The vapour so transferred is pumped away and the controlled nitrogen pressure reapplied to the trap side of the two-way tap. A liquid-nitrogen-filled Dewar vessel is then placed around the trap.

When steady-state conditions have been reached the two-way tap is operated and the energy input to the boiler, the temperature difference between the boiler bath and the vapour above the surface of the boiling liquid (as measured on DT), and the temperature of the boiler bath are monitored. The two-way tap operates a micro-switch connected to a timer, which measures the time interval during which the vapour is diverted. Liquid is introduced to the boiler from the burette at the same rate as it is removed by the boiling process in order to keep a constant level in the boiler. After a period of 100 to 300 s the two-way tap is returned to its initial position and the burette tap is closed. The sample, of mass about 10 g, is allowed to condense in the trap. After about 10 min the nitrogen gas is pumped away and the sample is collected in the cooled trap. The taps on the trap are closed and the trap removed and weighed.

The enthalpy of vaporization is determined by dividing the power input to the boiler by the amount of substance collected. The following corrections, each of which contribute less than 0.2 per cent of the measured quantity, may have to be applied.

(i) A correction to the amount of substance collected equivalent to the amount of vapour in the space caused by the net change in sample volume in the boiler during the run.

(ii) A correction due to the quantity of heat carried into the boiler by liquid from the burette, which enters the boiler at a temperature about 0.1 K higher than the boiler contents.

(iii) A correction due to the enthalpy change in the boiler arising from the dependence of the temperature profile in the boiler upon the liquid level, which may change during the experiment.

(iv) A correction to the energy due to any change in temperature of boiler and contents during the experiment.

These corrections are explained in greater detail in ref. 4.

6 Heat Capacity

The heat capacity of the vapour is measured for at least two pressures at a series of temperatures. The pressure controller maintains the pressure at the desired value and the boiler bath is controlled at about 0.1 K above the temperature of the liquid. The calorimeter bath is controlled at a temperature about 5 K below that of the required heat capacity measurement so that for a temperature rise of 10 K at the heater the vapour will have the required mean temperature. (For temperature ranges in which C_p is strongly temperature-dependent a smaller temperature rise is used.) Measurements are carried out at four flow rates (in the range 0.001 to 0·005 mol s^{-1}) with the calorimeter heater both energized and unenergized. At the first of the four selected flow rates measurements are made on thermometers T1, T2, T3, T4, T5, and the thermocouple DT. Measurements are repeated until a steady state has been reached as indicated by the consistency of the temperature differences $(T_2 - T_1)$ and $(T_3 - T_1)$. The controlled pressure in the system and the height of the liquid in the return tube to the boiler, which indicates the pressure drop in the system, are recorded. Similar measurements are then made at three other flow rates (selected by the potential applied to the boiler). The calorimeter heater is then activated and measurements at each flow rate are then repeated, and now the energy input to the boiler and calorimeter are also recorded.

For any flow rate the temperature rise at T2 (or T3) due to the energy input to the calorimeter can now be calculated as the difference between the $(T_2 - T_1)$ values for the energized and non-energized states. This method compensates for any Joule–Thomson cooling of the vapour due

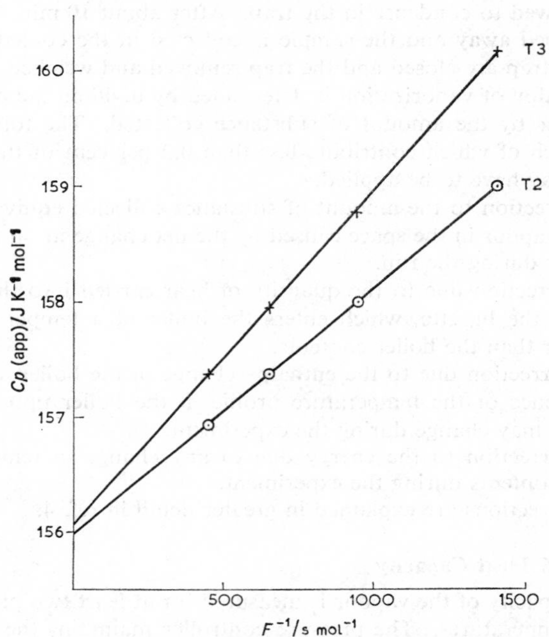

Figure 3 *Plot of* C_p(app) vs. *the reciprocal of the molar flow rate F for* Et_2O *at a temperature of* 450.02 K *and a pressure of* 101 kPa

to the pressure drop through the calorimeter. A further small correction may be necessary due to the Joule–Thomson cooling being slightly less at the higher temperature of the energized experiment than for the non-energized one. A correction arising from the heat carried into the boiler by the returning liquid must be applied, and it may also be necessary to allow for the effect of the small change in pressure in the calorimeter at various flow rates.

After application of these corrections the apparent heat capacity is calculated by dividing the calorimeter heater power by the product of the flow rate and the temperature rise. At infinite flow rate the vapour would not have time to lose heat so at this point C_p(app) becomes equal to C_p. Figure 3 shows the result of plotting C_p(app) against F^{-1} for diethyl ether[14] at 450.02 K and 101 kPa. For the curve marked T2 the slope

[14] J. F. Counsell, D. A. Lee, and J. F. Martin, *J. Chem. Soc.* (*A*), 1971, 313.

k was found to be 0.00218 J K^{-1} s^{-1}, and omission of the third term on the right-hand side of equation (8) would affect the calculated value of C_p by only 0.006 per cent. This term, $k^2/(2C_p F^2)$, becomes larger at higher temperatures (as k is approximately proportional to T^3) and is also larger for substances of low heat capacity.

7 Analysis of Heat Capacity and Enthalpy of Vaporization Measurements

The Clapeyron equation (9) relates the enthalpy of vaporization ΔH_v to the difference between the gaseous and liquid volumes of the substance $(V_G - V_L)$ and the first differential with respect to temperature of the vapour pressure (dp/dT).

$$(V_G - V_L) = (\Delta H_v/T)/(dp/dT). \tag{9}$$

If the vapour under consideration is assumed to obey an equation of the form:

$$pV_G = RT + B'p, \tag{10}$$

and higher virial coefficients are assumed to have no effect, then

$$B' = \{(\Delta H_v/T)/(dp/dT)\} + V_L - (RT/p), \tag{11}$$

Thus, it is possible to calculate B' for temperatures at which the enthalpies of vaporization are known so long as an accurate vapour pressure equation is available. The greatest contribution to any error in the calculation usually comes from the (dp/dT) term, as a 0.1 per cent uncertainty in this term, would lead to an uncertainty of 0.00003 m^3 mol^{-1} in B' for calculations at atmospheric pressure and normal temperatures.

Equation (12) correlates the pressure dependence of heat capacity with the second differential of the molar volume of the vapour with respect to temperature. The derivation of this equation is given in the Appendix.

$$(\partial C_p/\partial p)_T = -T(\partial^2 V_G/\partial T^2)_p. \tag{12}$$

Combining equation (10) with equation (12) leads to a relation between C_p and the second differential with respect to temperature of B'.

$$C_p = C_p^\circ - pT(d^2B'/dT^2). \tag{13}$$

If B' is assumed to have a temperature dependence of the form

$$-B' = b_1 + b_2 T \exp(b_3/T), \tag{14}$$

then

$$C_p = C_p^\circ + (b_2 b_3^2/T^2) \exp(b_3/T)\, p. \tag{15}$$

Equation (15) can be used to correlate a set of vapour heat capacities; see, for example, the treatment of observations on diethyl ether in reference 14. The heat capacities at rounded temperatures were obtained by linear interpolation over small temperature ranges (< 0.04 K). Values of $C_p(p_n, T) - C_p(p_m, T)$ at each rounded temperature for all possible pairs of pressure values (p_n, p_m) were calculated and then fitted to equation (16) by the method of least squares.

$$C_p(p_n, T) - C_p(p_m, T) = (b_2 b_3^2/T^2)\exp(b_3/T)(p_n - p_m). \quad (16)$$

Values of b_2 and b_3 were obtained and values of C_p° were calculated from each C_p value. These were then found to be fitted by an equation having linear temperature-dependence and the equation (17), which represents the data in the measured range 35 to 101 kPa and 310 to 450 K, was obtained.

$$C_p/(\mathrm{J\,K^{-1}\,mol^{-1}}) = 48.2374 + 0.237\,499(T/\mathrm{K}) + \{0.11331/(T/\mathrm{K})^2\}\exp\{1150/(T/\mathrm{K})\}(p/\mathrm{Pa}). \quad (17)$$

The largest difference between the observed and calculated values of the heat capacity is slightly less than 0.1 per cent of the heat capacity, and the root mean square value of all the differences is 0.06 per cent. The remaining coefficient b_1 in equation (14) can now be determined by using b_2 and b_3 from equation (17) and the values of B' calculated from equation (11).

The equation of state (18) may be used as an alternative to equation (10).

$$pV_G = RT(1 + B/V_G + \ldots). \quad (18)$$

This leads to the equation:

$$C_p = C_p^\circ - pT(\mathrm{d}^2 B/\mathrm{d}T^2) + 2B(\mathrm{d}^2 B/\mathrm{d}T^2)(p^2/R)(1 - 2Bp/RT). \quad (19)$$

This equation contains a small quadratic pressure-dependent term which can make a contribution equal to about 5 per cent of the effect of the linear pressure-dependent term.

If the coefficients b_1 and b_2 in equation (14) are positive, as they are for most substances, the form of equation (14) does not allow positive values of B'. Since at sufficiently high temperatures B' should become positive, an equation containing more terms than equation (14) should be used for results measured over long temperature ranges.

8 Standard Substances

When constructing and testing apparatus it is useful to examine the performance of the equipment by the use of one or more standard substances which have been employed by many workers in making similar measurements. For enthalpies of vaporization water [15, 16] and benzene [11, 17, 18] are suitable standard test substances. The high stability and ease of boiling of the fluorocarbons may lead to one of them, *e.g.* hexafluorobenzene,[19] being used as a standard in the future.

[15] N. S. Osborne, H. F. Stimson, and D. C. Ginnings, *J. Res. Nat. Bur. Stand.*, 1939, **23**, 261.
[16] J. P. McCullough, R. E. Pennington, and G. Waddington, *J. Amer. Chem. Soc.*, 1952, **74**, 4439.
[17] D. W. Scott, G. Waddington, J. C. Smith, and H. M. Huffman, *J. Chem. Phys.*, 1947, **15**, 565.
[18] E. F. Fiock, D. C. Ginnings, and W. B. Holton, *J. Res. Nat. Bur. Stand.*, 1931, **6**, 881.
[19] J. F. Counsell, J. H. S. Green, J. L. Hales, and J. F. Martin, *Trans. Faraday Soc.*, 1965, **61**, 212.

Standard substances for heat capacity measurements are benzene,[11, 17] water,[16] and carbon disulphide.[13] Water and carbon disulphide are structurally simple substances and thus it is possible to compare accurate calculated heat capacities with the measured values. McCullough and Waddington [4] discuss fully the employment of standard substances in heat capacity measurements and the reasons governing their choice.

9 Experimental Results

Alcohols and Ketones.—Heat capacities and enthalpies of vaporization have been published in the last decade on ethanol,[12] the propanols,[11, 20, 21] the butanols, [12, 22–24] and pentan-1-ol.[12] Isothermal C_p *vs.* p data for these compounds are highly non-linear so the simple form of equation (13) is not applicable. In refs. 11 and 25 the theory was developed that the vapour consisted of an equilibrium mixture of monomeric, dimeric, and r-meric forms, and that the equation of state (20) was applicable.

$$pV_G = RT + Bp + Dp^{r-1}. \qquad (20)$$

If ΔS_2 and ΔH_2 are respectively the entropy and enthalpy of dissociation of dimer to momomer, ΔS_r and ΔH_r are the corresponding quantities for r-mer to monomer, and b is the co-volume then the following equations apply.

$$B = b - RT\exp\{-(\Delta S_2/R) + (\Delta H_2/RT)\}, \qquad (21)$$

$$D = -(r-1)RT\exp\{-(\Delta S_r/R) + (\Delta H_r/RT)\}, \qquad (22)$$

$$C_p(p,T) = C_p^\circ(T) + a(T)p + c(T)p^{r-1}, \qquad (23)$$

$$a(T) = (\Delta H_2^2/RT^2)\exp\{-(\Delta S_2/R) + (\Delta H_2/RT)\}, \qquad (24)$$

$$c(T) = (\Delta H_r^2/RT^2)\exp\{-(\Delta S_r/R) + (\Delta H_r/RT)\}. \qquad (25)$$

The use of these equations with $r = 4$ allows representation of the experimental results to within an accuracy of about 0.2 per cent. In view of the temperature range of the measurements (*e.g.* for ethanol, 330 to 475 K) and therefore the improbability that ΔH or ΔS will be constant, it is noteworthy that this method gives excellent representation.

Berman [26] considered the theory of the formation of dimers and tetramers as applied to five of these alcohols and methanol. He concluded that both cyclic and linear dimers exist and that higher polymers are present only in a cyclic form. This conclusion is supported by the greater consistency of the ΔH_4 values among the alcohols than the ΔH_2 values. When comparing calculated enthalpies it must be remembered that they have high standard errors due to the short temperature range of the measurements and the

[20] N. S. Berman, C. W. Larkam, and J. J. McKetta, *J. Chem. and Eng. Data*, 1964, **9**, 218.
[21] J. F. Mathews and J. J. McKetta, *J. Phys. Chem.*, 1961, **65**, 758.
[22] N. S. Berman and J. J. McKetta, *J. Phys. Chem.*, 1962, **66**, 1444.
[23] E. T. Beynon and J. J. McKetta, *J. Phys. Chem.*, 1963, **67**, 2761.
[24] J. F. Counsell, J. L. Hales, and J. F. Martin, *Trans. Faraday Soc.*, 1965, **61**, 1869.
[25] W. Weltner and K. S. Pitzer, *J. Amer. Chem. Soc.*, 1951, **73**, 2606.
[26] N. S. Berman, *J. Amer. Inst. Chem. Eng.*, 1968, **14**, 497.

nature of the function being fitted.[12] Tucker, Farnham, and Christian [27] have proposed a new model for the association of methanol vapour. From *p*, *V*, *T* measurements at 15, 25, and 35 °C they consider that the vapour is composed of monomers, trimers, and octomers. Strömsöe, Rönne, and Lydersen [28] have measured the vapour heat capacities of eleven lower alcohols at atmospheric pressure between their boiling temperatures and 600 K. They derived an equation to represent their results for all the alcohols and claim that it may be used for extrapolation to higher alcohols.

Hales, Lees, and Ruxton [29] measured the vapour heat capacities and enthalpies of vaporization of diethyl ketone, ethyl propyl ketone, and methyl isopropyl ketone. Nickerson, Kobe, and McKetta [30] made measurements on methyl ethyl ketone and methyl propyl ketone, and Pennington and Kobe [31] made measurements on acetone. Hales, Lees, and Ruxton suggest that enough results are now available to calculate the ideal heat capacities of higher ketones in the temperature range 350 to 500 K.

Fluorine Compounds.—In the last ten years measurements have been made on a number of fluorine-containing benzene derivatives. 4-Fluorotoluene was studied by Scott *et al.*[32] and a vibrational assignment consistent with the calorimetric results was obtained; 1,2-difluorobenzene was similarly studied.[33] Benzene derivatives richer in fluorine were studied at the National Physical Laboratory; thus, vapour heat capacities and enthalpies of vaporization were measured for hexafluorobenzene,[19] pentafluorobenzene,[34] pentafluorochlorobenzene,[35] and 2,3,4,5,6-pentafluorotoluene.[36]

Sulphur and Nitrogen Compounds.—In the last ten years, staff at the Bureau of Mines, Bartlesville, U.S.A., have continued with their measurements on some of the nitrogen constituents of petroleum. This series of investigations started with measurements on pyridine [37] and has since included 2-methylpyridine,[38] 3-methylpyridine,[39] and pyrrole.[40] Organic sulphur

27 E. E. Tucker, S. B. Farnham, and S. D. Christian, *J. Phys. Chem.*, 1969, **73**, 3820.
28 E. Strömsöe, H. G. Rönne, and A. L. Lydersen, *J. Chem. and Eng. Data*, 1970, **15**, 286.
29 J. L. Hales, E. B. Lees, and D. J. Ruxton, *Trans. Faraday Soc.*, 1967, **63**, 1876.
30 J. K. Nickerson, K. A. Kobe, and J. J. McKetta, *J. Phys. Chem.*, 1961, **65**, 1037.
31 R. E. Pennington and K. A. Kobe, *J. Amer. Chem. Soc.*, 1957, **79**, 30.
32 D. W. Scott, J. F. Messerly, S. S. Todd, I. A. Hossenlopp, D. R. Douslin, and J. P. McCullough, *J. Chem. Phys.*, 1962, **37**, 867.
33 D. W. Scott, J. F. Messerly, S. S. Todd, I. A. Hossenlopp, A. Osborn, and J. P. McCullough, *J. Chem. Phys.*, 1963, **38**, 532.
34 J. F. Counsell, J. L. Hales, and J. F. Martin, *J. Chem. Soc.* (*A*), 1968, 2042.
35 R. J. L. Andon, J. F. Counsell, J. L. Hales, E. B. Lees, and J. F. Martin, *J. Chem. Soc.* (*A*), 1968, 2357.
36 J. F. Counsell, J. L. Hales, E. B. Lees, and J. F. Martin, *J. Chem. Soc.* (*A*), 1968, 2994.
37 J. P. McCullough, D. R. Douslin, J. F. Messerly, I. A. Hossenlopp, T. C. Kincheloe, and G. Waddington, *J. Amer. Chem. Soc.*, 1957, **79**, 4289.
38 D. W. Scott, W. N. Hubbard, J. F. Messerly, S. S. Todd, I. A. Hossenlopp, W. D. Good, D. R. Douslin, and J. P. McCullough, *J. Phys. Chem.*, 1963, **67**, 680.
39 D. W. Scott, W. D. Good, G. B. Guthrie, S. S. Todd, I. A. Hossenlopp, A. G. Osborn, and J. P. McCullough, *J. Phys. Chem.*, 1963, **67**, 685.
40 D. W. Scott, W. T. Berg, I. A. Hossenlopp, W. N. Hubbard, J. F. Messerly, S. S. Todd, D. R. Douslin, J. P. McCullough, and G. Waddington, *J. Phys. Chem.*, 1967, **71**, 2263.

compounds have also been studied and measurements have been made on cyclopentanethiol,[41] 3,3-dimethyl-2-thiabutane,[42] carbon disulphide,[13] 2-methyl-2-butanethiol,[43] and 1-pentanethiol.[44] Special attention was paid to the use of carbon disulphide as a standard substance for vapour flow calorimetry.

Appendix

Equation (12) is a relation between thermodynamic quantities which are not obviously connected. Its derivation can start from two more fundamental equations:

$$C_p = T(\partial S/\partial T)_p, \tag{i}$$

$$\mathrm{d}G = V\,\mathrm{d}p - S\,\mathrm{d}T. \tag{ii}$$

Equation (ii) leads to equations (iii) and (iv):

$$(\partial G/\partial p)_T = V, \tag{iii}$$

$$(\partial G/\partial T)_p = -S. \tag{iv}$$

Equations (iii) and (iv) can both be used to calculate $(\partial^2 G/\partial p\,\partial T)$:

$$(\partial^2 G/\partial p\,\partial T) = (\partial V/\partial T)_p = -(\partial S/\partial p)_T. \tag{v}$$

Equation (vi) is obtained by differentiating equation (v) with respect to T:

$$(\partial^2 V/\partial T^2)_p = -(\partial^2 S/\partial p\,\partial T). \tag{vi}$$

Differentiating equation (i) with respect to p at constant T leads to equation (vii):

$$(\partial C_p/\partial p)_T = T(\partial^2 S/\partial p\,\partial T). \tag{vii}$$

Comparison of equation (vi) with equation (vii) thus leads to equation (12).

41 W. T. Berg, D. W. Scott, W. N. Hubbard, S. S. Todd, J. F. Messerly, I. A. Hossenlopp, A. Osborn, D. R. Douslin, and J. P. McCullough, *J. Phys. Chem.*, 1961, **65**, 1425.

42 D. W. Scott, W. D. Good, S. S. Todd, J. F. Messerly, W. T. Berg, I. A. Hossenlopp, J. L. Lacina, A. Osborn, and J. P. McCullough, *J. Chem. Phys.*, 1962, **36**, 406.

43 D. W. Scott, D. R. Douslin, H. L. Finke, W. N. Hubbard, J. F. Messerly, I. A. Hossenlopp, and J. P. McCullough, *J. Phys. Chem.*, 1962, **66**, 1334.

44 H. L. Finke, I. A. Hossenlopp, and W. T. Berg, *J. Phys. Chem.*, 1965, **69**, 3030.

7
Vapour Pressures

BY D. AMBROSE

1 Introduction

This Chapter deals with the vapour + liquid equilibrium of pure substances over the range of pressure 0.1 kPa to their critical pressures – which seldom exceed 5000 kPa. It therefore excludes consideration of all methods for the determination of very low vapour pressures, such as that of Knudsen (also excluded are consideration of measurements at temperatures greater than 700 K, and more than passing reference to measurements at temperatures below ambient), and it deals primarily with the determination at high accuracy of the vapour pressures of organic liquids, and with the subsequent treatment of the results obtained. The subject has been well covered in two chapters in Volume 1 of the series 'Physical Methods of Organic Chemistry',[1,2] and matters dealt with in those two chapters are included in the pages which follow only when they are necessary for the development of the discussion, or when they are matters the writer wishes to emphasize. Another source of relevant information is the book by Hála, Pick, Fried, and Vilím.[3]

For convenience, discussion of vapour-pressure determination is presented in two sections, one dealing with a low-pressure range, 0.1 to 200 kPa, and the other with a high-pressure range, 200 kPa to the critical pressure. These are preceded by a section on the measurement of pressure in the range 0.1 to 5000 kPa, and are followed by one on equations for the representation of the variation of vapour pressure with temperature. The second variable, temperature, is not discussed because information on its measurement is available elsewhere: two chapters in books,[4,5] and the series of symposia

1 W. Swiętosławski and J. R. Anderson, in 'Physical Methods of Organic Chemistry', ed. A. Weissberger, Interscience, New York, 3rd edn., 1959, vol. 1, part 1, ch. VIII, p. 357.

2 G. W. Thomson, in 'Physical Methods of Organic Chemistry', ed. A. Weissberger, Interscience, New York, 3rd edn., 1959, vol. 1, part 1, ch. IX, p. 401.

3 E. Hála, J. Pick, V. Fried, and O. Vilím, 'Vapour–Liquid Equilibrium' (trans. G. Standart), Pergamon, Oxford, 2nd edn., 1968.

4 J. M. Sturtevant, in 'Physical Methods of Organic Chemistry', ed. A. Weissberger, Interscience, New York, 3rd edn., 1959, vol. 1, part 1, ch. VI, p. 259.

5 H. F. Stimson, D. R. Lovejoy, and J. R. Clement, in 'Experimental Thermodynamics', ed. J. P. McCullough and D. W. Scott, Butterworths, London, 1968, vol. 1, ch. 2, p. 15.

organized by the American Institute of Physics[6] are recommended; more recent developments, particularly directed to automatic recording of the observations, are discussed in the chapters in this Report by Head (Chapter 3) and by Martin (Chapter 4). Reference to the current temperature scale, the International Practical Temperature Scale of 1968, will be found on p. 163. Primary standardization in the United Kingdom for measurements of pressure and temperature and calibrations of high grade instruments are carried out at the National Physical Laboratory. Advice on making measurements of pressure and temperature may be sought from this Laboratory and information about the types of calibration undertaken will be available in pamphlets now in course of preparation.

2 Units and Conversion Factors

The unit of pressure used in the preceding paragraph is that of the International System of Units (SI), the newton per square metre, now known as the pascal, Pa. Conversion of pressures expressed in one unit to values expressed in another unit is frequently required, and the following conversion factors will be found useful:

$$1 \text{ standard atmosphere} = 101\,325 \text{ Pa exactly} = 1.013\,25 \text{ bar exactly}$$
$$= 760.000 \text{ mmHg} = 1.033\,227 \text{ kgf cm}^{-2} = 14.695\,95 \text{ lbf in}^{-2};$$
$$1 \text{ mmHg} = 133.3224 \text{ Pa}; \quad 1 \text{ Pa} = 7.500\,62 \times 10^{-3} \text{ mmHg};$$
$$1 \text{ kgf cm}^{-2} = 98\,066.5 \text{ Pa} = 735.559 \text{ mmHg};$$
$$1 \text{ lbf in}^{-2} = 6894.76 \text{ Pa} = 51.7149 \text{ mmHg};$$
$$1 \text{ kg} = 2.204\,62 \text{ lb}; \quad 1 \text{ lb} = 0.453\,592 \text{ kg}; \quad 1 \text{ in} = 2.54 \text{ cm exactly}.$$

3 Measurement of Pressure

In a later section the advantage of comparative ebulliometry for vapour-pressure measurement will be stressed because the method does not require that the pressure be directly measured. Discussion of pressure measurement is apposite, however, because the comparative method is not always applicable and, from the point of view of thermodynamic measurements in general, pressures frequently must be measured with accuracy – in this discussion we shall be concerned with techniques of the precision and accuracy which can reasonably be achieved in a laboratory engaged in this type of work, not with the ultimate limits sought in a standards laboratory.

Besides the familiar Bourdon gauge, numerous instruments for the measurement of pressures have been described which are based on elastic deformation either of a diaphragm (of glass,[7] of metal,[8] or of a plastic[9]),

[6] 'Temperature, Its Measurement and Control in Science and Industry', Reinhold, New York, 1941; vol. 2, ed. H. C. Wolfe, Reinhold, New York, 1955; vol. 3, Parts 1, 2, and 3, ed. C. M. Herzfield, Reinhold, New York, 1962.
[7] F. H. Spedding and J. L. Dye, *J. Phys. Chem.*, 1955, **59**, 581.
[8] M. Waxman and W. T. Chen, *J. Res. Nat. Bur. Stand., Sect. C*, 1965, **69**, 27.
[9] J. D. E. Beynon and R. B. Cairns, *J. Sci. Instr.*, 1964, **41**, 111.

or of a curved tube (in the form, for example, of the glass spoon-gauge,[10] or of a glass helix[11]), or of a metal bellows[12] (the references quoted are only a selection from those which can be found). The type of construction varies according to the pressures at which the instrument is to be used (and these range from very low pressures[9] to beyond 10 MPa[8]) and the complexity (including that of any electronic or mechanical devices for sensing the movement caused by change in the pressure) varies greatly. These instruments can be made sensitive, but all need calibration, and their accuracy depends upon the absence of effects due to hysteresis; it will often be preferable to use them as null instruments which isolate the experimental system from the instrument with which the pressure is measured (and to which the pressure is transmitted by admission of a gas to the connecting tubing or, at high pressures, by means of oil).

Many commercial gauges based on elastic deformation are available, and some are of good accuracy. Most of them incorporate a diaphragm, but one of the most successful types is that in which the sensitive element is a helical quartz tube. A variety of tubes can be obtained to cover different ranges of pressure and temperature. Although the remarks in the preceding paragraph apply to these gauges as well as to others, nevertheless the good elastic properties of quartz allow it to be used as a measuring instrument, and there are several reports of its use in this way for vapour-pressure measurement, for example, of deuterium and hydrogen sulphides.[13]

The Mercury Manometer.—For pressures not exceeding 130 kPa the pressure-measuring instrument will usually be a mercury manometer, by means of which the pressure is related to the fundamental quantities of length (*i.e.* the height of the mercury column) and the density of the mercury. In amplification of what follows, one introductory[14] and two comprehensive[15,16] articles on the principles of barometry will repay study.

The simplest instrument for measurement of pressure with moderate accuracy by determination of the height of a mercury column is the fixed cistern (Kew type) barometer; models are available on which heights from a nominal zero up to 1000 mm can be read. The error associated with the reading, however (and with that on the normal Fortin barometer used in the laboratory), is not less than ± 0.15 mm because of the effect of capillarity in the relatively narrow tube, and because of the uncertainties in determining the height of the mercury meniscus above its level in the cistern, and in the temperature. Reduction of the error requires refinement

10 W. D. Machin, *Canad. J. Chem.*, 1967, **45**, 1904.
11 H. Blend, *Rev. Sci. Instr.*, 1967, **38**, 1527.
12 R. Hackam, W. E. Austin, and R. D. Thomas, *J. Sci. Instr.*, 1965, **42**, 344.
13 E. C. W. Clarke and D. N. Glew, *Canad. J. Chem.*, 1970, **48**, 764.
14 'Measurement of Pressure with the Mercury Barometer', National Physical Laboratory, Notes on Applied Science No. 9, H.M.S.O., 1962.
15 F. A. Gould, 'Barometers and Manometers', in 'Dictionary of Applied Physics', ed. R. Glazebrook, Macmillan, London, 1923, vol. 3, p. 140.
16 W. G. Brombacher, D. P. Johnson, and J. L. Cross, 'Mercury Barometers and Manometers', NBS Monograph 8, U.S. Department of Commerce, Washington, D.C., 1964.

in the technique. It is usual in the laboratory to construct the manometer from a U-tube of glass (or of steel with glass windows for observation) and to determine the height of the mercury column by means of an independent instrument, the cathetometer – used preferably as a vertical comparator in the way described below.

Capillarity causes depression of the mercury meniscus, and tables are available relating the capillary depression to the height of the meniscus (*i.e.* the height of the centre of the meniscus above the line of contact with the manometer tube) and the bore of the tube. It is to be noted, however, that different authorities give slightly different values for the same conditions; for example, the depression in a tube of 10 mm bore when the meniscus height is 1 mm is stated as 0.34 mm in the table given by Thomson [2] and as 0.322 mm in the table computed at the NPL [17] which is given in Kaye and Laby.[18] The latter figure is also stated to be dependent on the surface tension of the mercury, which is affected by the presence of trace impurities at the surface; it is most commonly taken to be 0.45 N m^{-1} and the corresponding values of the depression for surface tensions of 0.4 and 0.5 N m^{-1} are 0.270 mm and 0.375 mm, respectively. Although a technique, employing two tubes of different diameters in parallel for each limb of the manometer, has been described [19] by which the surface tension may be determined simultaneously with the pressure measurement, this has not been widely used, and in general the surface tension will not be known.

One meniscus will normally have reached its position of rest after rising, and the other after falling, so that the meniscus heights differ, and it may be desirable to attempt to restore symmetry by tapping or vibrating the manometer. If a mercury manometer is used in a static method for vapour-pressure measurement and one meniscus is exposed to the vapour of the material under study, its behaviour may differ from that of the other which is exposed to a gas or is under vacuum. There is therefore some uncertainty in the corrections to be applied, and the best practice is to work as far as possible with two identically shaped menisci so that the uncertainty is approximately the same in both limbs of the manometer and becomes insignificant when the height of the column is obtained by difference (this is also desirable in that any bias inherent in the method used for locating the positions of the menisci will tend to be cancelled).

In planning an experiment it is best to make the diameter of the manometer tube sufficiently large for any residual uncertainties to be negligible, and study of the tables of capillary depression suggests for precise work a diameter of not less than 20 mm; Thomas and Cross [20] have described a manometer for measurement of differential pressures up to 100 mmHg in

[17] F. A. Gould and T. Vickers, *J. Sci. Instr.*, 1952, **29**, 85.

[18] 'Tables of Physical and Chemical Constants' (originally compiled by G. W. C. Kaye and T. H. Laby), Longmans, London, 13th edn., 1966.

[19] J. Kistemaker, *Physica*, 1945, **11**, 270, 277.

[20] A. M. Thomas and J. L. Cross, *J. Vac. Sci. Tech.*, 1967, **4**, 1; NBS Technical Note 420, U.S. Department of Commerce, Washington, D.C., 1967.

which the tubes are 38 mm in diameter, and they estimate that the difference in capillary depression between two well-shaped menisci in these tubes will be less than 3×10^{-4} mm. A commercial instrument of the same type (Hass Instrument Corporation) incorporates tubes 90 mm in diameter. Such manometers must be mounted on vibration-free supports if their potential accuracy is to be attained.

There will be little point in taking to the limit the precautions outlined for eliminating capillary effects if the height of the mercury column is measured by means of the scale of a cathetometer, because even the best-designed cathetometers suffer from the disadvantages that the readings obtained depend upon the levelling of the telescope (and at a distance of 1 m it is not difficult to develop an error of 0.1 mm by faulty levelling) and upon the mechanical perfection of the bar (*i.e.* its freedom from bend and twist) and of the telescope carriage. These disadvantages may be eliminated if the cathetometer is used as a vertical comparator by which the position of the meniscus is related to an accurate (calibrated) scale beside the manometer. The scale is suspended (so that it does not bow) close to the manometer in such a position that refocussing of the telescope is not required when it is swung from scale to meniscus and *vice versa*, and the telescope carriage is provided with a micrometer screw and graduated thimble for adjustment of the height of the telescope. After sight has been taken on the meniscus and the micrometer reading has been noted, the bar is swung through a small angle so that the telescope points at the scale and a second sight is taken on a convenient graduation; the difference between the two readings of the micrometer added to (or subtracted from) the scale reading gives the height of the meniscus in relation to the scale. For determination of one manometric height in this way, four observations must be made, but the expected overall error will be less than that arising from the two observations required when the measurements are made by means of the scale of the cathetometer because of cancellation of the largest errors. It is a convenience if the comparator is provided with two telescopes so that a separate telescope can be used for each meniscus when the manometric height is greater than about 200 mm, and it is desirable for the two limbs of the manometer to be arranged on the same vertical axis,[21, 22] as is customary with standard barometers,[23] because this arrangement makes the readings easier and also makes the cancellation of errors (the 'sine errors' [15, 16]) more exact. A useful idea for the construction of such manometers – metal plates sealed to glass tubing of precision bore by means of O-rings – has been suggested by Darby,[24] but the Reporter does not like the detailed design given because the mercury menisci are distorted by tubes passing through them.

21 (*a*) H. C. Brown, M. D. Taylor, and M. Gerstein, *J. Amer. Chem. Soc.*, 1944, **66**, **431**; (*b*) D. Ambrose, *Trans. Faraday Soc.*, 1956, **52**, 772.
22 F. Akerboom and H. H. van Mal, *J. Phys.*, (*E*), 1968, **1**, 689.
23 J. E. Sears and J. S. Clark, *Proc. Roy. Soc.*, 1933, **139A**, 130.
24 J. F. G. Darby, *J. Phys.*, (*E*), 1969, **2**, 1103.

Wherever it is located, the scale must be vertical because the error (the 'cosine error') arising from tilting of the scale at, for example, a height of 760 mm amounts to 0.1 mm for a 1° departure from verticality; there is little difficulty in reducing this to insignificance with the aid of a plumb-line. An artificial horizon, consisting of a liquid in wide glass tubes joined by flexible tubing, will be found useful in setting up a comparator: the comparator is first set vertically and the telescope is set horizontally by means of the spirit levels incorporated on the instrument; the two limbs of the artificial horizon are then set up so that they subtend an angle embracing that which will be used for the readings, and fine adjustment of the comparator is made until sight can be taken on the two menisci when the telescope is swung from one side to the other without significant vertical movement of the telescope.

An error of about 0.1 mm for a height of 760 mm will also arise if the temperature of the mercury column is incorrect by 1 K and, besides checks for error in the thermometer, precautions must be taken against the existence of temperature gradients – there may be vertical gradients in the room (in the Reporter's laboratory these can be as much as 1 K over a height of 1 m), and temperature changes may be caused by lamps used for illuminating the menisci (which should be switched on therefore only when observations are actually being made) and by the observer's body. The last is the reason why, on a Fortin barometer, the thermometer should be read before the mercury height [14] (because the thermometer, with its small thermal capacity, will respond to temperature changes more quickly than will the more massive mercury charge) and why it is recommended that the cathetometer (comparator), when used, should not be closer than about 1 m to the manometer tube (which, of course, makes the levelling errors larger than they would be if the instrument were closer to the tube, but close approach is undesirable as it makes the arc through which the telescope must be swung inconveniently large). Good mercury thermometers are satisfactory for measuring the temperature of the mercury column, but their lagging times should be increased so that they are approximately as long as that of the manometer, *e.g.* by immersing the bulbs in mercury-filled tubes of the same diameter as the manometer. If any permanent temperature gradients exist in the room it may be desirable to enclose the manometer in a box and reduce the gradient by blowing air through it; the temperature should be taken as the mean value from a number of thermometers appropriately disposed.

When sight is taken on the meniscus, the simplest optical system is similar to that used in the Fortin or Kew barometer: the meniscus is illuminated from the back through a diffusing screen, and adjustable blackened opaque screens in front and behind are set about 1 mm above the meniscus so that light does not fall on its top surface and a sharply defined silhouette is seen.[21] A simple mechanical device for this operation, in which the source of illumination and the screen are attached to an endless belt, has been described by Freeman.[25] In a more complex optical system, which is used in one of the NPL primary barometers,[23] a real image of an illuminated crosswire is formed above the centre of the meniscus where it is effectively planar; this image and its reflection are observed through the telescope, and the position of the meniscus is then exactly midway between the two (Sears and Clark [23] discuss the justification for this statement in detail). A manometer in which this method is used, described by Rowlinson and his co-workers,[26] was made by machining slots through a stainless steel bar 1 m long and 5 cm by 3.5 cm in cross-section, to which polished plate-glass strips were then attached at the front and back surfaces by epoxy resin. The mercury columns are 3.5 cm square in section and are bounded on two sides by steel surfaces and on two sides by glass. The effect of vibration is eliminated by suspending the manometer from a system of springs fixed to the wall so that the whole assembly has a time period of about 0.5 s. The manometer is in a room kept at a controlled temperature, and is itself maintained at a temperature constant to 0.01 K by water passing through channels attached to the two sides of the manometer. The reproducibility claimed in pressure measurement is 0.002 kPa.

If the mercury is confined, not behind plate glass, but in glass tubing, the optical quality of the tubing must be considered; Cawood and Paterson [27] checked this by taking sight on a stretched horizontal wire and comparing the vertical setting for the telescope with that obtained when the glass tube was interposed vertically, at various points along its length, in the line of sight. They frequently found measurable displacement of the image and recommended use of tubing of not more than 0.3 mm wall thickness. An alternative method of checking the optical quality is to lay the tube horizontally and to measure, by means of a travelling microscope, the apparent diameter of a steel ball at various points along the length of the tube. Although a check of this type is still necessary, it appears probable that the glass tubing available today is of better quality than could be obtained in 1933, and there should be little difficulty in obtaining pieces fairly free from irregularities.

For measurement of low pressures where the height of the mercury column is within the compass of standard micrometer screws, the positions of the menisci may be determined by observation of the contact of probes attached to micrometers, the spindles of which are sealed to the apparatus by means of O-rings.[20, 28] When the mercury surface is located by contact in this way, the design of the probes is important if sticking of the mercury to the probe (or a 'jump' in the mercury surface as the probe approaches)

[25] M. P. Freeman, *Rev. Sci. Instr.*, 1957, **28**, 59.

[26] R. D. Weir, I. Wynn Jones, J. S. Rowlinson, and G. Saville, *Trans. Faraday Soc.*, 1967, **63**, 1320.

[27] W. Cawood and H. S. Paterson, *Trans. Faraday Soc.*, 1933, **29**, 514.

[28] V. B. Gerard, *J. Sci. Instr.*, 1966, **43**, 194.

is to be avoided. Thomas and Cross [20] recommend conically shaped probes, with an included angle of not more than 30°, finished at the point with a radius of 0.08 mm or less, whereas Stillman [29] found a 1 mm diameter steel pin with a rounded end satisfactory—it is probable that the condition of the surface of the probe, *i.e.* that it should be polished and free from pitting when examined under moderate magnification, is more important than the exact profile to which it is made. Contact of the probe with the mercury is most easily observed through a lens (with diffused back illumination), and, with care, coincidence of the probe and its reflection in the mercury can be fixed to the nearest 0.001 mm. It was pointed out many years ago,[29] in connection with the Fortin barometer, that setting a mercury surface to a pointer (and this is equally applicable to the reverse operation of setting a probe to the surface) is simplified if a scale of alternate black and white horizontal lines, each about 0.5 mm wide, is placed behind the meniscus in such a position that the observer can see the reflection of the scale in the mercury surface: then contact of the pointer with the mercury destroys the planarity of the surface and causes distortion of the pattern of the reflection, a change to which the eye is very sensitive. The advantage of an aid such as this is probably not so much that a single reading can be made more accurately as that, by reducing fatigue, it will make a series of readings more accurate. A refined optical procedure for setting the level to a point, claimed to be reproducible to ± 0.0002 mm, has been described in detail by Bottomley.[30]

Moser and his co-workers [31] have used a method in which the end of a sharp-edged metal tube just touches the mercury, and the planarity of the meniscus so formed in the tube is examined by a special optical system, but the method does not seem to have been used generally in manometry. Contact between a metal probe and the mercury may be detected by completion of an electrical circuit. The current allowed to pass must be very small, and visual observation is also necessary so that adjustment just before contact may be made at a sufficiently slow rate. Good performance has been claimed if the contact is allowed to complete a circuit of low potential difference (1.5 V) [32] but use of a sensitive electronic relay requiring a current of only a few μA or less is to be preferred.

Electronic methods of detecting the level of the mercury can be made very sensitive, but as they must depend upon the balancing of the electronic circuits it may be difficult to ensure the absolute accuracy of the measurement: examples of manometers incorporating such devices are those described by Los and Morrison [33] (based on measurement of capacitance)

[29] M. H. Stillman, *Bull. Bur. Stand.*, 1914, **10**, no. 214.
[30] G. A. Bottomley, *J. Sci. Instr.*, 1958, **35**, 254.
[31] H. Moser, 'Temperature, Its Measurement and Control in Science and Industry', ed. H. C. Wolfe, Reinhold, New York, 1955, vol. 2, p. 103; H. Moser, J. Otto, and W. Thomas, *Z. Physik*, 1957, **147**, 59.
[32] L. Akobjanoff, *Rev. Sci. Instr.*, 1952, **23**, 447.
[33] J. M. Los and J. A. Morrison, *Rev. Sci. Instr.*, 1951, **22**, 805.

and by Jasper and Miller [34] (based on measurement of mutual inductance). A photoelectric method is described by Farquharson and Kermicle.[35]

So far, measurement of height by simple means only – scales and micrometer screws – has been considered because these are the most generally useful. In Akobjanoff's manometer,[32] rods of calibrated length are used, but as described it is only intended for the measurement of one fixed pressure. Other methods (some of which, such as the use of gauge blocks,[16] may be more exact) are not generally practicable. One method which would be attractive is measurement of the angle of rotation of a screw, and Maslach [36] has described a manometer, for measurement of low pressures with a column of oil, in which a screw is used in this way. However, a screw for the measurement of a height of 1 m with sufficient precision is very expensive, and the instrument incorporating it will require expert mechanical design.

The theoretical advantages gained from the use of wide tubes and the best procedures for determination of the height of the column will be nullified if the menisci are subject to vibration: some sites may be unsuitable unless elaborate precautions are taken, because of periodic vibrations caused by motors which set up large standing waves in the mercury surface. The manometer and its comparator must be set upon firm supports on a floor strong enough for no relative displacement of the parts of the assembly to be caused by the observer's movements. On the other hand, the manometer–comparator assembly has an advantage over the instrument incorporating a micrometer in that with the former the mercury is relatively isolated from the observer as he adjusts the telescope, whereas the latter must necessarily be touched for adjustment.

One limb of the manometer must be under vacuum (except in a null-manometer used in the static method) and it is best for pumping to be continuous, or at least for provision to be made for periodic repumping. If there is any possibility of leakage (*e.g.* through the O-ring seal of the spindle on a micrometer used as described above) a vacuum gauge should be mounted in the pumping line close to the manometer. If only degassing and slow creep of air into the vacuum space because of movement of the mercury are to be expected, an alternative, simpler than continuous pumping, is incorporation of a gas trap above the vacuum space [37] into which any air may be transferred by tilting the manometer; when the manometer is restored to its upright position the air is separated from the vacuum by a mercury-filled U.

The equation relating pressure to the height of a column of liquid is

$$p/\text{Pa} = \{\rho(t)/\text{kg m}^{-3}\}(g/\text{m s}^{-2})\{h(t)/\text{m}\}, \quad (1)$$

where p is the pressure, $\rho(t)$ and $h(t)$ are the density of the liquid and the

[34] J. J. Jasper and G. B. Miller, *J. Phys. Chem.*, 1955, **59**, 441.
[35] J. Farquharson and H. A. Kermicle, *Rev. Sci. Instr.*, 1957, **28**, 324.
[36] G. J. Maslach, *Rev. Sci. Instr.*, 1952, **23**, 367.
[37] E. L. Harrington, *Rev. Sci. Instr.*, 1949, **20**, 368.

height of the column respectively at the Celsius temperature t at which the observation is made, and g is the local value of the acceleration of free fall.

If the height of the column has been measured by means of a scale adjusted to be correct at 20 °C (the temperature at which scales in general, including those on cathetometers and micrometer screws, are standardized), the observed height $h'(t)$ is related to the true height $h(t)$ by the relation

$$h(t)/\mathrm{m} = \{h'(t)/\mathrm{m}\}\,[1 + (\alpha\,\mathrm{K})\,\{(t'/^{\circ}\mathrm{C}) - 20\}], \qquad (2)$$

where t' is the Celsius temperature and α is the linear coefficient of expansion of the scale.

An expression for the variation in the density of mercury with temperature and pressure is given in the International Practical Temperature Scale of 1968 (IPTS-68),[38] and this is appropriate for use in the most accurate barometry. At pressures up to about 100 kPa the density of mercury varies approximately linearly with pressure, but the effect of compression is negligible – the coefficient is about $4 \times 10^{-11}\,\mathrm{Pa}^{-1}$ – in relation to the measurement of pressures up to and somewhat above atmospheric, and for work of ordinary accuracy a linear relation between density and temperature, independent of pressure, may be assumed. Then

$$\rho(t)/\mathrm{kg\,m^{-3}} = \{\rho(20\,^{\circ}\mathrm{C})/\mathrm{kg\,m^{-3}}\}\,[1 + (\beta\,\mathrm{K})\,\{(t/^{\circ}\mathrm{C}) - 20\}]^{-1}, \qquad (3)$$

where $\rho(t)$ is the density at the Celsius temperature t and β is the coefficient of cubical expansion of mercury; and

$$p/\mathrm{Pa} = \frac{\{\rho(20\,^{\circ}\mathrm{C})/\mathrm{kg\,m^{-3}}\}\,(g/\mathrm{m\,s^{-2}})\,\{h'(t)/\mathrm{m}\}\,[1 + (\alpha\,\mathrm{K})\,\{(t'/^{\circ}\mathrm{C}) - 20\}]}{1 + (\beta\,\mathrm{K})\,\{(t/^{\circ}\mathrm{C}) - 20\}}. \qquad (4)$$

The effect of expansion of the scale is small in comparison with that of the mercury, and normally the error is negligible if the temperature of the scale is taken to be the same as that of the mercury, $t' = t$.

The appropriate values of the coefficients of expansion for use in the above equations and in equation (5) are

$$\alpha(\text{brass}) = 1.84 \times 10^{-5}\,\mathrm{K}^{-1}, \quad \alpha(\text{steel}) = 1.1 \times 10^{-5}\,\mathrm{K}^{-1},$$

$$\text{and } \beta = 1.818 \times 10^{-4}\,\mathrm{K}^{-1},$$

and the density of pure mercury at 20 °C (IPTS-68) is

$$13.545\,87 \times 10^{3}\,\mathrm{kg\,m^{-3}}.$$ [39]

Pressures are frequently expressed in terms of the millimetre of mercury, mmHg, which is the pressure exerted under conditions of standard acceleration of free fall by a column 1 mm high of mercury at 0 °C, where by 'mercury at 0 °C' is meant a hypothetical fluid having an invariable

[38] 'The International Practical Temperature Scale of 1968', H.M.S.O., London, 1969; *Metrologia*, 1969, **5**, 35.

[39] M. V. Chattle, 'The Density of Mercury over the Temperature Range 0 °C to 40 °C', National Physical Laboratory Report Qu 9, 1970.

density of exactly 13.5951×10^3 kg m^{-3} (this figure is intended as an unchanging conventional value and represents the density of pure mercury at 0 °C under a pressure of 101 325 Pa as closely as is necessary).[40] When the observed height of a mercury column has been measured by means of a scale adjusted to be correct at 20 °C, the pressure may be obtained from the expression:

$$p/\text{mmHg} = \frac{13.545\,87\,(g/\text{m s}^{-2})\,\{h'(t)/\text{mm}\}\,[1 + (\alpha\,\text{K})\{(t'/°\text{C}) - 20\}]}{13.595\,1\,(g_n/\text{m s}^{-2})\,[1 + (\beta\,\text{K})\{(t/°\text{C}) - 20\}]}, \quad (5)$$

where g_n is the standard value of the acceleration of free fall, 9.806 65 m s^{-2}. It will be found by substitution of the values of α and β that for many purposes the ratio:

$$[1 + (\alpha\,\text{K})\{(t'/°\text{C}) - 20]\,/[1 + (\beta\,\text{K})\{(t/°\text{C}) - 20\}],$$

which appears in equation (5), and also in equation (4), may be simplified in the range $10\ °\text{C} < t = t' < 30\ °\text{C}$ to $[1 - 1.634 \times 10^{-4}\{(t/°\text{C}) - 20\}]$ for a brass scale, or $[1 - 1.708 \times 10^{-4}\{(t/°\text{C}) - 20\}]$ for a steel scale.

Pressures may be obtained from readings of Fortin and Kew type barometers by means of tables of corrections:[40, 41] the appropriate correction is subtracted from a reading to give the height of the column at 0 °C and this height is then corrected again to take account of the local value of the acceleration due to gravity. The corrections are calculated on the basis that the instrument gives a true reading of the height when it is at 0 °C, *i.e.* that the scale is adjusted to be correct at 0 °C. Significant errors may arise if these tabulated corrections are applied to measurements made with a scale adjusted for 20 °C, *e.g.* for a reading of 760 mm made at normal room temperatures, the value of the pressure obtained by means of the tabulated Fortin correction will be nearly 0.3 mmHg greater than that calculated by means of equation (5). The correction for a 20 °C brass scale may be obtained by entering the Fortin table 2.3 K above the actual temperature of the barometer, or from an empirical formula which is rarely in error by as much as 0.01 mm,

$$\Delta h/\text{mm} = \{(t/°\text{C}) + 2.5\}\,\{h'(t)/\text{mm}\}/6200, \quad (6)$$

where Δh is the correction to be subtracted from the observed height to obtain $h(0\ °\text{C})$.[42] Tables are also included in references 40 and 41 for the gravity correction, but their use to obtain p/mmHg is not so simple as

[40] 'Barometer Conventions and Tables', BS 2520, British Standards Institution, London, 1967.

[41] 'Smithsonian Physical Tables', Smithsonian Institute, City of Washington, 9th edn., 1954.

[42] D. Ambrose and P. H. Bigg, *J. Sci. Instr.*, 1956, **33**, 126.

direct multiplication of $h(0\ °C)$ by the ratio g/g_n. The temperature-correction tables apply to brass scales and their use, or use of the method of equation (6), with steel scales is more complicated – it will probably be easier to use equation (4) or (5) to obtain the pressure directly from readings made with steel scales. In any case, with the widespread availability of desk calculating machines and computers, the need for simple subtractive corrections which can be evaluated with, at most, the aid of a slide rule is passing.

The effect of the variation from place to place in the acceleration of free fall is significant in all accurate measurements. For example, at Teddington the local value of g, corrected to the site of the Fundamental Gravity Station [43] is 9.811 817 7 m s^{-2} and therefore the ratio $g/g_n = 9.811\ 82/9.806\ 65 = 1.000\ 53$: in the most accurate work account must be taken of the variation in g with location in the laboratory, which may amount to several parts in 10^6. When a measured value of g is not available, recourse must be had to tables in which the value is related to latitude, and this value must then be adjusted for altitude.[16, 40, 41] The value so obtained is calculated from the general shape of the earth, and takes no account of anomalies due to local variations in the earth's structure, which may amount to as much as 1 part in 10^4 in parts of the U.S.A. and the British Isles (corresponding to about 0.05 mmHg at atmospheric pressure). For the British Isles, advice on the value of g appropriate to any locality may be sought from the Institute of Geological Sciences, Exhibition Road, London, S.W.7.

The Bourdon Gauge.—The usual upper limit for the height of a mercury manometer is 1 m; higher pressures may be determined by using several such manometers in series [21] and, for calibration purposes, longer mercury manometers and multiple manometers capable of measuring pressures up to 20 MPa and more have been built.[44, 45] However, for practical measurements the method becomes inconvenient when the total height of the columns exceeds 2 to 3 m, and the instruments generally used at higher pressures are Bourdon gauges or pressure balances. Bourdon gauges are rapid in operation and have frequently been used for vapour-pressure measurement, but they must be calibrated (at the working level, sometimes against a more accurate standard Bourdon gauge, but finally and more accurately against a pressure balance) and, although large ones (of nominal diameter 40 cm) are available, they are of limited accuracy and sensitivity. They are not suitable for making measurements of high precision, but would be the best choice for measurements on compounds which are liable to decompose at elevated temperatures, when speed becomes more important than accuracy.

[43] 'Units and Standards of Measurement Employed at the National Physical Laboratory', Inst. Mechanics, H.M.S.O., London, 1967.

[44] J. R. Roebuck and H. W. Ibser, *Rev. Sci. Instr.*, 1954, **25**, 46.

[45] K. E. Bett and D. M. Newitt, 'Report of Symposium on the Physics and Chemistry of High Pressures', June 1962, Soc. Chem. Ind., London, 1963, p. 99.

The Pressure Balance (Piston Gauge).—The pressure balance was originally used as an instrument for the measurement of high pressures. In the past twenty years its performance and the theory underlying its operation have been refined, and it is not only accepted as a primary standard in its original application, but is also taking a place in applications hitherto reserved for the mercury column, *i.e.* in the measurement of pressures below 130 kPa. The pressure balance is unlikely to supersede the mercury column completely, but for many purposes it is more suitable than the latter because, to give one reason, of its relative portability.

The pressure balance consists of a vertical steel piston sliding in a closely fitting cylinder. A pressure applied to the lower end of the piston is balanced by loading the piston with weights, and the value of the pressure is calculated from the effective area of the piston and the force applied to it. The piston must be as truly cylindrical as possible; it is lapped into the cylinder and must fit well so that it slides smoothly with a low rate of leakage of the fluid used for transmitting the pressure. Yet, however well made the assembly, there must be a clearance between the piston and the cylinder, and the effective area lies between the geometrical area of the piston and that of the cylinder. The effective area increases with increase in the applied pressure because of elastic distortion of the piston and cylinder, but this effect does not normally become significant at pressures below 1 MPa at the accuracy with which vapour pressures in the higher range can be measured (from 1 in 10^3 to 1 in 10^4). Manufacturers normally specify the values (and limits of error) for the components of their balances, *i.e.* values of the effective area of the piston and of the masses of the weights, and this information may be adequate in many instances. For primary standardizing instruments the practice at NPL is to compute the effective area of the piston–cylinder unit from a series of very accurate diametral measurements of the piston and cylinder: if the two components are truly cylindrical the effective area of the unit is the mean of that of the piston and that of the cylinder. Figures obtained in this way are found to be in close agreement with those obtained indirectly by balancing the unit against a mercury column.[46] Calibration of a working instrument is carried out by balancing it against a standard instrument.

The effective area of a piston gauge has frequently been obtained by using it to measure the vapour pressure of carbon dioxide at 0 °C or at 0.010 °C, the values of which are well established.[47] Carbon dioxide is relatively easily prepared in a state of high purity, and a simple apparatus has been described for the calibration experiment.[48] Its value, however, probably lies not so much in its use for calibration of the balance, since this can be done more easily and economically in a standards laboratory, as in its use in the opposite sense to demonstrate that the technique adopted for

46 R. S. Dadson, R. G. P. Greig, and A. Horner, *Metrologia*, 1965, **1**, 55.
47 R. G. P. Greig and R. S. Dadson, *Brit. J. Appl. Phys.*, 1966, **17**, 1633.
48 J. L. Edwards and D. P. Johnson, *J. Res. Nat. Bur. Stand., Sect. C*, 1968, **72**, 27.

the measurement of pressure in experiments of this type is satisfactory. In the procedure described for the use of the apparatus, the carbon dioxide itself is used as the pressure transmitting fluid: this may not always be appropriate, but the apparatus can easily be adapted for oil-transmission by interposition of a suitable null gauge between the carbon dioxide and the oil. The method adopted is a dynamic one in which the liquid is boiled and the temperature of the reflux liquid is measured, and the apparatus has the advantage that the final purification can be done *in situ* since low-boiling impurities can be bled off from the condenser during the course of the experiment. (Other apparatus for purification of carbon dioxide by distillation, in which a fractionating column is incorporated, has also been described.[49, 50])

Movement of the piston is usually limited by stops to a distance of 2 to 3 mm and it will be balanced near the middle of its traverse: as movement will disturb the equilibrium of the vapour-pressure system, it may be desirable to modify the instrument so as to limit the traverse still further.[47] The condition of balance will usually be determined most easily from the null manometer used to separate the oil-filled part of the system from that containing the substance under investigation, but a means of observing the position of the piston is useful. For example, either a capacitance meter may be used or a real image of a pointer, fixed at a convenient place on the moving assembly of piston and weights, may be projected on to a graduated screen with a magnification of three or four. The effect of friction on the movement is made as small as possible by spinning the piston or by causing the piston or cylinder to oscillate so that the two mating surfaces are always in motion relative to one another. Any mechanical device which will produce this motion may impart also a spurious vertical force to the piston with the result that the balance weight is too large or too small. It is not difficult, however, to ensure and check that any bias caused in this way is insignificant in relation to a particular experiment, and for physico-chemical measurements (where the major problem may be the achievement and maintenance of equilibrium conditions in the experimental system) mechanical spinning of the piston is preferable to hand spinning because the impulse from the latter is almost certain to contain a vertical component which will disturb the system. Some instruments are sold with spinning devices; others may be adapted by fitting vanes, against which air is blown, or more complex mechanical arrangements.[51, 52]

Piston gauges were originally developed for the measurement of pressures above atmospheric (the pressure measured is the 'gauge' pressure – for the 'absolute' value the atmospheric pressure must be added to the value

[49] A. I. M. Keulemans, 'Gas Chromatography', Reinhold, New York, 2nd edn., 1959, p. 227.
[50] M. Wilkins and J. D. Wilson, AERE Report 3244, 1960.
[51] C. H. Meyers and R. S. Jessup, *J. Res. Nat. Bur. Stand.*, 1931, **6**, 1061.
[52] J. F. Pearce, *J. Sci. Instr.*, 1952, **29**, 18.

obtained from the piston gauge) and the pressure was transmitted, and the piston lubricated, by an oil such as liquid paraffin. With such instruments it is difficult to measure with accuracy pressures below 400 to 500 kPa. In recent years, piston gauges (often of somewhat larger diameter than those used with oil) have been developed which will operate with a gas as transmitting fluid. As the gas is of much lower viscosity than the oil, the piston can be made to spin with a lighter load, and pressures much closer to atmospheric can be measured. If the whole gauge is placed in an evacuated chamber, then the value obtained is absolute and pressures below atmospheric may be measured; this arrangement, however, has the practical disadvantage that the load on the piston cannot be changed easily and the experimental conditions must be adjusted to fit the pressure instead of *vice versa*: this may be inconvenient in vapour-pressure measurement. Pressures down to about 2 kPa may be measured with a gas-operated pressure balance used under vacuum, the limit being determined by the total mass of the piston and a load which has sufficient inertia for free rotation to be maintained over a fair period of time. A method developed in the NPL, in which two pressure balances are used to determine a pressure by difference, allows the determination of much lower pressures: the method may not be directly usable in vapour-pressure experiments, but it will provide accurate calibrations of other instruments such as elastic diaphragm or helix gauges for the measurement of pressures down to, and beyond, the lower limit covered in this review.[53]

The force F exerted by the total mass m_w of the piston and its load (the *weights*) is given by

$$F/\mathrm{N} = \{(m_w - m_a)/\mathrm{kg}\}\,(g/\mathrm{m\,s^{-2}}) + x/\mathrm{kg\,m\,s^{-2}}, \qquad (7)$$

where g is the local value of the acceleration of free fall, m_a is the mass of air displaced by the weights, and x is a correction applicable to oil-operated balances for the effects of surface tension (*i.e.* of the ring of oil seeping round the piston) and of any buoyancy of the piston in the gauge fluid. These latter effects are discussed in detail by Cross.[54] The pressure p is then given by

$$p/\mathrm{Pa} = (F/\mathrm{N})/(A/\mathrm{m}^2), \qquad (8)$$

where A is the effective area of the piston corrected for the effects of temperature and the applied pressure. For a steel piston, the superficial coefficient of thermal expansion (twice the linear coefficient) is $2.2 \times 10^{-5}\ \mathrm{K}^{-1}$, and a change of 4 to 5 K from the temperature of calibration is needed to cause a change of 1 in 10^4 in A (and p). The pressure coefficient depends upon the construction of the piston–cylinder assembly, and can be ascertained only from the manufacturers of the gauge or by

[53] R. S. Dadson, 'A New Development in the Accurate Measurement of Low Gas Pressures', National Physical Laboratory Report MC 6, 1970.

[54] J. L. Cross, 'Reduction of Data for Piston Gauge Pressure Measurements', NBS Monograph 65, U.S. Department of Commerce, Washington, D.C., 1964.

expert calibration. A typical value for the pressure coefficient of an oil-operated gauge is 3×10^{-12} Pa^{-1}: it is therefore not significant in the measurements now being considered.

The problem of buoyancy in air, *i.e.* the factor $(m_w - m_a)$, is confusing. In normal weighing, both arms of the balance are subject to buoyancy, and the correction required for determination of mass (for which tables are available[55, 56]) is dependent upon the difference between the densities of the weights and of the object weighed. In a weighing for the purpose of calibration, the gauge weights and the standardizing weights are likely to be of similar densities and the buoyancy differences (usually less than 1 part in 10^5) will be so small that they can be neglected. In such cases the mass of a gauge weight may be taken to be the same as that of the standard weights it balances, and this would be the meaning in any NPL report worded to the effect that the values given were obtained without the application of buoyancy corrections. However, standard weights of mass m exert the force mg in the gravitational field *in vacuo* and not in air, and for accurate measurements a buoyancy correction is therefore required when weights are used on a pressure balance in order to balance a force. The mass of air m_a in equation (7) can be taken as that displaced by the standard weights equivalent to the weights in use or by the weights themselves (because the difference is so small), and $m_a = m_w \rho_a/\rho_w$ where ρ_a and ρ_w are the densities respectively of air and of the weights. For weights of density 8.0×10^3 to 8.4×10^3 kg m^{-3} the ratio ρ_a/ρ_w may be taken as 1/7000. If the balance is operated under vacuum so that the weights are not buoyed by the atmosphere, $m_a = 0$, and the force is correctly obtained as $m_w g$ even though the value for m_w is obtained from calibration weighings in air.

4 Hydrostatic Heads

The value of the pressure calculated in one of the ways just described is the pressure at some reference point in the instrument, *e.g.* at the level of the mercury in the reservoir of a Fortin barometer, or at the base of the piston in a piston gauge. If, when a barometric measurement is made, the height of the reservoir differs from that of the point at which the pressure needs to be known, a correction for the hydrostatic head of air may be required: at atmospheric pressure, for example, a difference in height of 1 m corresponds to about 12 Pa. Tables are available for making barometric corrections, but in apparatus for vapour-pressure measurement the situation is complicated by the fact that some parts of the system may contain a gas (probably nitrogen or helium) and some parts will contain the vapour (which is of different density); the corrections are small and may be calculated by treating both gas and vapour as ideal gases, *i.e.* the

[55] 'Balances, Weights and Precise Laboratory Weighing', National Physical Laboratory, Notes on Applied Science No. 7, H.M.S.O., 1954.

[56] 'Handbook of Chemistry', ed. N. A. Lange and G. M. Forker, McGraw-Hill, New York, 10th edn., 1961, p. 1848.

density ρ of each gas or vapour is given by the expression:

$$\rho/\text{kg m}^{-3} = (p/\text{Pa})(M_m/\text{kg mol}^{-1})(R/\text{J K}^{-1}\text{mol}^{-1})^{-1}(T/\text{K})^{-1}, \quad (9)$$

where p is the pressure, M_m is the molar mass, T is the thermodynamic temperature, and $R = 8.3143\ \text{J K}^{-1}\text{mol}^{-1}$. As an example we consider ebulliometry in apparatus of the type shown in Figure 3 on p. 244. The substance condenses at the pressure p prevailing at the condenser, but at the point where the temperature is measured the pressure is greater than p because of the head of vapour, of height h, between the condenser and the thermometer.[57] If the method is used in the comparative manner there will be two sets of apparatus which we will assume are arranged with the two condensers at the same level so that no correction need be made for a head of the buffer gas. Then, if we indicate the two substances (the material under study and the standard) by subscripts A and B, we have

$$p = p_A - \rho_A g h_A = p_B - \rho_B g h_B. \quad (10)$$

If we substitute in this expression the values for ρ_A and ρ_B from equation (9) we get

$$p_A/p_B = (1 - gh_B M_B/RT_B)/(1 - gh_A M_A/RT_A), \quad (11)$$

where T_A and T_B are the two boiling temperatures for the pressure $p \approx p_A \approx p_B$. The right-hand side of equation (11) may be taken as approximately a constant, the static-head correction factor, which may conveniently be evaluated from the normal boiling temperatures of the two substances. If we consider two boilers in which the vapour heads are 0.5 m, one containing water and the other benzene, and substitute in equation (11) the appropriate values of the molar masses, 0.018 02 kg mol^{-1} and 0.078 11 kg mol^{-1} respectively, and normal boiling temperatures 353.2 K and 373.2 K, we get for the static head correction factor the value 1.0001, which corresponds to a temperature shift of about 0.003 K at atmospheric pressure. If for benzene we substitute mercury (molar mass, 0.2006 kg mol^{-1}; normal boiling temperature 629.8 K) the static-head correction factor becomes 1.0002 and this corresponds to a temperature shift of 0.01 K.

With systems containing liquid columns, for example, a U-tube containing mercury used as a null manometer which may be only approximately balanced, or a piston gauge in which oil is used, the hydrostatic heads – evaluated according to equation (1) – of all the fluids must be added or subtracted appropriately. It may be most straightforward if all fluid heads are first converted into the corresponding heights of one fluid (most usually mercury) by multiplication by the inverse ratio of their densities at

[57] J. A. Beattie, M. Benedict, and B. E. Blaisdell, *Proc. Amer. Acad. Arts Sci.*, 1936, **71**, 327; J. A. Beattie and B. E. Blaisdell, *Proc. Amer. Acad. Arts Sci.*, 1936, **71**, 361; J. A. Beattie, B. E. Blaisdell, and J. Kaminsky, *Proc. Amer. Acad. Arts Sci.*, 1936, **71**, 395.

the appropriate temperatures, and then the final sum (which may include an atmospheric barometric reading also) is converted by equation (1).

It is implied in the preceding paragraph that the densities of liquids are unaffected by pressure, whereas in reality liquids are compressible, as has already been mentioned in relation to the measurement of pressures around atmospheric with a column of mercury. At high pressures, the compressibility of mercury may have to be taken into account (but it will be of greater importance in the measurement of volumes than of pressures) and for this purpose the work of Bett, Weale, and Newitt [58] should be consulted. The density of mercury at atmospheric pressure over the range of temperature − 20 to + 300 °C was given by Bigg [59] and the figures have recently been recomputed in the range 0 to 40 °C on the basis of the IPTS-68 for precise barometric work.[39]

5 Vapour-pressure Measurement in the Range 0.1 to 200 kPa

The Static Method.—In the static method for vapour-pressure measurement, a sample of the substance is confined in a vessel attached to a manometer, and the pressure exerted by the sample, when it is maintained at a fixed temperature, is measured. The key points about the method are as follows:

(*i*) The measured pressure will be too high if air is present, and it is an advantage if the apparatus allows repeated degassing of the liquid after the start of the experiment.

(*ii*) The measured pressure will be too high if any low-boiling impurity is present, and too low if any high-boiling impurity is present. Absence of impurities to the degree required will be demonstrated if the pressure can be measured with different vapour volumes (so that the conditions of the sample range from as near to the dew volume to as near to the bubble volume as possible, *i.e.* from it being nearly all in the vapour phase to being nearly all in the liquid phase, *cf.* p. 251) and is found to be unchanged. An equivalent method is to distil the sample and to verify that the vapour pressures of the first and last fractions are the same. This is conveniently done isothermally in the vapour pressure apparatus (as has been described for gases [60]) so that there is no possibility of contamination of the fractions during transfer from one apparatus to another.

The last paragraph has been written on the assumption that no azeotrope is formed with impurities present. If one is formed and the sample being examined is of the azeotropic composition (as is possible if purification has been carried out by fractional distillation) the presence of impurity will not be revealed by this test. Although azeotropic composition changes with temperature, and consequently the sample might not pass the test if it were

58 K. E. Bett, K. E. Weale, and D. M. Newitt, *Brit. J. Appl. Phys.*, 1954, **5**, 243.
59 P. H. Bigg, *Brit. J. Appl. Phys.*, 1964, **15**, 1111.
60 M. Shepherd, *J. Res. Nat. Bur. Stand.*, 1934, **12**, 185.

repeated at another temperature, the change in composition is small, and the sensitivity of detection of an azeotropic-forming impurity will usually be low.

(*iii*) For work above room temperature, precautions must be taken to ensure that the sample does not condense in cool parts of the apparatus with the result that the pressure is lowered, and one side of the manometer must therefore be kept at an elevated temperature – at least as high as the temperature of measurement. Measurements are now seldom made by means of a mercury manometer kept at the temperature of the sample, except at low pressures (say, up to 30 kPa), because it is preferable to have the precise measuring instrument at room temperature, and the sample is normally confined by a null manometer (either of the diaphragm type, or a short U-tube containing mercury) in the thermostat; the pressure is transmitted by gas from the null manometer to the main measuring manometer.

(*iv*) The method is called static, but it is static only to the extent that the system is isolated from external disturbances. It may be possible to maintain a thermostat sufficiently constant for changes in its temperature to be insignificant, but volume changes are unavoidable when the pressure is balanced, because manometers operate by movement, either of the mercury surface or of the diaphragm. Compression of the vapour leads to condensation and a rise in the temperature of the sample, expansion to evaporation and a fall in temperature. These excursions must be reduced as far as possible by careful balancing with progressively smaller adjustments over a period of time if equilibrium is to be approached. The difficulty of reaching and preserving equilibrium will be increased if a mercury manometer of very large diameter is used; the volume change of a diaphragm gauge may also be appreciable, and in this respect the quartz helix gauge may be advantageous because the volume change required to operate it is very small.

Degassing of the sample is frequently carried out by freezing it, pumping away the non-condensable gases, allowing it to melt, and then repeating the sequence of operations, the number of repetitions being determined by the operator's experience or exhaustion: if vapour pressure measurements are made at one temperature after each sequence the value obtained after an initial fall should eventually become constant at the true value. It is the Reporter's belief that procedures for degassing which include distillation or sublimation are always more effective than that just described. For example, in one apparatus [61] the frozen sample is allowed to sublime on to a cold finger above the sample bulb while the system is pumped; the authors recommend that conditions (*i.e.* the temperatures of the sample and of the cold finger) should be adjusted so that a sample of 40 cm^3 takes from 1 to 2 h to sublime, and the sublimate remains solid at all times – because if the sublimate melted they found that some of the gas redissolved. One sublimation is normally adequate but, if it is not, the cold finger is

[61] T. N. Bell, E. L. Cussler, K. R. Harris, C. N. Pepela, and P. J. Dunlop, *J. Phys. Chem.*, 1968, **72**, 4693; K. R. Harris and P. J. Dunlop, *J. Chem. Thermodynamics*, 1970, **2**, 805.

allowed to warm so that the sample runs back into the sample bulb and is then ready for repetition of the process. Loss from the sample may be reduced if the pumping line passes through a conventional cold trap from which any material carried over and condensed may be distilled back when the pumping has been discontinued.

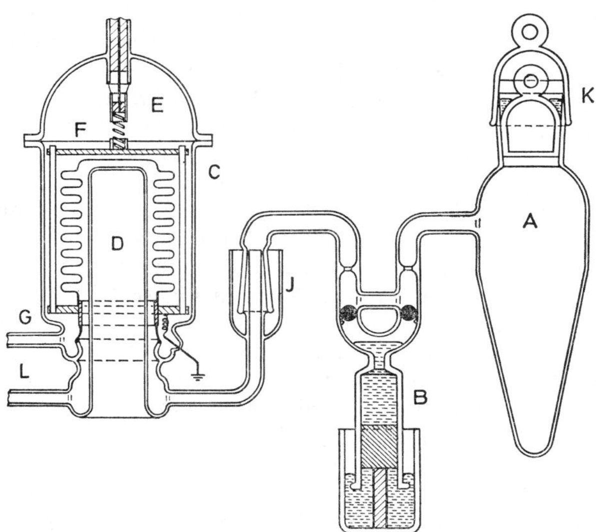

Figure 1 *Apparatus for measurement of vapour pressure by the static method.* A, *sample vessel;* B, *mercury cut-off;* C, *null manometer;* D, *hollow glass finger* (*to reduce internal volume and limit movement of bellows*); E, *outer chamber of manometer;* F, *metal plate;* G, *connection to measuring manometer;* J, *mercury-sealed joint;* K, *mercury-sealed stopper and cap;* L, *connection to vacuum* (via *a second mercury cut-off*)
(Reproduced by permission from *J. Chem. and Eng. Data*, 1967, **12**, 326)

In another apparatus [62] the sample bulb incorporates a condenser, and immediately above this a tap, so that the sample may be degassed by boiling under reflux at atmospheric pressure; if the condenser is then allowed to warm, the rising vapour expels the remaining air from the sample container, and the tap is shut (with simultaneous discontinuance of the heating) at the moment the condensing ring reaches it. However, it does not seem that this method will lead to elimination of the last trace of air as easily as the other method just described, and it does not avoid the need for vacuum pumps in the apparatus.

An example of the use of the static method is provided by the work of Cruickshank and Cutler [63] whose apparatus, used in the temperature range 25 to 75 °C, is shown in Figure 1. A is the sample container, and this is

[62] R. R. Davison, W. H. Smith, jun., and K. W. Chun, *Amer. Inst. Chem. Engineers J.*, 1967, **13**, 590.
[63] A. J. B. Cruickshank and A. J. B. Cutler, *J. Chem. and Eng. Data*, 1967, **12**, 326.

connected *via* the mercury cut-off B to the null manometer C. This manometer is a laboratory-made instrument depending on the extension of a metal bellows, and movement of the end of the bellows is detected by determination of the change in capacitance of the capacitor formed by the bellows and plate F. A problem arises here which was not mentioned in the earlier section on pressure measurement: even if the zero of an elastic gauge stays constant at one temperature it will almost certainly change when the temperature changes. Provision must therefore be made so that the sample can be isolated (*e.g.* by means of the cut-off B) and the two sides of the null manometer can be connected together for determination of the zero under all experimental conditions. If the sample in A is frozen, the absence of air can be checked by repumping the apparatus with B open. Other authors [64, 65] have recommended use of a stirrer in the sample vessel as an aid to reaching equilibrium, but this is probably more necessary when mixtures rather than single substances are being studied. A particular point to note about the apparatus shown in Figure 1 is that mercury is present with the sample and contributes to the pressure exerted on one side of the null manometer but not on the other: the vapour pressure of mercury at the experimental temperature must therefore be subtracted from the measured pressure to obtain the vapour pressure of the sample (*cf.* p. 251), whereas if a mercury-filled U-tube had been used as null manometer there would have been mercury vapour on both sides and no correction would have been required. The mercury-vapour-pressure correction at temperatures up to 75 °C is small, but for low vapour pressures measured at higher temperatures the presence of mercury becomes objectionable because its vapour pressure is likely to be a significant proportion of the total pressure, and use of a metal bellows- or diaphragm-sealed valve in place of the cut-off B would be preferable.

Apparatus has been described by Spedding and Dye [7] for measurement of the vapour pressure of mercury between 520 and 630 K, which would be suitable for measurements on other substances of low vapour pressure: the liquid in the thermostat was a fused salt mixture (sodium nitrate + potassium nitrate) and the null manometer was of all-glass construction, the sensitive element being a thin glass diaphragm to which was fused a platinum wire and bead able to make and break contact with a similar bead held rigidly in relation to the body of the gauge. An appreciable zero correction was determined in initial calibration experiments over the full range of temperature before the mercury was put in the apparatus.

An example has already been given of the use of the quartz helix gauge, not as a null instrument but for direct measurement of vapour pressures of samples at temperatures below ambient.[13] Singh and Benson [65] and Fowler, Trump, and Vogler [66] have used this type for samples at temperatures above

64 P. W. Allen, D. H. Everett, and M. F. Penney, *Proc. Roy. Soc.*, 1952, **212A**, 149.
65 Jaswant Singh and G. C. Benson, *Canad. J. Chem.*, 1968, **46**, 1249.
66 L. Fowler, W. N. Trump, and C. E. Vogler, *J. Chem. and Eng. Data*, 1968, **13**, 209.

ambient, and have prevented condensation of the sample by heating the gauge to a fixed temperature higher than their highest experimental temperature, *i.e.* they used two thermostats, one for the sample and one for the gauge. This arrangement requires that the tube connecting the sample vessel and the gauge be also heated above the experimental temperature.

For measurements on aniline at the low end of the pressure range Röck [67] has described a simple apparatus. The degassed sample was sealed in a U-tube with two vertical limbs 12 cm high joined by a horizontal tube of 2 mm bore. The two limbs were surrounded by jackets through which liquids were circulated, one at the temperature at which the vapour pressure was required and one at a temperature at which the vapour pressure was very low. The vapour-pressure difference between the two temperatures was therefore obtained in terms of the height of a column of the aniline itself. The vapour pressure at the lower temperature, which was a small correction to be added to the difference, had in this investigation been measured by another method.

The discussion of piston gauges already given was limited to consideration of standard instruments. Two special designs, used for measuring vapour pressures at temperatures near ambient, should also be mentioned. The first, used by Back and Betts [68] for measurement of the vapour pressure of hydroxylamine between − 12 and + 70 °C, was constructed from a glass syringe of 5 cm^3 capacity. The plunger was cut to a length of about 20 mm (mass 3 g) and was suspended from a torsion balance modified to allow its operation under vacuum. The vapour was admitted to the barrel of the syringe below the plunger, and the torsion wire was rotated to balance the force on the piston, the effect of friction being reduced by manual tapping. (This is not strictly a static method because there was a constant stream of vapour flowing past the piston to be pumped away.) The second special design is much more complex and only its principle of operation can be described here; for details of the construction the original paper by Douslin and Osborn should be consulted.[69] In this instrument the mass of the piston is fixed, and the force it exerts is varied by tilting the assembly of piston and cylinder. The cylinder is constantly rotated, but because the piston is weighted eccentrically, it is always moving about its axis relative to the cylinder. The force exerted is determined by precise measurement of the angle of tilt. The piston is lightly lubricated and the film of oil provides a seal between the lower end of the cylinder which communicates with the vessel containing the sample, and the upper end which is under vacuum. Pressures up to 5 kPa may be measured, and the instrument could be used like any other pressure gauge for vapour pressures

[67] H. Röck, *Z. phys. Chem.*, 1955, **4**, 242.
[68] R. A. Back and J. Betts, *Canad. J. Chem.*, 1965, **43**, 2157.
[69] D. R. Douslin and A. Osborn, *J. Sci. Instr.*, 1965, **42**, 369.

at temperatures above ambient if a suitable null manometer were used to confine the sample.

Ebulliometric Methods.—It is sometimes suggested that methods of vapour-pressure measurement which depend on boiling are less accurate than static methods, presumably because at one time the temperature measured was that of the boiling liquid and, since a liquid must be to some extent superheated for it to boil, it is difficult to ensure that the temperature measured is that truly corresponding to the prevailing pressure. The problem remains in experiments on multi-component mixtures where the composition of the vapour differs from that of the liquid, and much effort has been devoted to its solution. For a single substance, however, this difficulty does not arise, and its condensation temperature, as distinct from its boiling temperature, is the correct one. If two ebulliometers are used – one containing the substance under investigation and one a standard substance (normally water) for which the relation of vapour pressure and temperature is accurately known, both boiling at the same pressure – two condensation temperatures may be determined, and from that of the standard the pressure may be calculated. In this way necessity for determination of the pressure may be avoided: a considerable advantage because temperatures can more easily be measured accurately than can pressures, and the same measuring equipment is used for both variables. This comparative method also helps to reduce any systematic errors caused by imperfections in the design of the apparatus or in the measuring equipment because they tend to cancel out, and there is the advantage that if measurements have been made on a series of substances by means of a single standard they can all easily be recalculated and brought up to date if later more accurate work shows that the values for the standard need amendment.

Extensive work on ebulliometry was carried out by Swiętosławski [70] who developed ebulliometers in which both the temperature of the boiling liquid (reduced as far as possible from its superheated condition) and of the condensing vapour are measured; the value of these ebulliometers for measurement of vapour pressures by the comparative method was demonstrated by Zmaczynski.[71] Rossini and his co-workers,[72, 73] however, in their work on hydrocarbons from petroleum, preferred to measure the condensation temperature only, and in the past 20 years workers in this field have divided into two groups: one, of those who follow Swiętosławski and measure two temperatures in each ebulliometer, and the other, of those who follow Rossini and measure only the condensation temperature.

[70] W. Swiętosławski, 'Ebulliometric Measurements', Reinhold, New York, 1945.
[71] A. Zmaczynski, *J. Chim. phys.*, 1930, **27**, 503.
[72] C. B. Willingham, W. J. Taylor, J. M. Pignocco, and F. D. Rossini, *J. Res. Nat. Bur. Stand.*, 1945, **35**, 219.
[73] F. D. Rossini, B. J. Mair, and A. J. Streiff, 'Hydrocarbons from Petroleum', Reinhold, New York, 1953.

In an ebulliometer of the Swiętosławski type, the difference between the temperatures of the boiling liquid and of the condensing vapour may indicate whether impurity is present, or whether the substance is decomposing in the conditions of the experiment. It does not give positive evidence of purity because an azeotrope may not be detected, and – to the extent that the determination of purity is now firmly based on other methods, in particular on chromatography and on the study of the solid-to-liquid phase change – the indications to be obtained in this way from the ebulliometer, although still useful, are less necessary. Furthermore, if the temperature-measuring equipment is sufficiently sensitive, the behaviour of the condensation temperature alone may reveal inadequacy of the sample. It is standard practice in the determinations made at the NPL for a small quantity of the sample to be distilled off before measurements are made: with a pure sample, change arising in the condensation temperature (which alone is measured) is barely detectable. If water is present in the sample the temperature record is usually unsteady (as, for example, it was during many measurements made on methanol, a substance from which it proved difficult to eliminate traces of water [74]).

A stringent test of the experiment (and therefore of the sample) is the fitting of an equation: if measurements have been made over the pressure range 10 to 200 kPa, then 95% of the points will usually lie within 0.005 K of the best curve passing through all the points, whereas the scatter if the sample is wet or otherwise impure may exceed 0.01 K. With high-boiling compounds the onset of decomposition at the upper end of the range will again be apparent from unsteadiness in the temperature record during the experiment, and later study during the curve-fitting procedure may suggest that the scatter of some of the topmost points is unduly large and that they should be rejected, even though they were apparently acceptable when the experiment was in progress. It seems, therefore, that in most instances there is little advantage to be gained by measuring the boiling temperature in addition to the condensation temperature, but that is the Reporter's personal opinion with which not all will agree.

In ebulliometric methods a second component is always present: the gas by which the pressure in the system is controlled. It is assumed (and the assumption is borne out in practice) that at low pressures this gas does not affect the behaviour of the substance under study, but at higher pressures the basis for the assumption is less secure. If the gas dissolves in the liquid, the boiling and condensation temperatures of the latter could be affected, and for this reason helium is in principle to be preferred as the gas because its solubility in all liquids is smaller than that of other gases. Equally it can be demonstrated, both theoretically and practically, that pressure applied to a liquid or solid will raise its vapour pressure; for example, the concentration of water in a compressed gas saturated with water vapour is greater

[74] D. Ambrose and C. H. S. Sprake, *J. Chem. Thermodynamics*, 1970, **2**, 631.

than the concentration corresponding to the normal equilibrium vapour pressure of water (or ice) at the particular temperature.[75, 76] However, in the pressure range now under consideration no differences have been found between values obtained by the static method and values obtained by ebulliometry that are to be attributed to these effects. Ambrose, Sprake, and Townsend[77] used the ebulliometric method for determination of the vapour pressure of benzene in the range 100 to 500 kPa and found no difference between the values obtained when helium was used as the buffer gas and those obtained when nitrogen was used, and all the values were in satisfactory agreement with an interpolation between values at lower pressures and values obtained by the static method at higher pressures. In the higher range, a metal ebulliometer has been described[78] for use in the comparative manner with water as standard at pressures up to the critical; at these pressures, however, there are few sets of results more precise than 1 part in 10^3, and it is not possible to say from any comparison whether the presence of helium or any other gas at a pressure of 1 MPa or more will have a significant effect on the measured condensation temperature.

The two approaches to ebulliometry (measurement of two temperatures, or of only one) are exemplified by the equipment in use at the U.S. Bureau of Mines, Bartlesville,[79] and at the National Physical Laboratory.[80] In the Bartlesville apparatus, two identical ebulliometers of the Swiętosławski type (Figure 2), one of which contains the sample and the other water, are each connected through their condensers to a buffer volume of about 16.5 l, and through a diffusion barrier and cold trap to a common line through which the pressure is adjusted. The boiler is arranged so that vapour rising from the boiling liquid pumps a mixture of liquid and vapour over the thermometer well E; the boiler and percolator tube J are not lagged, and the surroundings are maintained at least 10 K below the boiling temperature, so that any superheating caused by the boiling can be dissipated during the passage of the mixture to the thermometer-well. The boiler heater-well is wrapped with glass thread to promote steady boiling, and the heater itself is immersed in the well in a silicone oil. The baffles I assist disengagement of the vapour from the liquid, and the vapour passes over the thermometer-well D, where the condensation temperature is obtained, before it is condensed and returned to the boiler. The purity and stability of the sample are considered to be adequate if the difference between the two temperatures does not exceed 0.005 K.

The special feature of the NPL apparatus for measurement of condensation temperatures is the boiler (Figure 3) in which four inverted cups

[75] T. J. Webster, *J. Soc. Chem. Ind.*, 1950, **69**, 343.

[76] S. Robin and B. Vodar, *J. Phys. Radium*, 1952, **13**, 264.

[77] D. Ambrose, C. H. S. Sprake, and R. Townsend, *J. Chem. Thermodynamics*, 1969, **1**, 499.

[78] G. D. Oliver and J. W. Grisard, *Rev. Sci. Instr.*, 1953, **24**, 204.

[79] A. G. Osborn and D. R. Douslin, *J. Chem. and Eng. Data*, 1966, **11**, 502.

[80] D. Ambrose, *J. Phys.*, (*E*), 1968, **1**, 41.

(about 18 mm in diameter and 40 mm high) are held so that their open ends just dip below the surface of the liquid and entrap, first gas and later, after the gas has diffused out, vapour. This arrangement ensures that there is at all times a vapour-to-liquid interface below the free liquid surface in

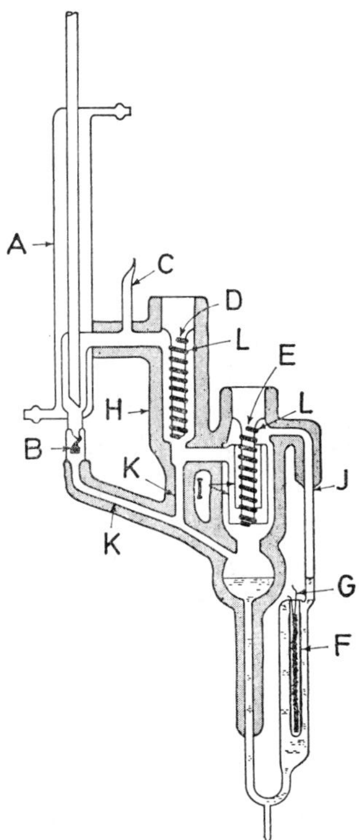

Figure 2 *Świętosławski-type ebulliometer.* A, *condenser;* B, *drop counter;* C, *filling tube;* D, *thermometer pocket* (*condensation temperature*)*;* E, *thermometer pocket* (*boiling-liquid temperature*)*;* F, *heater pocket* (*wrapped with glass thread*)*;* G, *heater;* H, *lagging;* I, *glass shields;* J, *percolator tube;* K, *condensate return tubes;* L, *glass helices*
(Reproduced by permission from *J. Chem. and Eng. Data*, 1966, **11**, 502)

the boiler. As heat is applied and the trapped vapour increases in volume it pushes down the liquid until bubbles can escape at one or more of the indentations in the rim of each cup. Waves are set up in the liquid, and regular smooth rhythmic boiling takes place from all the bubble caps at pressures down to 1 kPa and below. The diameter of the caps is such that they cover about half of the cross-sectional area of the boiler. It is essential that the caps do not dip far below the surface of the liquid and the boiler is shaped as shown so that the level of the liquid is less affected by the size of the sample (about 25 cm^3) than it would be if the outer tube closely fitted the bubble caps. Two identical boilers are connected through reflux condensers and cold traps to a common main in which currently the pressure is controlled by a Texas Instrument Company controller. The jacket G is

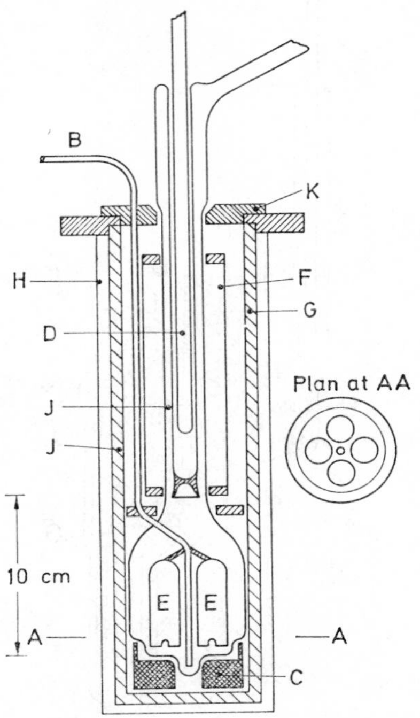

Figure 3 *Boiler for measurement of condensation temperature.* AA, *liquid level;* B, *filling tube;* C, *heater;* D, *thermometer pocket;* E, *bubble caps;* F, *radiation shield;* G, *heated jacket;* H, *outer canister;* J,J, *differential thermocouples;* K, *Sindanyo plate and lid*
[Reproduced by permission from *J. Phys.*, (*E*), 1968, **1**, 41]

heated electrically and controlled automatically so that its temperature, as determined by differential thermocouples located at J,J, is between 1 and 5 K below that of the boiler. It has been found in the use and development of this apparatus that the design of any drop counter fitted below the condenser is of some importance: first, and obviously, that it must not restrict the passage available for the flow of vapour to the condenser and secondly, that the drops produced should be small, and should run down the wall of the boiler rather than fall straight back into the boiling liquid, if they are not to cause irregularity in the temperature record. It is also doubtful whether the radiation shield F as shown in Figure 3 is useful; at low temperatures (for example, those measured in the water boiler over the whole range of pressure) its presence or absence appears to make no difference to the results, whereas at high temperatures (for example, those measured in a boiler containing mercury [81]) the apparatus as shown did not give satisfactory results and it was necessary to place radiation shields inside the boiler.[57] (The use of internal radiation shields in ebulliometric equipment for establishing the temperature scale, *e.g.* for the boiling temperature of sulphur, is a standard practice which no longer receives comment.[82] It is therefore of interest to note here that Mueller and Burgess investigated the design of radiation shields for this purpose and recorded [83] in 1919 that temperatures measured at the normal boiling temperature of sulphur might be 0.2 K low if unsuitable radiation shields were used.)

Temperatures in both the apparatuses just described are measured by means of platinum resistance thermometers used with high quality bridges, and a sensitivity of 0.001 K or better is attained. The results obtained in the two laboratories are in close agreement; for example, the boiling temperature of hexafluorobenzene at 101.325 kPa calculated from the results reported from Bartlesville [69] is 353.398 K while the most recent NPL value is 353.401 K;[84] another value obtained in the NPL with a different sample of hexafluorobenzene in an earlier design of apparatus [84] is 353.406 K (these temperatures are expressed on IPTS–68 and the numbers given for the two values already published [69, 85] have been obtained by recomputation in NPL on the new scale). At the precision attained in these measurements, differences of more than 0.01 K between the values of the normal boiling temperatures for one substance obtained by different workers are likely to be due to differences in the purities of the samples used—it is purity which is probably the limiting factor in the accuracy attainable in boiling-temperature measurements. This example of highly accurate work at two specialized laboratories is not intended to give the impression that the method is applicable only when the most refined equipment is available. Results obtained with simpler temperature-measuring equipment should be satisfactory, although necessarily of lower precision.

The upper limit of pressure at which this type of apparatus may be used is fixed by considerations of safety, but the lower limit is fixed by one of several factors: (*i*) the tendency of the boiling liquids, particularly the water, to bump; (*ii*) if water (as is to be preferred) is used as the standard it cannot be refluxed below a pressure of around 2 to 3 kPa because the

[81] D. Ambrose and C. H. S. Sprake, *J. Chem. Thermodynamics*, 1972, **4**, 603.
[82] C. R. Barber, 'The Calibration of Thermometers', National Physical Laboratory, H.M.S.O., London, 1971.
[83] E. F. Mueller and H. A. Burgess, *J. Amer. Chem. Soc.*, 1919, **41**, 745.
[84] D. Ambrose, unpublished results.
[85] J. F. Counsell, J. H. S. Green, J. L. Hales, and J. F. Martin, *Trans. Faraday Soc.*, 1965, **61**, 212.

condenser temperature cannot be taken below 0 °C; (*iii*) when the pressure of the gas is reduced below 10 kPa diffusion of the sample is little hindered and a volatile sample is liable to move from the boiler to the cold trap; (*iv*) there must be a pressure drop in any dynamic system, and at some pressure this will become significant in relation to the equilibrium vapour pressure – if the boil-up rate is reduced to reduce the pressure drop the rising vapour may not carry enough heat to control the temperature of the thermometer and its reading will not then be meaningful. Of these factors (*i*), the onset of bumping, has usually occurred at pressures of 15 kPa or more; this problem has been eliminated in the NPL apparatus, smooth boiling being obtained with all substances tried, and (*ii*) and (*iii*) have become the limiting factors. It is not yet clear at what pressure the final limit of (*iv*) is reached, but determinations of the vapour pressure of mercury by this method down to a pressure of 0.2 kPa (the pressure being measured not comparatively by means of a standard, but directly with the aid of a manometer with micrometer probes such as has been described earlier in this Report) appear to be satisfactory.[81]

Satisfactory measurements at pressures down to the limit set for this Report and beyond have been described by Thomas, Smith, and Meatyard [86] who used a method based on that of Ramsay and Young [2] in which the temperature of the liquid, the evaporation temperature, is measured. A glass-fibre skirt tied to the bulb of a mercury thermometer dips into the liquid in a wide tube which is immersed in a heated bath. The temperature of the bath is raised slowly and, when it is just above that at which the vapour pressure of the liquid equals the pressure set in the system, the liquid, which rises up the skirt by capillary action, evaporates from around the thermometer and is cooled from its slightly superheated condition to its true boiling temperature. At a fixed pressure the reading of the thermometer is stated to be independent of the bath temperature over a range of 30 to 40 K. Evaporation from a wick in this way may provide another satisfactory method of obtaining smooth boiling generally applicable in ebulliometry (and in this connection the boiler described by Wormald, for another purpose, is of interest [87]).

Other Methods.—The isoteniscope [2] is a form of apparatus for making static experiments, but it has an affinity with ebulliometry in that the sample is repeatedly degassed by boiling. If two isoteniscopes, one of which contains water and each of which is in a separately controlled bath, are both balanced against the same pressure by adjustment of the temperatures of the baths, measurements may be made in the comparative manner. The Reporter has used this method but did not succeed in making its precision approach that easily attainable by comparative ebulliometry.[84] It does, however, have the advantage that no more than 2 cm^3 of each liquid are

[86] L. H. Thomas, H. Smith, and R. Meatyard, *J. Phys.* (*E*), 1968, **2**, 1119.
[87] C. J. Wormald, *J. Sci. Instr.*, 1965, **42**, 794.

required. Goncharov and Karapet'yants [88] have reported use of an isoteniscope containing a small sample of the substance under study in conjunction with an ebulliometer containing a standard; this combination should be easier to control than one comprising two isoteniscopes, but it lacks the symmetry between the halves of the apparatus which is a valuable feature of the comparative method as described up to this point.

Of the methods available for determination of the boiling temperatures of very small samples, refinement of the one first proposed by Siwoloboff [2] seems to offer good possibility of achieving fair accuracy. A sealed capillary tube is inverted in a small quantity of the liquid in a test tube to form a focus for the formation of bubbles. The test tube is heated in a stirred bath to the boiling temperature of the sample, and after a period of boiling during which the air trapped in the capillary tube is displaced by vapour, the temperature at which boiling just ceases, *i.e.* the temperature at which the liquid begins to be drawn into the capillary tube, is determined. Sunner and Magnusson [89] described apparatus based on this method with which they were able to detect differences in condensation temperatures of 0.001 K between closely related samples, *e.g.* between successive fractions obtained by distillation. For the actual temperature, as distinct from temperature differences between samples, the method suffers from the disadvantage that bubble formation is affected to an unknown extent by surface tension, but determinations can be made with samples as small as 0.1 cm^3 and it is unlikely that any other method operating at that scale is more accurate.

Kemme and Kreps described the application of differential thermal analysis, another method requiring only a small sample, for the measurement of the vapour pressures of aliphatic alcohols and alkyl chlorides: samples ranging in size from 4 to 12 mm^3 were mixed with about 40 mg of 100 μm glass beads, and an equal mass of beads formed the inert reference material. After the sample block had been placed in the analyser, the pressure in the chamber was adjusted to the required value and the temperature of the oven was raised at about 10 K min^{-1}. As soon as the instrument trace showed that evaporation was taking place, the temperature of the sample was determined. Three replicate determinations at one pressure could be made in 30 to 40 min and repeated tests with water evaporating at the pressure corresponding to 70 °C showed that the accuracy to be expected was 0.1 K. An advantage of the method is that the trace from the instrument may give indication of the presence of impurity, or of decomposition of the sample at elevated temperatures.[90]

The transpiration method is adequately discussed by Thomson,[2] but there has been one recent development: the use of a flame-ionization

[88] A. K. Goncharov and M. Kh. Karapet'yants, *Trudy Mosk. Khim.-Tekhnol. Inst.*, 1969, No. 62, 17.
[89] S. Sunner and N. Magnusson, *Acta Chem. Scand.*, 1950, **4**, 1464.
[90] H. R. Kemme and S. I. Kreps, *J. Chem. and Eng. Data*, 1969, **14**, 98.

detector for determining the concentration of the sample in the carrier gas; the method is applicable for vapour pressures of 1 kPa and below.[91] Saturators (designed for use with aqueous solutions of salts) containing mechanically driven drums dipping into the liquid and having a particularly low pressure drop are described by Smith, Combs, and Googin.[92]

6 Vapour-pressure Measurement in the Range 200 kPa to the Critical Pressure

Application of the ebulliometric method at elevated pressures has already been discussed. The method has been used [93] only occasionally, and most vapour-pressure measurements in the range now considered have been made by the static method in apparatus which is, in principle, the same as that used by Andrews for his work on the isotherms of carbon dioxide, and subsequently developed by Hannay and by Ramsay and Young.[94] The sample is confined over mercury in a thick-walled glass tube sealed at its upper end, and the tube is attached to a steel U-tube containing mercury; the pressure is transmitted through the mercury to the pressure gauge and a piston on a screw allows the level of the mercury, and hence the volume occupied by the sample, to be adjusted. The experimental tube is heated by means of a vapour jacket. (Alternatively, the glass tube itself may be bent into the form of a U, often then known as a Cailletet tube, and the apparatus is heated by immersion in a normal thermostat bath.[95])

The differences between the original designs and that used today, for example by Kay,[96] lie in the details of the methods used for sealing the glass tube to the steel U-tube, in the methods of measuring pressure (by Bourdon or piston gauge in place of the tube containing air on which Boyle's-law measurements were made by the early experimenters), in the availability of glass tubing of uniform bore which simplifies and improves the volumetric measurements often made at the same time, and, of course, in greatly improved methods of temperature measurement. Kay's apparatus has also always included in the glass tube a steel ball which may be moved up and down by a magnet in order to stir the sample. The glass tube is sealed to the metal U-tube by means of a union incorporating an unsupported-area seal of the type shown in Figure 4.[97] An O-ring seal was used by Connolly and Kandalic [98] whose apparatus is illustrated in Figure 5. Air was removed from this apparatus by freezing the sample in the tip of

91 (*a*) F. T. Eggertsen, E. E. Seibert, and F. H. Stross, *Analyt. Chem.*, 1969, **41**, 1175; (*b*) A. Franck, *Chem.-Ztg.*, 1969, **93**, 668.
92 H. A. Smith, R. L. Combs, and J. M. Googin, *J. Phys. Chem.*, 1954, **58**, 997.
93 G. D. Oliver, H. T. Milton, and J. W. Grisard, *J. Amer. Chem. Soc.*, 1953, **75**, 2827.
94 E. P. Flint, *Chem. and Ind.*, 1968, 1618.
95 H. S. Booth and C. F. Swinehart, *J. Amer. Chem. Soc.*, 1935, **57**, 1337.
96 W. B. Kay, *J. Amer. Chem. Soc.*, 1946, **68**, 1336; 1947, **69**, 1273; W. B. Kay and G. M. Rambosek, *Ind. and Eng. Chem.*, 1953, **45**, 221.
97 D. Ambrose, B. E. Broderick, and R. Townsend, *J. Chem. Soc.* (*A*), 1967, 633.
98 J. F. Connolly and G. A. Kandalic, *Phys. Fluids*, 1960, **3**, 463.

the glass tube with liquid nitrogen so that the tube could be evacuated, and the sample was kept frozen while the assembly of tube and valve was mounted as shown. The rest of the apparatus had already been charged with mercury, which was propelled into the tube after the valve had been opened. Air may also be removed by rather a complex series of operations in which the sample is boiled in the tube.[96]

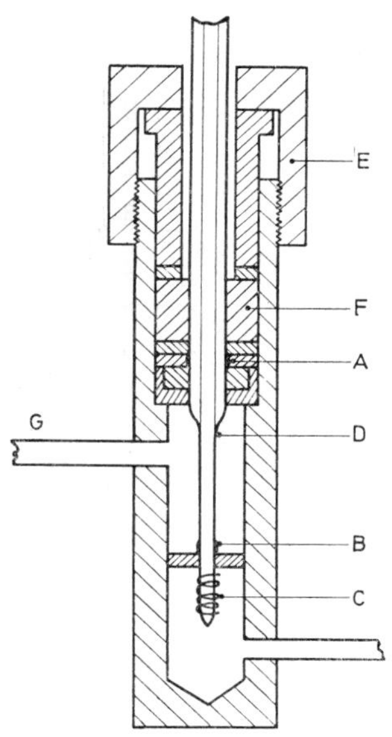

Figure 4 *Union for joining a sealed glass tube to metal aparatus and breaking* in situ. A,B, *swellings on glass tube;* C, *Nichrome spring;* D, *cut in glass;* E, *nut;* F, *elastomer sealing ring;* G, *connection to valve and vacuum system*

The union shown in Figure 4 allows a sealed tube containing a degassed sample to be broken *in situ*. The swelling of the glass tube at A prevents it from being ejected under pressure, and another swelling at B serves to locate a washer which is a loose fit in the outer steel tube and is held in position by the Nichrome spring C. Before assembly, a cut is made in the glass at D and the mercury surface is lowered to the level of C; the tube is put in position and the nut E is tightened to compress the elastomer sealing ring F. The space over the mercury is evacuated through valve G and the valve is then closed. The mercury is allowed to fill the evacuated space (any residual air is trapped under the seal) and the tube is broken at D by gentle rocking in the seal.

Another apparatus for static measurements is that originally developed for determination of the p, V, T behaviour of water, but later used for other substances by Beattie [99] and, more recently, by Douslin.[100] The principle of the method is the same, *i.e.* the volume occupied by a fixed mass of the substance is varied by means of a screw pressing a plunger into a reservoir containing mercury so that the variation of pressure with volume at a series of fixed temperatures may be found. However, the apparatus is

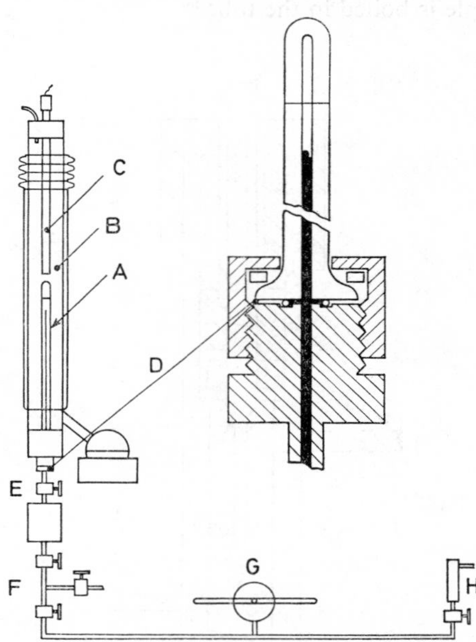

Figure 5 *Apparatus for measurement of vapour pressures in the higher range by the static method.* A, *experimental tube;* B, *vapour jacket;* C, *thermometer;* D, O-*ring-sealed glass-to-metal union;* E, *charging valve;* F, *valves;* G, *mercury injector;* H, *mercury–oil interface*
(Reproduced by permission from *Phys. Fluids*, 1960, **3**, 463)

made wholly of metal and the volumetric measurements depend upon calibration of the screw and not upon visual observation of the position of the mercury. The size of the sample container is no longer limited by the strength of glass tubing and a larger sample may be used: for this reason the results obtained with this apparatus may probably (but by no means necessarily) be of higher accuracy. The sample is contained in the metal bomb in a glass liner, and the experimental procedure allows the sample to be degassed and sealed in this liner which is then broken *in situ* after assembly and evacuation of the whole apparatus.

[99] J. A. Beattie, *Proc. Amer. Acad. Arts Sci.*, 1934, **69**, 389; J. A. Beattie and D. G. Edwards, *J. Amer. Chem. Soc.*, 1948, **70**, 382.
[100] D. R. Douslin, R. T. Moore, J. P. Dawson, and G. Waddington, *J. Amer. Chem. Soc.*, 1958, **80**, 2031.

A prime requisite in all static methods, as has already been emphasized in the discussion of measurements in the low-pressure range, is that all air must be eliminated from the sample. With the apparatus just described, this is confirmed by checking the constancy of the vapour pressure measured over the full range from the dew volume to the bubble volume in the course of the volumetric measurements.

A correction must be made for the effect of the mercury present, generally by subtraction of the vapour pressure of mercury from the measured value of the pressure (*cf.* p. 238). However, as has already been mentioned in the discussion of ebulliometry, vapour pressure is increased by external pressure, and the increase is in fact greater than that calculated by simple thermodynamic considerations, *i.e.* by the Poynting equation.[101] Rowlinson and his co-workers [102] have determined the equilibrium concentrations and rates of diffusion of mercury in several compressed gases, but such experiments have not yet shown how a more accurate correction in the general case is to be determined. A theoretical discussion of the enhancement of the mercury vapour pressure in dense gases, in which the experimental evidence is reviewed, has been published by Haar and Sengers.[103] The practical situation may also be complicated by doubt whether the vapour is saturated with mercury or not. Young [104] (who had no stirrer in his apparatus) pointed out that the rate of diffusion of mercury is very slow through the layer of liquid covering it when the sample is in two phases, and that when observations were made in a series in which the volume was successively allowed to expand, the concentration of mercury could be expected to be low, whereas in a series in which the volume was made successively to contract, the concentration could approach saturation. In Kay's apparatus where the sample is stirred, or in Beattie's apparatus where the ratio of the diameter to the height of the volume occupied by the sample is much greater, saturation must be reached more quickly. Until a more well-founded method is proposed the best course in making these measurements seems to be to aim at saturation and to correct by subtraction of the mercury vapour pressure adjusted, when appropriate, by application of the Poynting equation. (A related question is whether the presence of mercury affects the value found for the critical temperature: some results of an investigation into this by Kay are reported by Kudchadker, Alani, and Zwolinski.[105]) Ambrose, Broderick, and Townsend [97] avoided the problems associated with the presence of mercury in the heated zone by heating only the upper end of the tube of apparatus otherwise generally

101 S. Glasstone, 'Thermodynamics for Chemists', van Nostrand, New York, 1947, p. 235.

102 W. B. Jepson and J. S. Rowlinson, *J. Chem. Phys.*, 1955, **23**, 1599; W. B. Jepson, M. J. Richardson, and J. S. Rowlinson, *Trans. Faraday Soc.*, 1957, **53**, 1586; M. J. Richardson and J. S. Rowlinson, *Trans. Faraday Soc.*, 1959, **55**, 1333; D. Stubley and J. S. Rowlinson, *Trans. Faraday Soc.*, 1961, **57**, 1275.

103 L. Haar and J. M. H. Levelt Sengers, *J. Chem. Phys.*, 1970, **52**, 5069.

104 S. Young, *J. Chim. phys.*, 1906, **4**, 425.

105 A. P. Kudchadker, G. H. Alani, and B. J. Zwolinski, *Chem. Rev.*, 1968, **68**, 659.

similar (but without the stirrer) to that used by Kay. This is a satisfactory procedure for vapour-pressure measurement, but precludes the possibility of making volumetric measurements at the same time.

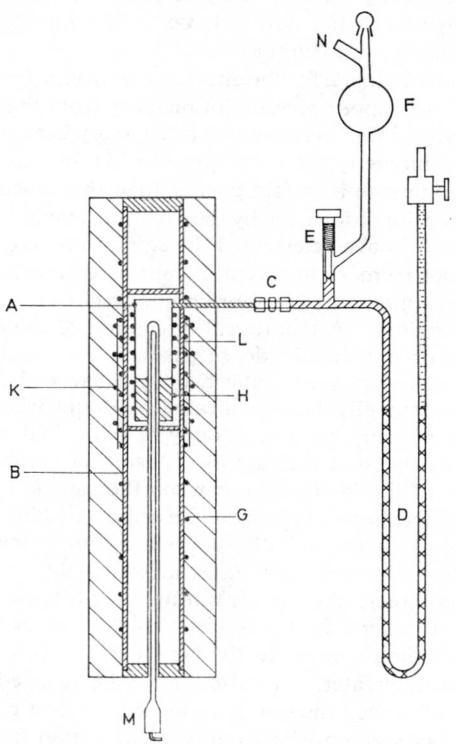

Figure 6 *Apparatus for determination of vapour pressure by dynamic method.* A, *Monel boiler provided with central thermometer pocket;* B, *aluminium tube-oven;* C, *coupling linking metal connecting tube to glass;* D, *mercury-filled manometer connected to oil-filled line leading to piston gauge;* E, *glass valve with PTFE stem* (*Fisher and Porter Company, U.S.A.*)*;* F, *reservoir for introduction of the sample;* G,H, *electrical heater windings;* K,L, *control resistor windings, insulated by suitable wrappings from* G *and* H*;* M, *platinum resistance thermometer;* N, *connection to vacuum system*

An apparatus from which air or other volatile impurities may be withdrawn after the start of the experiment is the calorimeter used at the National Bureau of Standards by Osborne, Stimson, and Ginnings [106] for determination of the heat capacity and vapour pressure of water from 100 °C up to its critical temperature, but the apparatus is very complex

[106] N. S. Osborne, H. F. Stimson, and D. C. Ginnings, *J. Res. Nat. Bur. Stand.*, 1939 **23**, 197.

and its use for any other substance has not been reported. A simpler apparatus (Figure 6) may be used for the determination of vapour pressure only:[97, 107] the method is a dynamic one and boiling takes place, but the presence of a second component, the gas normally used to control the pressure in ebulliometry, is avoided. The metal vessel A, suitable for operation at pressures up to 5 MPa, contains the sample and is provided with a pocket for the thermometer which reaches up into the vapour space; the vessel A is surrounded by the jacket B and the two are provided with heater windings (G and H) and resistance temperature-sensors (K and L). Heat applied to the liquid in A causes it to evaporate, and the pressure and temperature to rise; the vapour condenses on the upper wall and the thermometer pocket, raising their temperatures, and a steady state is reached when the temperature of A exceeds that of B by an amount sufficient for transfer to B of the heat supplied to A. The tube C, which passes outside the heated enclosure, fills with condensed liquid and this transmits the pressure to the mercury-filled U-tube D (from where the line leads to a piston gauge). The size of sample used is sufficient to fill the connecting tubing as shown and to fill about one third of A at room temperature. After the start of the experiment, when the temperature has been raised sufficiently for the pressure in the system to exceed the atmospheric pressure, air initially present may easily be moved by operation of the connected hydraulic system to the space below the valve E and bled off; in a similar manner volatile decomposition products may be removed – a valuable advantage (which is also possessed by the ebulliometric method) over some of the types of apparatus described earlier in this section from which satisfactory results cannot be obtained if there is any decomposition of the sample.

7 The Critical Point

The vapour-pressure curve terminates at the critical point. The definition and identification of this point are of interest both practically and theoretically, and the critical region has been the subject of much study because at no time in the past hundred years has the behaviour of fluids in the critical region been free of anomalies according to the then current theories. No attempt is made to survey the subject here and for further study the reader is directed to accounts of the early difficulties,[108, 109] of the practical investigations up to the present time,[105] and of modern theoretical ideas about the critical state,[110] and to a book by Rowlinson.[111] It is clear now that the grosser anomalies once thought to occur in the critical region were anomalies only because there was a lack of understanding of the effect of

[107] D. Ambrose and C. H. S. Sprake, *J. Chem. Soc.* (*A*), 1971, 1263.
[108] A. L. Clark, *Chem. Rev.*, 1938, **23**, 1.
[109] O. Maass, *Chem. Rev.*, 1938, **23**, 17.
[110] J. V. Sengers and A. L. Sengers, *Chem. Eng. News*, 10 June 1968, **46**, 104.
[111] J. S. Rowlinson, 'Liquids and Liquid Mixtures', Butterworths, London, 2nd edn., 1969.

gravity – which causes a measurable change in density over the height of a column of fluid at temperatures near the critical because the fluid is at this point highly compressible – and of the long time required for equilibrium to be reached when the two co-existing phases have almost the same properties.

When compilations of recorded critical temperatures[105, 112] are studied it is found that agreement between the results of different investigations on one substance closer than 0.1 K is the exception rather than the rule. Even for carbon dioxide, a substance easily purified and with a critical temperature most conveniently situated for exact measurement, we find the values obtained by careful investigators range from 30.96 to 31.10 °C; if we eliminate the highest and lowest of the values (which also happen to be the oldest) the range still amounts to 0.07 K. The different values have been obtained from p, V, T measurements, from density studies, by simple visual observation, and by more refined optical studies; we may therefore hazard the estimate that a range of about 0.1 K in the values of the critical temperature obtained for any one substance may arise from the use of different experimental methods (in effect, from differing definitions of the critical point) and that variation over and above 0.1 K (if not due to faulty thermometry) will be due to differences in the purities of samples. If that is so, clearly from the point of view of measurement the latter is more important than the former and no argument about the correct method for determining the critical point is of significance if in fact the experiment is performed with an impure material. Alternatively, we may say that the value of the critical temperature of a stable substance is a sensitive check on the purity of the sample and, indeed, its use in this way as a check has been suggested.[113]

The determination of the critical temperature is most simply carried out in glass apparatus by observation of the disappearance of the meniscus between the liquid and vapour phases, and the pressure exerted at this temperature is the critical pressure. Alternatively, the critical pressure may be determined by extrapolation of a suitable equation to the critical temperature: for an accurate value the observations must be taken close to the critical temperature. When metal apparatus is used, the critical temperature can be determined by analysis of the inflections of the isotherms, *i.e.* it is the temperature at which $(\partial p/\partial V)_T = 0$ and $(\partial^2 p/\partial V^2)_T = 0$.

8 Vapour-pressure Equations

In the discussion which follows, symbols have the following meanings:

p	vapour pressure
p_c	critical pressure
$p_r = p/p_c$	reduced vapour pressure
T	thermodynamic temperature

[112] K. A. Kobe and R. E. Lynn, *Chem. Rev.*, 1953, **52**, 117.
[113] R. Fischer and T. Reichel, *Mikrochemie*, 1943, **31**, 102.

T_c	critical temperature
T_b	normal boiling temperature (*i.e.* at a pressure of 101.325 kPa)
$T_r = T/T_c$	reduced temperature
$A, B, C, \ldots$ $a, b, c, \ldots$	constants appropriate to a particular equation; as used here A is invariably positive, B invariably negative
T_{max}, T_{min}	maximum and minimum temperatures in equation (27) — Chebyshev equations
$E(x)$	Chebyshev polynomial — Chebyshev equations
$a_0, a_1, a_2, \ldots$	coefficients in equation (26) — Chebyshev equations
$b_0, b_1, b_2, \ldots$	coefficients in equation (29) — Chebyshev equations

In the section of his treatise dealing with equations relating vapour pressure to temperature, Partington[114] listed more than fifty formulae which have been used since Dalton first suggested that the vapour pressure was proportional to some power of the temperature. Few of them are of more than historical interest, and today almost all are related to the simple approximate form:

$$\log p = A + B/T. \quad (12)$$

This equation is derived theoretically in textbooks on the basis of assumptions valid at temperatures up to the normal boiling temperature – constancy of the enthalpy of evaporation, behaviour of the vapour as an ideal gas, and neglect of the molar volume of the liquid – from the thermodynamically exact Clapeyron equation. In practice the equation is a better representation of the vapour-pressure line than simple theory would lead us to expect because, although at higher temperatures the justifying assumptions are invalid, the quantities relating to them change in such a way that the ratio $\Delta H/(z_G - z_L) = \Delta H/\Delta z$, *i.e.* $-RB\ln 10$, where B is defined by equation (12), remains more nearly constant as the temperature is changed than do the separate quantities ΔH, z_G, and z_L: in the expression just given ΔH is the molar enthalpy of evaporation, $z_G = pV_G/RT$, and $z_L = pV_L/RT$, where V_G and V_L are the molar volumes of the vapour and liquid, respectively, at the temperature T.

If $\log p$ is plotted against $1/T$ an approximately straight line is obtained. The curvature in the line may be most clearly seen on plots of the type of Figure 7, where $\Delta = \{\log_{10} p_{obs} - (A + B/T)\}$ has been plotted by the Reporter against T_r for three substances, the constants A and B having in each instance been adjusted so that the residuals are zero at $T_r = 0.7$ and 1.0. Curve I (2,2,4-trimethylpentane) is typical for non-associated substances boiling above room temperature; there is reversal of curvature with the point of inflection at $T_r = 0.85$. Curve II (oxygen) is of the same general shape but inflects at a lower reduced temperature. Curve III (n-butyl alcohol) is typical of associated substances such as the C_3 and C_4 alcohols

[114] J. R. Partington, 'An Advanced Treatise on Physical Chemistry', Longmans, London, 1951, vol. 2, p. 265.

for which equation (12) is a poorer representation of the vapour pressure than it is for other types of substance; there is no reversal but only a diminution in the curvature at $T_r \approx 0.85$. (It is of interest to note that the curve for water lies between curves I and II, and does show reversal.)

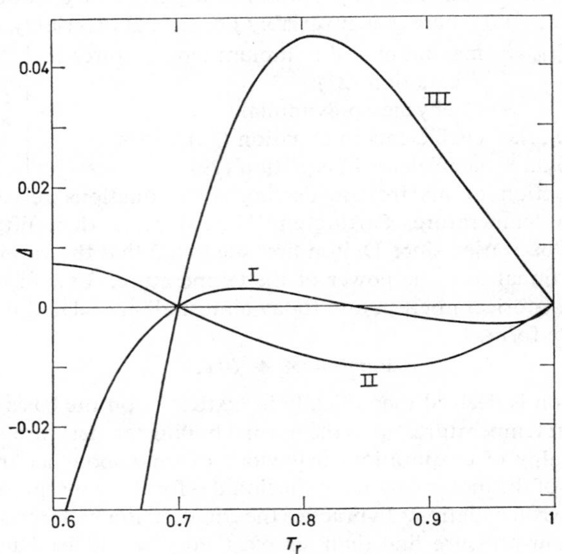

Figure 7 *Plot of* $\Delta = \{\log_{10} p_{obs} - (A + B/T)\}$, *equation* (7), *against reduced temperature* T_r *for* 2,2,4-*trimethylpentane* (I); *oxygen* (II); *and n-butyl alcohol* (III)

At temperatures less than $T_r = 0.75$, the equation most widely used today is that due to Antoine, which the Reporter now prefers in the form

$$\log p = A + B/(T + C), \tag{13}$$

with thermodynamic temperature instead of Celsius temperature as has been customary. In equation (13) the temperature of equation (12) is reduced by the amount C (which is usually between -30 and -120 K), and adjustment in this way allows data of all but the highest accuracy from the triple point to $T_r = 0.75$ to be represented. One attraction of this equation, apart from the excellence of its fit, is the ease with which it may be inverted so as to obtain T from p, whereas all the other equations to be discussed, with the exception of equation (14), require iterative methods. Despite the almost universal availability of computers, which eliminate all the labour from iteration, those who still keep slide rules and logarithm tables on their desks will continue to find use for results expressed by means of the Antoine equation as well as by means of other equations which may be more accurate. Another advantage of the equation is that its fit is relatively insensitive to change in the value assigned to C, and rules are available which allow *a priori* selection of this constant [2, 115, 116] so that an equation of fair accuracy can be set up from only two points.

A different method of adjustment of the temperature is adopted in the equation suggested by Henglein:

$$\log p = A + B/T^n, \qquad (14)$$

where $n \approx 1.5$ (the range in the values of n for different substances is about $\pm$ 0.2). The fit of this equation is comparable with that of equation (13),[117] but it appears to have no advantages in comparison with equation (13) and is not widely used.

Another frequently used three-constant equation is that used by Rankine and by Kirchhoff:[115]

$$\log p = A + B/T + C \log T. \qquad (15)$$

This equation can be given a better theoretical justification [2, 115] than can equation (13), but it is the Reporter's experience that its fit to accurate experimental points is seldom as good as that of equation (13), and it therefore seems questionable whether it should be preferred (as it often is) for extrapolation outside the experimental range.

Application of equations (13), (14), and (15) cannot be extended satisfactorily up to the critical point because, with only three constants, these equations cannot be made to reproduce the reversal of curvature which occurs at $T_r = 0.85$ with the result that, for non-associated substances, the critical pressure is about 8% higher than the value calculated from an equation (13) fitted at temperatures below $T_r = 0.75$.[118] One of the most successful equations for application over the whole liquid range was proposed by Cox.[119] If equation (12) is rewritten in the form:

$$\log(p/\text{atm}) = A + B(T/\text{K})^{-1},$$

since $p = 1$ atm when $T = T_b$, the constants are related by $-B = AT_b$, and we can put

$$\log(p/\text{atm}) = A(1 - T_b/T). \qquad (16)$$

If A is evaluated from observed values of $\{\log(p/\text{atm})\}(1 - T_b/T)^{-1}$, it is found not to be a constant, but to vary with temperature, passing through a minimum at $T_r \approx 0.85$. Cox expressed this variation by putting

$$\log A = \log A_c + E(1 - T_r)(F - T_r), \qquad (17)$$

where A_c is the value of A at the critical point, and $F \approx 0.85$. Equation (16) used with A not as a constant but as a variable in accordance with equation (17) usually reproduces observed vapour pressures within about 0.5%

[115] G. W. Thomson, *Chem. Rev.*, 1946, **38**, 1.
[116] R. R. Dreisbach, 'Vapour Pressure–Temperature Data for Organic Compounds', The Dow Chemical Company, Midland, Michigan, 2nd edn., 1946.
[117] J. Cornelissen and H. I. Waterman, *Chem. Eng. Sci.*, 1956, **5**, 141.
[118] G. W. Thomson, *Chem. Eng. Progr.*, 1949, **45**, 153.
[119] E. R. Cox, *Ind. and Eng. Chem.*, 1936, **28**, 613.

(except for associated substances); the two equations are useful also for the purpose of estimation because F may be assigned the value 0.85, and the other constants can then be calculated from three values of the vapour pressure: at the critical temperature, the normal boiling temperature, and one other.

If the critical temperature and pressure are not known, equation (17) may be written as

$$\log A = a + bT + cT^2. \qquad (18)$$

Over short ranges, equations (16) and (18) can be adjusted to fit the best results within the precision of the measurements, and the range of high quality fit can be increased if more terms are added to equation (18). For example, terms in T^3 and T^4 were used by Douslin, Harrison, and Moore for the representation of their accurate measurements on hexafluorobenzene.[120]

Another method of obtaining an improved fit over a long range is to add terms to equation (12) as follows:

$$\log p = A + B/T + CT + DT^2, \text{ (ref. 121)}, \qquad (19)$$

$$\log p = A + B/T + C\log T + DT^6, \text{ (ref. 122)}, \qquad (20)$$

$$\log p = A + B/T + C\log T + Dp/T^2, \text{ (ref. 123)}. \qquad (21)$$

Of these three, equation (21) probably gives the best fit for the complete liquid range (within 0.2% of the pressure), but it has the drawback that iteration is required for its solution for p from T as well as for T from p. Its application, particularly in reduced form, has been extensively discussed by Thodos and his co-workers [124] and also by Miller.[125]

More complex modifications of equation (12), which were devised by Martin [126] for representation of the vapour pressure of water (the data for which are probably the most accurate known), are the following:

$$\log p = A + B/T + C\log T + DT + \{E(b - T)/bT\}\log(b - T), \qquad (22)$$

$$\log p = A + B/T + C\log T + DT + ET^2 + F\log(b - T), \qquad (23)$$

where $b = (T_c + 8\text{ K})$. Good though their fit is, these forms present difficulties even to the computer for evaluation of the constants and, at the

120 D. R. Douslin, R. H. Harrison, and R. T. Moore, *J. Chem. Thermodynamics*, 1969, **1**, 305.

121 C. S. Cragoe, 'International Critical Tables', McGraw-Hill, New York and London, 1928, vol. III, p. 228.

122 R. Plank and L. Riedel, *Ing.-Arch.*, 1948, **16**, 255.

123 A. A. Frost and D. R. Kalkwarf, *J. Chem. Phys.*, 1953, **21**, 264.

124 E. G. Reynes and G. Thodos, *Amer. Inst. Chem. Engineers J.*, 1962, **8**, 357; *Ind. and Eng. Chem. (Fundamentals)*, 1962, **1**, 127; L. I. Stiel and G. Thodos, *Canad. J. Chem. Eng.*, 1962, **40**, 253.

125 D. G. Miller, *Ind. and Eng. Chem. (Fundamentals)*, 1963, **2**, 78.

126 J. J. Martin, R. M. Kapoor, and R. D. Shinn, in 'Proc. 1958 Dechema Congress', Frankfurt, 1958, p. 46; J. J. Martin, 'Thermodynamic and Transport Properties of Gases, Liquids and Solids'. American Society for Mechanical Engineering, New York, 1959, p. 110.

NPL, a modified form has been used:

$$\log p = A + B/T + CT + DT^2 + ET^3 + F\log(b - T). \quad (24)$$

The fit of equations (22), (23), and (24) is generally superior to that of equation (21).

Although equations (22) and (23) are empirical, they were devised by careful analysis of the shape of the vapour-pressure curve. An alternative approach is to eschew such subtleties, and to rely on the ability of computers to solve large, but straightforward, matrices. For example, the vapour-pressure curve of oxygen may be represented by a power series:

$$\log p = a + bT + CT^2 + dT^3 + eT^4 + fT^5 + gT^6 + hT^7. \quad (25)$$

The data for oxygen are probably second only to those for water in accuracy, and, with appropriate values for the constants, this equation represents them from the triple point to the critical point within the precision of the measurements.[127]

Chebyshev Equations.—The disadvantage from the point of view of fitting by computer of all the more complex equations listed so far is that they are ill-conditioned: the addition of a term can result in large changes of the previously calculated coefficients. If a term in $\log T$ is present, the normal fitting procedure may collapse, because for $T_{max} < 2T_{min}$, a usual situation, $\log T$ can be adequately expressed by a combination of terms in T^{-1}, T^0, T^1, and T^2, the terms which are already explicitly present in the equation. A second disadvantage is that the number of terms is predetermined so that the equations cannot be expanded or curtailed to suit the range and accuracy of the data to be fitted. The representation of vapour pressure by means of Chebyshev polynomials [128] overcomes these difficulties, and a satisfactory way of treating vapour pressures with the aid of a computer is by means of the equation

$$T\log p = a_0/2 + a_1E_1(x) + a_2E_2(x) + \ldots + a_sE_s(x) + \ldots + a_nE_n(x), \quad (26)$$

where $E_s(x)$ is the Chebyshev polynomial in x of degree s, and x is a function of temperature, varying between the limits 1 and -1, defined as

$$x = \{2T - (T_{max} + T_{min})\}/(T_{max} - T_{min}), \quad (27)$$

where T_{max} and T_{min} are temperatures respectively just above and just below the extreme temperatures of the measured values. The first few Chebyshev polynomials are

$$E_0(x) = 1; \quad E_1(x) = x; \quad E_2(x) = 2x^2 - 1;$$
$$E_3(x) = 4x^3 - 3x; \; E_4(x) = 8x^4 - 8x^2 + 1,$$

[127] J. G. Hust and R. B. Stewart, 'A Vapour Pressure Equation for Oxygen', NBS Report 8753, U.S. Department of Commerce, Washington, D.C., 1965.

[128] D. Ambrose, J. F. Counsell, and A. J. Davenport, *J. Chem. Thermodynamics*, 1970, **2**, 283.

and they are related by the recurrence relation:

$$E_{s+1}(x) - 2xE_s(x) + E_{s-1}(x) = 0. \tag{28}$$

Any polynomial in the series can therefore be calculated from its two predecessors, and the series can be easily summed and differentiated. To sum the series the coefficients $b_n, b_{n+1}, \dots b_s, \dots b_0$ are formed from $b_s = 2xb_{s+1} - b_{s+2} + a_s$ (where $b_{n+1} = b_{n+2} = 0$), and then

$$T \log p = (b_0 - b_2)/2. \tag{29}$$

Equation (29) may be extrapolated outside the experimental range – limitation of the value of x is required only for the fitting procedure. The fitting of experimental results by orthogonal polynomials and their transformation to the Chebyshev form by computer is straightforward; the experimental points should be evenly spaced throughout the temperature range, but in practice little difficulty arises if this condition is not fully met.

If, besides the coefficients, the program is arranged so that tables of dp/dT and of the residuals between the observed and calculated values are produced, inspection of these allows choice of the appropriate order to be made. In making this choice account must be taken of the sums of squares of the residuals, the magnitude of the residuals, and of the behaviour of dp/dT. Experience with data of high precision quickly shows that in general, for a range of temperature up to about 100 K, third order is appropriate; for a range of 100 to 150 K, fourth order may be needed; and for the whole liquid range fifth or higher order (depending upon the precision of the data) will be required – for oxygen, tenth order,[128] and for water, eleventh order [129] equations have been proposed. On the other hand, if extrapolation is required, a third-order equation (or one of higher order truncated) is preferable to one with more coefficients. The usual requirement will be for extrapolation of the curve obtained in the atmospheric range to higher or lower pressures, and it appears that the third-order equation, which is equivalent to equation (19), based on observations covering a 50 to 100 K range of temperature may be extrapolated upwards by 50 K or more before the error reaches 1% of the pressure or 0.5 K, and downwards by the same amount before the error reaches 2% of the pressure or 0.25 K (these figures are estimates, the correctness of which will depend on, among other things, the accuracy of the data on which the curve is based; the equivalence between the figures given for pressure and temperature is only approximate). The value of the critical pressure obtained by extrapolation of a third-order equation may be in error by 10%, whereas higher-order equations extrapolated so far may give not just errors but nonsensical results, for example, negative values for dp/dT.

The resemblance of equations (19)—(24) to equation (12) is largely illusory: the additional terms are not merely correcting terms but make a

[129] M. R. Gibson and E. A. Bruges, *J. Mech. Eng. Sci.*, 1967, **9**, 24.

major contribution to the final value, and the value of the constant B therefore cannot be related to the enthalpy of evaporation as it is in equation (12). If we consider, for example, the following equation, in the form of equation (19), for pentafluorobenzene,

$$\log_{10}(p/\text{kPa}) = 13.116\,06 - 2621.879(T/\text{K})^{-1} - 1.461\,537 \times 10^{-2}(T/\text{K}) + 1.118\,355 \times 10^{-5}(T/\text{K})^2,$$

and evaluate it for the normal boiling temperature, 358.891 K (IPTS-68), we get

$$\log_{10}(p/\text{kPa}) = 13.116\,06 - 7.305\,50 - 5.245\,32 + 1.440\,47 = 2.005\,71$$

(for $p/\text{kPa} = 101.325$, $\log_{10}(p/\text{kPa}) = 2.005\,717$). Omission of any term will give a nonsensical result, but in the corresponding Chebyshev equation all terms other than the first two have only small effects. For $T_{\text{min}} = 321$ K and $T_{\text{max}} = 368$ K, the coefficients are $a_0 = 1236.899$, $a_1 = 165.264$, $a_2 = -0.844$, and $a_3 = 0.036$; then

$$(T/\text{K})\log_{10}(p/\text{kPa}) = (b_0 - b_2)/2 \approx a_0/2 + a_1 x, \qquad (30)$$

and evaluation of this for $T = 358.891$ K gives $\log_{10}(p/\text{kPa}) = 2.005\,22$. Addition of the terms including a_2 and a_3 results in $\log_{10}(p/\text{kPa})$ changing to 2.005 80 and 2.005 71, respectively.

The property of Chebyshev series just demonstrated assists the calculation of T from p because the approximate values of T and $\Delta \log p/\Delta T$ obtainable directly from equation (30) simplify the iteration required for evaluation of the full equation. Equation (30) is in effect an alternative way of writing equation (12), and to the extent that equation (12) is an adequate expression for the vapour-pressure curve, the coefficients a_0 and a_1 of equation (30) are related to the enthalpy of evaporation ΔH by

$$\Delta H = (R \ln 10)\left\{\frac{(T_{\text{max}} + T_{\text{min}})a_1}{T_{\text{max}} - T_{\text{min}}} - \frac{a_0}{2}\right\}. \qquad (31)$$

They are also related to the constants of equation (12) by

$$a_0 = 2B + (T_{\text{max}} + T_{\text{min}})A, \qquad (32)$$

and

$$a_1 = (T_{\text{max}} - T_{\text{min}})A/2. \qquad (33)$$

These close relationships with equation (12) provide grounds for choice of the form of equation (26). Merely to obtain a good fit (at order 4 and higher), $\log p$ may well be substituted for $T \log p$ as the variable on the left-hand side, *cf.* equation (25).

Theoretically Based Equations.—Although semi-theoretical derivations can be produced for some vapour-pressure equations, for example equation (12), these are valid at fairly low pressures only, and there are no theoretically based equations which relate vapour pressure to temperature over the whole range from the triple point to the critical point. Equations can be written in terms of other thermodynamic properties[130–132] which allow the consistency between vapour-pressure and heat-capacity measurements to be checked, but they do not bring us any closer to a theoretical equation because the manner of variation with temperature of the related properties is no better known than is that of the vapour pressure itself. Greatest uncertainty in the manner of variation of all the properties arises in the critical region, the theoretical aspects of which are currently of much interest.[133] It has been suggested that thermodynamic properties here are non-analytic, and that a vapour-pressure equation should contain terms such as $(T_c - T)\log(T_c - T)$ or $(T_c - T)^\varepsilon$, where $1 < \varepsilon < 2$, both of which result in the value of d^2p/dT^2 becoming infinite at T_c.[134, 135] Equations containing these terms have been tried by Goodwin[136] for oxygen and for nitrogen, and by Sengers *et al.*[137] for carbon dioxide. The second form was found the better for the precise oxygen data. These are well represented by

$$\log p = a + bT + cT^2 + dT^3 + eT(1 - T)^\varepsilon, \tag{34}$$

where $\varepsilon = 1.633$ (as presented in the paper[136] the temperature is scaled so that the variable is not T but is a number ranging from 0 to 1 as T ranges from the triple-point temperature to the critical temperature). For carbon dioxide over a restricted range, 4 °C to the critical temperature 30.95 °C, the fit of the equation,

$$p/p_c = a + b\Delta T + c\Delta T^2 + d\Delta T^\varepsilon, \tag{35}$$

is markedly superior to that of the simple power series

$$p/p_c = a + b\Delta T + c\Delta T^2 + d\Delta T^3, \tag{36}$$

where $\Delta T = (T_c - T)/T_c$ and $\varepsilon \approx 1.8$.[137]

However, although these equations have theoretical support, and equation (34) has fewer coefficients than have equation (25) or the tenth-order Chebyshev equation,[128] its fit to the data for oxygen is not as good and there are systematic deviations which the polynomials do not exhibit. It therefore seems questionable whether derived thermodynamic quantities which depend upon the differential coefficients dp/dT or d^2p/dT^2 are certain to be more accurate when calculated from equation (34) than they are when calculated from equation (25) or the Chebyshev equation.

Generalized and Predictive Equations.—Equations have so far been considered for fitting to experimentally observed values, but they will sometimes

130 T. B. Douglas, A. F. Ball, and D. C. Ginnings, *J. Res. Nat. Bur. Stand.*, 1951, **46**, 334.
131 A. J. Gottschal and A. E. Korvezee, *Rec. Trav. chim.*, 1953, **72**, 465.
132 J. W. Armitage and P. Gray, *J. Chem. Soc.*, 1963, 1807; P. Gray and P. G. Wright, *Trans. Faraday Soc.*, 1963, **59**, 1.
133 J. V. Sengers and A. L. Sengers, *J. Res. Nat. Bur. Stand.*, 1968, **46**, 104.
134 C. N. Yang and C. P. Yang, *Phys. Rev. Letters*, 1964, **13**, 303.
135 M. E. Fisher, *J. Math. Phys.*, 1964, **5**, 944.
136 R. D. Goodwin, *J. Res. Nat. Bur. Stand., Sect. A*, 1969, **73**, 487.
137 J. M. H. L. Sengers, W. T. Chen, and M. Vincent-Massoni, *Bull. Amer. Phys. Soc., Ser. II*, 1969, **14**, 593.

be needed for the estimation of values for substances for which the data are too sparse or of too low a quality for a multi-constant fitted equation to be useful. Reference has already been made to the value of equations (13) and (16) in these circumstances; for further guidance three sources may be consulted: the investigation by Miller [138] of a variety of predictive equations; work on the use of a reference substance by Othmer;[139] and the book by Reid and Sherwood.[140] Waring [141] has discussed the thermodynamic requirements in a long-range vapour-pressure equation, and in particular has shown that the quantity $\Delta H/\Delta z$, where $z = pV/RT$, passes through a minimum at $T_r = 0.85$ [this is putting in another form what is implied by equation (17)] and that the first, second, and third derivatives of $\Delta H/\Delta z$ with respect to temperature are positive or negative in particular parts of the reduced temperature scale; these points may be of importance if an equation is to be extrapolated.

The principle of corresponding states was transformed from its position as a theoretical curiosity of little practical value – because of its poor predictive performance – through the introduction by Pitzer and his co-workers [142] of a third parameter, the acentric factor ω. This parameter is defined by $-\log p_r = 1 + \omega$ for $T_r = 0.7$, and for simple spherical molecules $\omega \approx 0$, *i.e.* $p \approx 0.1p_c$ for $T = 0.7T_c$. If T_c, p_c, and T_b (or, better, p for $T_r = 0.7$) of a substance are known the value of ω may be calculated, and vapour pressures obtained from the tables given by Pitzer *et al.* relating p to T_r and ω are seldom in error by more than 2% in the range $0.7 < T_r < 1.0$ (that is, for non-associated substances). Riedel [143] independently proposed a similar scheme, with a third parameter α_k defined in a different way but related to ω, and it is to be noted that his tables are advantageous in that the argument in temperature is at smaller intervals so that, within the accuracy to be expected of such schemes, interpolation is not required. In principle, it is possible by trial and error to fit a set of points to the table even if T_c and p_c are not known and to make an estimate of these two properties and of α_k (or ω).[144] At temperatures below $T_r = 0.7$ or T_b (whichever point has been used for definition of ω), discrepancies between observed and tabulated values begin to increase and they may rise to 20% or more at the triple-point temperature. Halm and Stiel [145] introduced yet a fourth parameter for application in this range and have greatly improved the correlation for associated substances such as alcohols.

[138] D. G. Miller, *Ind. and Eng. Chem.*, 1964, **56**, 46; *J. Phys. Chem.*, 1964, **68**, 1399; 1965, **69**, 3209.

[139] D. F. Othmer and Erl-Sheen Yu, *Ind. and Eng. Chem.*, 1968, **60**, 22.

[140] R. C. Reid and T. K. Sherwood, 'The Properties of Gases and Liquids, Their Estimation and Correlation', McGraw-Hill, New York, 2nd edn., 1966.

[141] W. Waring, *Ind. and Eng. Chem.*, 1954, **46**, 762.

[142] K. S. Pitzer, D. Z. Lippmann, R. F. Curl, C. M. Huggins, and D. E. Petersen, *J. Amer. Chem. Soc.*, 1955, **77**, 3433.

[143] L. Riedel, *Chem.-Ing.-Tech.*, 1954, **26**, 83.

[144] L. Riedel, *Chem.-Ing.-Tech.*, 1955, **27**, 475.

[145] R. L. Halm and L. I. Stiel, *Amer. Inst. Chem. Engineers J.*, 1967, **13**, 351.

Fitting of Vapour-pressure Equations.—Useful advice on the graphical fitting of equations and simple evaluation of constants from selected points is given by Thomson,[2, 115] by Hála *et al.*,[3] and by Miller.[146] Least-squares treatment of the Antoine equation is described by Willingham *et al.*[72] The matter will not be pursued here since it is assumed that most fitting will be done today by computer, and the procedure cannot be covered usefully in a short account. One question which does need discussion however is that of weighting.

If data from different observers are being combined, it will be appropriate for the correlator to make an assessment of the relative accuracies of the several sets and weight them accordingly, and sometimes, on the basis of the magnitude of their residuals or of evidence from measurements of other properties, to reject whole sets or individual observations. However, when data covering a long range from one observer are being considered, weighting may still be needed on various counts (and in any case is unavoidable by the nature of the equations).

(*i*) If, in the fitting procedure, residuals in $\log p$ are minimized, it is implied that the errors in measurement of pressure are a constant fraction of the pressure, and this is probably not true if pressures below 10 kPa are included: low-pressure points should not have the same weight as those near to atmospheric pressure. If residuals in $T \log p$ are minimized the points at higher pressures are given greater weight than those at lower pressures, in comparison with what they are given in the $\log p$ method, but the effect of T in the term is not sufficient to cover large changes in the fractional error in p. It has been suggested that residuals in p rather than in $\log p$ should be minimized,[66] *i.e.* that the error in measurement of pressure is constant irrespective of the pressure and, therefore, that the higher points are relatively more accurate than the lower, which may or may not be correct in any particular instance. Choice of one method or another is a choice of the way in which the points are weighted.

(*ii*) The distribution of points may be uneven: if there is any region with a larger number of points than others this region has more effect than the others in determining the curve, and this may or may not be desirable.

(*iii*) If a very wide range has been covered, different methods of differing accuracies may have been used in different regions. For example, at the NPL ebulliometric measurements, *i.e.* the points at pressures up to 200 kPa, are considered to be of better accuracy than those obtained at higher pressures up to the critical pressure: a much better fit overall is obtained if the former are given a weight of 10 and the latter a weight of only 1 than if they are all treated as being of equal weight. When the points are given weights of 10 and 1, the resulting fit of a fifth- or higher-order Chebyshev equation to the ebulliometric points is almost as good as the fit of the third-order equation fitted to the ebulliometric points only; and in

[146] D. G. Miller, *Ind. and Eng. Chem.* (*Fundamentals*), 1963, **2**, 68.

comparison with an equally weighted fit the price paid in deterioration of the fit to the higher points is not greater than their estimated experimental error. Hust and McCarty have presented a formula for calculation of weighting factors in accordance with estimated errors in p and T.[147]

(*iv*) It may be desirable to make the equation fit best in a particular region, for example, at the normal boiling temperature, and the appropriate points can then be given an arbitrarily high weight to force the equation through them.

In simple least-squares treatments one variable – in this connection, usually T – is considered to be free of error so that all the error is attributed to the other variable, pressure. This assumption may be so nearly true as to be justified, but in the comparative ebulliometric method where the measurement is of two temperatures, it is clearly incorrect to attribute all the error to one temperature, that of the standard, from which the pressure is calculated on the assumption that we know the relation of vapour pressure and temperature of the standard, water, very exactly. The rigorous statistical treatment allowing for errors in both variables is much more complex (and troublesome to apply even with a computer), and if the equation obtained by the simple treatment reproduces the data within their estimated precision (as it does), there seems little point in adopting a more exact treatment when its results offer no advantages. As has been implied above, the equation obtained depends upon the judgement of the correlator, and in most instances the effect of his decisions will be far greater than changes deriving from statistical rigour.

9 Conclusion

In conclusion, the Reporter suggests to those who measure vapour pressures that, before publishing a set of values, they ought always to study the consistency of their measurements, both internally and with any others which have been published, because the job of the correlator who, in due course, will seek to produce a table of best values (or its equivalent equation) will then be made easier and his table of values will be more reliable.

As has already been discussed, the precise mathematical description of the vapour-pressure curve is complex but, nevertheless, the curve is a smooth one. In Figure 8 the Reporter has plotted (against reduced temperature) the vapour pressure of water (curve I) from its triple point to its critical point as $\Delta = (\log_{10} p - \log_{10} p_{\text{Ant.}})$ where $\log_{10} p_{\text{Ant.}}$ has been obtained from an Antoine equation fitted in the range $0.51 < T_r < 0.59$ (22 to 129 kPa). This equation is not suitable for the exact representation of the vapour pressure of water but, at the sensitivity given by the scale adopted for the graph, the curve representing its actual behaviour meets the Antoine line $\Delta = 0$ smoothly. Curve II, plotted in a similar way for benzene, is typical of non-associated substances (for which the Antoine

[147] J. G. Hust and R. D. McCarty, *Cryogenics*, 1967, **7**, 200.

equation does provide a satisfactory representation over a wide range in the region of atmospheric pressure), and here a much more sensitive scale would be required to reveal anything other than a steady rise from the Antoine line at increasing temperatures. If observations plotted in this way do not join smoothly with the Antoine line representing the atmospheric range, or fail to meet it at all, then the measurements should be scrutinized carefully before they are published; either they are in error, or the measurements on which the Antoine equation has been based are in error. The

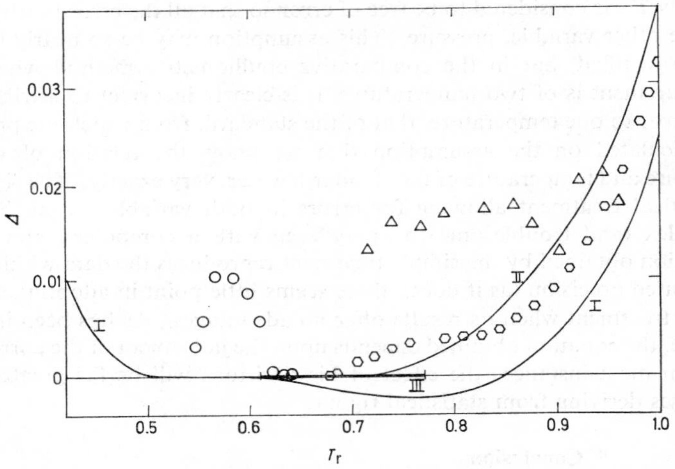

Figure 8 *Plot of* Δ = ($\log_{10} p$ − $\log_{10} p_{\text{Ant.}}$) *against reduced temperature* T_r *for water* (I), *benzene* (II), *and other compounds* (III; A, ⬭; B, △; C, ○)

points for compound A represent some measurements subjected to this test; their disposition in the range $0.85 < T_r < 1.0$ is satisfactory, but the discontinuity in slope at their junction with the zero-line suggests some fault at the lower temperatures. All the points for compound B appear of doubtful validity. (For both sets, the zero-line is based upon measurements of high reliability.)

Such dogmatism is not appropriate in respect of low-pressure measurements because the fractional errors in such measurements may be large despite care by the experimenter. Nevertheless, each stable phase is characterized by only one vapour-pressure curve, and if two methods have been used in one investigation for different ranges of pressure, the two sets of results should be consistent with each other. For example, a set of ebulliometric measurements of high precision for compound C, fitted by an Antoine curve, is represented by the line III, and static measurements made at lower temperatures by the same investigator are represented by the points indicated in the caption. The Reporter doubts the correctness of the latter observations, and suggests that they call for comment and an assessment by the author of the relative weight to be given to the values obtained by the different methods.

I wish to acknowledge the benefit I have received in the preparation of this Report from discussions with Dr. J. D. Cox, Mr. R. S. Dadson, Dr. E. F. G. Herington, and Mr. J. B. Rands, and from correspondence with Dr. G. W. Thomson.

8
Statistical Methods for Calculating Thermodynamic Functions

BY S. G. FRANKISS AND J. H. S. GREEN

1 Introduction

'The calculation of the thermodynamic properties of simple gases from a knowledge of the structure and vibrational frequencies of their molecules is now a well-established and important method of obtaining data useful in engineering as well as scientific problems.' A review [1] of the present subject written thirty years ago opened with this remark which remains equally valid today, and the intervening years have seen the extension of such calculations to a very wide range of compounds, and increasingly to molecules more complex than simple gases. The aim of the present chapter is to present the procedures involved in the calculation of the thermodynamic properties of different types of molecules in the ideal gas state, with examples of the treatment of only a relatively few specific compounds.

Since all the thermodynamic functions can be expressed in terms of the partition function involving the molecular energy levels, then 'clearly, if we know the energy levels as functions of volume and temperature, we can calculate all of thermodynamics. This is very satisfying. At the present time we need not measure thermodynamic properties experimentally if we have the necessary absorption spectra.'[2] In broad terms, the procedures for treating simple molecules were available by 1940, whereas those for molecules exhibiting internal rotations and other special features were worked out during 1935–50. Subsequently, the explicit treatment of larger and more-complex molecules has been made possible by considerably increased knowledge of molecular energy levels. This has in turn stemmed from new and improved spectroscopic techniques which may therefore be summarized briefly.

Microwave spectroscopy has provided reliable structural data for many molecules, values for the energy barriers to internal rotation and, more recently for a few molecules, rather detailed information concerning the dependence of potential energy on internal rotation. The developments of far-i.r. spectroscopy have enabled the direct observation of vibrational

[1] E. B. Wilson, *Chem. Rev.*, 1940, **27**, 17.
[2] H. Eyring, D. Henderson, B. J. Stover, and E. M. Eyring, 'Statistical Mechanics and Dynamics', Wiley, New York, 1964.

fundamentals giving rise to absorptions at long wavelengths, and whose position it was previously often necessary to infer. Additionally, the observation of torsional transitions in the far-i.r. region has been used to derive much information on energy barriers to internal rotation. Further improvements in vibrational spectroscopy have come from the use of laser sources for Raman spectroscopy, enabling more complete data, and more reliable polarization data, to be obtained. Investigations by n.m.r. spectroscopy have provided information on rotational isomerism, conformational equilibria, and restricted internal rotation for the liquid state. Improvements in electron diffraction by the sector technique have made this method of structural determination complementary to microwave spectroscopy for several molecules. Information on rotational isomerism and internal rotation is being provided increasingly by measurements of ultrasonic relaxation. For some molecules, investigations of electronic spectra under high resolution have afforded important information on the associated vibrational transitions. Electronic spectra are, of course, essential for determining the spectroscopic terms, and therefore the degeneracies, of the ground and higher states, as well as the energy separations of the latter, for species for which thermodynamic properties at high temperatures are required. Finally, and most recently, the technique of inelastic neutron scattering has begun to afford information on barriers to internal rotation, particularly for methyl groups. Neutron scattering is not limited by the usual optical selection rules involving dipoles or polarizabilities and therefore can afford information on vibrations which are forbidden for optical techniques.

Along with improved experimental techniques have gone advances in the interpretation of spectra and the results for large numbers of molecules have now been analysed to provide, for example, complete vibrational assignments. Calculations of molecular vibrational frequencies have advanced considerably through the facility of high-speed digital computers. Such force-constant calculations remain beset with some inevitable limitations, but they can provide more reliable vibrational assignments than were possible solely by the use of selection rules and regularities among structurally related molecules. They are of particular importance in yielding values for fundamentals forbidden by selection rules in the observed spectra, or otherwise unobserved; they are also important as guides to interpreting the spectra of compounds existing as mixtures of rotational isomers, by identifying the vibrations arising from the individual components.

At the same time, improvements in the accuracy of measured thermodynamic quantities – in particular, the determination of entropy by low-temperature calorimetry and of heat capacity by vapour-flow calorimetry – have enabled significance to be attached to more complete calculations of these quantities using more sophisticated molecular models. The calculations themselves, of course, readily allow extension of the thermodynamic data over a temperature range inaccessible to experimental measurement.

All these advances, and the many new insights into molecular structure and energetics, have made possible detailed treatments for a large number of compounds of scientific significance or industrial importance. For organic compounds, the results for simpler, 'key' molecules then provide the essential basis for the creation, by incremental methods, of tables of chemical thermodynamic properties for scores of homologous series containing thousands of compounds. The practical justification lies in the provision of these important data. But at the same time, and from a more fundamental standpoint, it should be recalled that a valid comparison of observed and calculated thermodynamic quantities for a molecule provides a unification of a very substantial amount of experimental thermochemical and spectroscopic information, interpreted on the basis of quantum and statistical mechanics. 'Indeed, the best empirical evidence for the truth of statistical mechanics lies in the agreement of calculated and observed thermodynamic properties.'[3] Additionally, such comparisons have required quantification of, and provided important data on, such important concepts as restricted internal rotation, and have led to new concepts such as pseudo-rotation.

In what follows, the computational methods for different types of molecules are presented with emphasis on the practical formulae rather than detailed derivations. It is convenient to consider firstly simple, 'rigid' molecules, that is, those without large vibrational amplitudes; then those exhibiting internal rotations, inversion motion, and ring motion. Finally, the details of some individual molecules are considered, as examples of the utilization of the methods.

All the formulae relate to the ideal gas state at the standard pressure of 101 325 Pa (1 atmosphere) and the thermodynamic functions comprise: the standard molar internal energy relative to the zero temperature energy $\{U^\ominus(T) - U^\ominus(0)\}$, the function $\{H^\ominus(T) - H^\ominus(0)\}/T$ where H denotes enthalpy, the function $\{G^\ominus(T) - H^\ominus(0)\}/T$ where G denotes Gibbs free energy, the entropy $S^\ominus(T)$, and the heat capacity at constant pressure $C_p^\ominus(T)$. It is, of course, the last two quantities which are to be compared with experimentally derived values.

2 Statistical Formulae for Thermodynamic Functions

The Partition Function.[4–7]—The basis for all the calculations is the evaluation of the partition function. For an assembly of N non-localized

[3] G. N. Lewis and M. Randall, 'Thermodynamics', revised K. S. Pitzer and L. Brewer, McGraw-Hill, New York, 2nd edn., 1961.

[4] R. H. Fowler and E. A. Guggenheim, 'Statistical Thermodynamics', C.U.P., Cambridge, 1939.

[5] J. E. Mayer and M. G. Mayer, 'Statistical Mechanics', Wiley, New York, 1940.

[6] J. G. Aston and J. J. Fritz, 'Thermodynamics and Statistical Thermodynamics', Wiley, New York, 1959.

[7] J. R. Partington, 'An advanced Treatise on Physical Chemistry', Longmans, Green, London, 1949, vol. 1.

molecules the system partition function Q is defined by

$$Q = q^N/N!, \tag{1}$$

where q is the molecular partition function given by the expression

$$q = \Sigma_i g_i \exp(-\varepsilon_i/kT), \tag{2}$$

that is, the sum of the Boltzmann factors over all the energy levels ε_i, the degeneracy or statistical weight of the ith level being g_i.

All the thermodynamic functions can be expressed in terms of the partition function and written in terms of q. They are (where L is Avogadro's constant)

$$U^{\ominus}(T) - U^{\ominus}(0) = LkT^2(\partial \ln q/\partial T)_V = LkTq'/q, \tag{3}$$

$$\{H^{\ominus}(T) - H^{\ominus}(0)\}/T = \{U^{\ominus}(T) - U^{\ominus}(0)\}/T + Lk, \tag{4}$$

$$\{G^{\ominus}(T) - H^{\ominus}(0)\}/T = -Lk\ln(q/L), \tag{5}$$

$$S^{\ominus}(T) = Lk\{1 + \ln(q/L) + T(\partial \ln q/\partial T)_V\}, \tag{6}$$

or

$$S^{\ominus}(T) = Lk\{1 + \ln(q/L) + q'/q\}, \tag{7}$$

$$C_p^{\ominus}(T) = C_V^{\ominus}(T) + Lk = (\partial U/\partial T)_V + Lk, \tag{8}$$

or

$$C_p^{\ominus}(T) = Lk\{1 + (q''/q) - (q'/q)^2\}. \tag{9}$$

The only completely general method of evaluating the partition function is by summation of the separately calculated terms in equation (2). If q cannot be expressed in a closed form, then for numerical evaluation of the thermodynamic functions equations (3), (7), and (9) are most convenient; they contain sums related to q and defined by

$$q' = T(\partial q/\partial T)_V = \Sigma_i g_i(\varepsilon_i/kT)\exp(-\varepsilon_i/kT), \tag{10}$$

$$q'' = (1/T^2)\{\partial^2 q/\partial(1/T)^2\}_V = \Sigma_i g_i(\varepsilon_i/kT)^2\exp(-\varepsilon_i/kT). \tag{11}$$

Translational Contribution.—Since for an ideal gas the energy of a molecule does not depend on its position in space, the total energy can be expressed as the sum of independent translational and internal energies

$$\varepsilon = \varepsilon_{\text{transl}} + \varepsilon_{\text{internal}}.$$

The molecular partition function can then be factored:

$$q = q_{\text{transl}} \times q_{\text{internal}},$$

and because of the dependence of the thermodynamic functions upon the logarithm of q, there are separate translational and internal contributions to each of them.

The translational partition function of an ideal gas occupying volume V is

$$q_{\text{transl}} = V(2\pi mkT)^{3/2}h^{-3}, \tag{12}$$

where m is the mass of each molecule. It follows directly, using equation (12) in equations (3), (4), and (8), that the translational contribution to $\{H^{\ominus}(T) - H^{\ominus}(0)\}/T$ and to $C_p^{\ominus}(T)$ is $5Lk/2$.

The partition function in equation (12) gives the translational contribution for all molecules, and the practical equations are given in Table 1.

Table 1 *Equations*[a] *for the translational, rigid-rotational, and harmonic oscillator contributions to the thermodynamic functions of molecules in the ideal-gas state and at the standard pressure* 101 325 Pa

Translational contributions (all molecules)

$$[\{H^{\ominus}(T) - H^{\ominus}(0)\}/T]/\mathrm{J\,K^{-1}\,mol^{-1}} = C_p^{\ominus}(T)/\mathrm{J\,K^{-1}\,mol^{-1}} = 20.7858,$$

$$-[\{G^{\ominus}(T) - H^{\ominus}(0)\}/T]/\mathrm{J\,K^{-1}\,mol^{-1}} = 28.7166\log_{10}(M/\mathrm{g\,mol^{-1}}) + 47.8610\log_{10}(T/\mathrm{K}) - 30.4716,$$

$$S^{\ominus}(T)/\mathrm{J\,K^{-1}\,mol^{-1}} = 28.7166\log_{10}(M/\mathrm{g\,mol^{-1}}) + 47.8610\log_{10}(T/\mathrm{K}) - 9.6858.$$

Rigid rotational contributions

(i) *Linear molecules*

$$[\{H^{\ominus}(T) - H^{\ominus}(0)\}/T]/\mathrm{J\,K^{-1}\,mol^{-1}} = C_p^{\ominus}(T)/\mathrm{J\,K^{-1}\,mol^{-1}} = 8.3143,$$

$$-[\{G^{\ominus}(T) - H^{\ominus}(0)\}/T]/\mathrm{J\,K^{-1}\,mol^{-1}} = 19.1444\log_{10}(I \times 10^{-39}/\mathrm{g\,cm^2}) + 19.1444\log_{10}(T/\mathrm{K}) - 19.1444\log_{10}\sigma - 11.5286,$$

$$S^{\ominus}(T)/\mathrm{J\,K^{-1}\,mol^{-1}} = 19.1444\log_{10}(I \times 10^{-39}/\mathrm{g\,cm^2}) + 19.1444\log_{10}(T/\mathrm{K}) - 19.1444\log_{10}\sigma - 3.2683.$$

(ii) *Non-linear molecules*

$$[\{H^{\ominus}(T) - H^{\ominus}(0)\}/T]/\mathrm{J\,K^{-1}\,mol^{-1}} = C_p^{\ominus}(T)/\mathrm{J\,K^{-1}\,mol^{-1}} = 12.4715,$$

$$-[\{G^{\ominus}(T) - H^{\ominus}(0)\}/T]/\mathrm{J\,K^{-1}\,mol^{-1}} = 9.7522\log_{10}(I_xI_yI_z \times 10^{117}/\mathrm{g^3\,cm^6}) + 28.7166\log_{10}(T/\mathrm{K}) - 19.1444\log_{10}\sigma - 12.6151,$$

$$S^{\ominus}(T)/\mathrm{J\,K^{-1}\,mol^{-1}} = 9.7522\log_{10}(I_xI_yI_z \times 10^{117}/\mathrm{g^3\,cm^6}) + 28.7166\log_{10}(T/\mathrm{K}) - 19.1444\log_{10}\sigma - 0.1436.$$

Harmonic oscillator contributions

$$u_i = 1.43879\tilde{\nu}_i/\mathrm{cm^{-1}} = \Theta_{\mathrm{vib}}/T,$$

$$[\{H^{\ominus}(T) - H^{\ominus}(0)\}/T]/\mathrm{J\,K^{-1}\,mol^{-1}} = 8.3143\Sigma_i u_i\{\exp(u_i) - 1\}^{-1},$$

$$-[\{G^{\ominus}(T) - H^{\ominus}(0)\}/T]/\mathrm{J\,K^{-1}\,mol^{-1}} = 8.3143\Sigma_i \ln\{1 - \exp(-u_i)\},$$

$$C_p^{\ominus}(T)/\mathrm{J\,K^{-1}\,mol^{-1}} = 8.3143\Sigma_i u_i^2\exp(u_i)\{\exp(u_i) - 1\}^{-2}.$$

[a] Physical constants from *Pure Appl. Chem.*, 1964, **9**, 453.

Internal Partition Function for Monatomic Gases.—For the present purpose, for the internal energy of a monatomic gas only the nuclear spin and electronic states need be considered. On the assumption that these energies are additive, the partition function can again be factored:

$$q_{\mathrm{internal}} = q_{\mathrm{nuclear\ spin}} \times q_{\mathrm{electronic}},$$

and the contributions to the thermodynamic functions are additive.

The nuclear spin partition function is given by $(2i + 1)$ where i is the nuclear spin quantum number, since the energy of nuclear orientations is very small compared with kT.[7] Thus, for the hydrogen atom (^{1}H), $i = 1/2$ and for the chlorine atom $i = 3/2$ giving nuclear spin contributions of 2 and 4, respectively, to the partition function. However, it is conventional to omit these factors from calculated entropies.

The electronic partition function is given by

$$q_{\text{electronic}} = \Sigma(2j + 1)\exp(- \varepsilon_{\text{el}}/kT),$$

where j is the quantum number determining the total electronic angular momentum, and the summation is over all the electronic states of the atom. For the ground state, ε_{el} is zero and the contribution is just the degeneracy $(2j + 1)$. The value of j is contained in the spectroscopic term symbol for the state, as the right-hand subscript. Thus, for hydrogen the ground state is a $^2S_{1/2}$ state and has a degeneracy of 2. The next three states are $^2P_{1/2}$, $^2S_{1/2}$, and $^3P_{3/2}$ with corresponding degeneracies of 2, 2, and 4; but since they lie at wavenumbers of approximately 82 260 cm^{-1} ($\triangleq$ 118 350 K) above the ground state their contribution to the thermodynamic functions is negligible. For the chlorine atom, the ground state is $^2P_{3/2}$ (with a degeneracy of 4) and the next highest level is $^2P_{1/2}$ (degeneracy of 2) at a wavenumber 881 cm^{-1} ($\triangleq$ 1268 K) above the ground state. So the electronic partition function becomes

$$q_{\text{electronic}} = 4 + 2\exp(- 1268\ \text{K}/T).$$

Using equations (9), (10), and (11) we find that this factor in the partition function contributes to $C_p^{\ominus}(T)/Lk$ an amount 0.128 at 300 K and 0.236 at 500 K. Using equations (7) and (10) we find that the contribution at 298.15 K to $S^{\ominus}/Lk$ is 1.42; the translational contribution is 18.44 so the electronic contribution is about 7% of the total. The electronic contributions must always be evaluated by summation; a further example may be found in reference 3. Much information on atomic states and energy levels has been critically evaluated and tabulated.[8]

Internal Partition Functions for Polyatomic Molecules.—The internal partition function for a polyatomic molecule comprises contributions from nuclear spin and electronic levels, and from rotational and vibrational degrees of freedom. On the assumption that the corresponding energies are additive and independent, these contributions can be factored, and the corresponding contributions to the thermodynamic functions are additive.

Nuclear Spin Contribution. The nuclear spin partition function is the product of the nuclear spin multiplicity $(2i_j + 1)$ for all the atoms in the molecule, where i_j is the nuclear spin of the jth atom. Since, apart from processes involving molecular hydrogen and its isotopes (and the transmutation of the elements), nuclear spins are conserved, this contribution is conventionally omitted from the total entropy leaving the 'practical' or 'virtual' entropy (these adjectives are frequently omitted also).

Rotational Contribution. For linear molecules, the rotational energy levels are given, to a good first approximation, by the expression

$$\varepsilon_{\text{rot}}/hc = BJ(J + 1),$$

[8] C. E. Moore, 'Atomic Energy Levels', National Bureau of Standards, Circular 467, Washington, D.C., 1949, 1952, 1958.

where J is the rotational quantum number and B is related to the moment of inertia I and the characteristic rotational temperature Θ_r by

$$B = h/8\pi^2Ic = k\Theta_r/hc.$$

It is convenient to define the dimensionless quantity:

$$y = hcB/kT = \Theta_r/T,$$

so that the rotational partition function can be written

$$q_{rot} = \Sigma_J(2J + 1)\exp\{-J(J + 1)y\}.$$

The summation for unsymmetrical molecules is over all J, and for symmetrical molecules is over even or odd values of J for *para-* or *ortho-*species. Since q_{rot} cannot be expressed in closed form, its evaluation as a sum with the thermodynamic functions calculated from equations (3)—(5), (7), and (9)—(11) is necessary, for example for hydrogen at moderate temperatures. However, if y is small, then the summation can be replaced by an integral and evaluated as $(\sigma y)^{-1}$ where σ is the symmetry number taking the value 2 or 1 for symmetrical or unsymmetrical molecules, respectively. The same result appears as the leading term in a power series for q_{rot} which may be derived:[5, 6]

$$q_{rot} = (\sigma y)^{-1}(1 + y/3 + y^2/15 + 4y^3/315 + y^4/315 + \ldots), \qquad (13)$$

which to the term in y^2 is accurate to 0.1% or better if $y < 0.3$. The rotational contributions to the thermodynamic functions are then

$$\{H^\ominus(T) - H^\ominus(0)\}/T = Lk(1 - y/3 - y^2/45 + \ldots),$$

$$-\{G^\ominus(T) - H^\ominus(0)\}/T = -Lk(\ln\sigma + \ln y - y/3 - y^2/90 + \ldots),$$

$$C_p^\ominus(T) = Lk(1 + y^2/45 + \ldots).$$

The corresponding practical equations are given in Table 1.

For non-linear, asymmetric top molecules the rotational partition function involves all three moments of inertia and the resulting expression is complex.[9] For most practical purposes the result reduces to the classical high-temperature approximation

$$q_{rot} = (8\pi^2kT)^{3/2}(\sigma h^3)^{-1}(\pi I_xI_yI_z)^{1/2}. \qquad (14)$$

Since only the product $I_xI_yI_z$ of the three principal moments of inertia is required, a simplified procedure can be used.[3, 10] The symmetry number σ is the number of indistinguishable positions into which the molecule can be turned by simple rigid rotations;[1] it is given directly by the point group of the molecule.[11] Equation (14) is valid for polyatomic molecules at or above room temperature. The rotational contributions to the thermodynamic functions follow from equation (14) using equations (3), (4), (5), and (8), and the practical equations are given in Table 1.

[9] K. F. Stripp and J. G. Kirkwood, *J. Chem. Phys.*, 1951, **19**, 1131.
[10] J. O. Hirschfelder, *J. Chem. Phys.*, 1940, **8**, 431.
[11] G. Herzberg 'Infrared and Raman spectra of Polyatomic Molecules', Van Nostrand, Princeton, 1945, p. 508.

Vibrational Contributions. A linear, or non-linear, molecule with n atoms has $(3n-5)$ or $(3n-6)$, respectively, vibrational degrees of freedom and that number of normal modes of vibration of wavenumber $\tilde{\nu}_i$. On the assumption of harmonic oscillators, which is always a good first approximation for the thermodynamic functions, the energy of each is independent, and this summation gives the vibrational energy of the molecule:

$$\varepsilon_{\text{vib}}/hc = \Sigma_i \tilde{\nu}_i (v_i + \tfrac{1}{2}),$$

where v_i is the vibrational quantum number, and each $v_i = 0, 1, 2, \ldots$. The vibrational partition function is the product of factors for each of the vibrations and if $u_i = hc\tilde{\nu}_i/kT$, each vibration contributes to the partition function a factor

$$q_{\text{vib}} = \Sigma_v \exp(-\ u_i v) = \{1 - \exp(-\ u)\}^{-1}.$$

Thus the vibrational contribution to the partition function can be written exactly in closed form. The resulting practical equations for the thermodynamic functions are given in Table 1. The total vibrational contributions will be the sum of the contributions from each of the vibrations; a vibration $\tilde{\nu}_i$ of degeneracy d_i will contribute d_i times the contribution of that $\tilde{\nu}_i$.

Several tabulations of the harmonic oscillator or Einstein functions are available;[1, 3, 5, 6, 12] the most reliable are those giving the values in dimensionless form and the most convenient for practical use are those of ref. 13.

Electronic Contribution. If a molecule has an electronic ground state of degeneracy g_0 and a first excited state at an energy ε_{el} with a degeneracy g_1, then the electronic partition function will be

$$q_{\text{el}} = g_0 + g_1 \exp(-\ \varepsilon_{\text{el}}/kT).$$

This contributes to the entropy, for example, an amount $Lk \ln g_0$ at low temperatures and $Lk \ln(g_0 + g_1)$ at high temperatures.[4, 14] The majority of polyatomic molecules of interest at moderate temperatures have a singly degenerate electronic ground state ($g_0 = 1$), and the higher electronic levels lie at sufficiently greater energies so that they make no significant contribution to the partition function.

For diatomic molecules the degeneracy of an electronic level is contained in the spectroscopic term symbol. States with total orbital angular momentum (in units of $h/2\pi$) of 0, 1, 2, 3, ... are denoted by Σ, Π, Δ, Φ, ..., respectively, with a left superscript of value $(2M + 1)$, the spin multiplicity, where M is the total electron spin quantum number. Then, the degeneracy of the level is given by $(2M + 1)$ for Σ states and $2(2M + 1)$ for Π, Δ, Φ, ... states. For example, the degeneracy of all ${}^1\Sigma$ states is 1; of the ${}^3\Sigma_g^-$, ${}^1\Delta_g$, and ${}^1\Sigma_g^+$ states of oxygen is 3, 2, and 1, respectively; and of the ${}^3\Phi$ state of cerium oxide is 6. For oxygen the two higher levels are at

[12] K. S. Pitzer, 'Quantum Chemistry', Prentice-Hall, Englewood Cliffs, 1953.

[13] J. Hilsenrath and G. C. Ziegler, 'Tables of Einstein Functions, Vibrational Contributions to the Thermodynamic Functions', National Bureau of Standards, Washington D.C., 1962.

[14] J. S. Rowlinson, 'The Perfect Gas', Pergamon Press, Oxford, 1963, p. 69.

7882 cm^{-1} ($\hat{=}$ 11 340 K) and 13 121 cm^{-1} ($\hat{=}$ 18 878 K) above the ground state, so that the electronic partition function can be written:

$$q_{el} = 3 + 2\exp(-\ 11\,340\ K/T) + \exp(-\ 18\,878\ K/T).$$

(Such an expression involves the assumption that the rotational and vibrational constants are identical for all three electronic states. To avoid such an assumption separate treatment of each of the states would be needed.)

Free radicals with an unpaired electron will have an electronic degeneracy of 2 arising from the resultant spin of 1/2. The same situation arises with NO_2, and $Lk \ln 2$ has therefore to be added to $S^{\ominus}(T)$ and $-\{G^{\ominus}(T) - H^{\ominus}(0)\}/T$ whereas $C_p^{\ominus}(T)$ and $\{H^{\ominus}(T) - H^{\ominus}(0)\}/T$ are unaffected.

Molecules of importance at high temperatures may have loosely bound electrons and low-lying electronic levels. Limitations remain with the calculation of thermodynamic properties of such molecules because of uncertainties concerning the electronic states. For example, in calculations of the thermodynamic functions for gaseous metal dioxides, the electronic contribution is the most difficult to evaluate.[15] Whilst MO_2 compounds of Group IV metals can be assumed to have $^1\Sigma$ ground states and no adjacent higher electronic states, for other dioxides appreciable electronic contributions to the partition function are likely, both from the multiplicity of the ground state and the presence of low-lying electronic states. Comparison has to be made with the M^{4+} ion or, lacking that, with an isoelectronic ion. Alternative comparisons have been made for other molecules,[16] and it may additionally be necessary to assume that an effective decrease in degeneracy arising from the separation of multiplet components is compensated by the presence of other low-lying electronic states.[17] For some compounds the resulting degeneracies can be estimated [16] correctly only within a factor of 2.

Higher Approximations.—The preceding equations give the thermodynamic functions on the assumptions of a rigid rotator and of harmonic oscillators. However, observed vibrational spectra show the departure of molecular vibrations from the harmonic law – an effect known as vibrational anharmonicity. A consequence of this anharmonicity is a dependence of the interatomic distance on the vibrational energy, an effect known as rotational–vibrational coupling. Thirdly, the centrifugal force of rotation stretches the molecule, increasing the moment of inertia and decreasing *B*. These effects give rise to small corrections to the thermodynamic functions and should be taken into account for accurate calculations.

[15] L. Brewer and G. M. Rosenblatt, *Chem. Rev.*, 1961, **61**, 257.

[16] P. Coppens, S. Smoes, and J. Drowart, *Trans. Faraday Soc.*, 1967, **63**, 2140; S. Smoes, P. Coppens, C. Bergman, and J. Drowart, *ibid.*, 1969, **65**, 682; C. Bergman, P. Coppens, J. Drowart, and S. Smoes, *ibid.*, 1970, **66**, 800; B. V. Fenochka, S. P. Gordienko, and V. V. Fesenko, *Russ. J. Phys. Chem.*, 1970, **44**, 1150.

[17] L. L. Ames and R. F. Barrow, *Proc. Phys. Soc.*, 1967, **90**, 869.

Diatomic Molecules. For diatomic molecules the correction terms can be evaluated explicitly, and in recent years have to an increasing extent been calculated [18, 19, 20] from improved spectroscopic data now available. The relevant equations [5] have been presented in equivalent but more convenient forms.[3, 18]

The vibrational anharmonicity gives a first additional term in the vibrational energy of $-\, x_e\tilde{\nu}_e(v + \frac{1}{2})^2 hc$. [The second term is the coefficient of $(v + \frac{1}{2})^3$ and is shown by observed spectra to be negligibly small.] Here, x_e is the (first) anharmonicity constant and $\tilde{\nu}_e$ is the equilibrium wavenumber, for small displacements. The energy of the vth vibrationally excited state is then given by

$$(\varepsilon - \varepsilon_0)/hc = \tilde{\nu}_0 v - x\tilde{\nu}_0 v(v - 1),$$

where $\tilde{\nu}_0$ is the observed wavenumber, $\tilde{\nu}_0 = \tilde{\nu}_e(1 - 2x_e)$, and where $x = x_e/(1 - 2x_e)$.

The coupling of rotation and vibration gives an additional term in the expression for $(\varepsilon - \varepsilon_0)/hc$ of $-\, \alpha(v + \frac{1}{2})J(J + 1)$ where α is the rotational–vibrational coupling constant. The effect of centrifugal stretching is expressed by a further term of $-\, DJ^2(J + 1)$ where D is the rotational stretching coefficient, which is related to the rotational constant for equilibrium separation, B_e, by $D = 4B_e^3/\tilde{\nu}_e^2$. If B_0 is the rotational constant for the first vibrationally excited state, then $B_0 = B_e - \alpha/2$.

Defining the quantities $u = hc\tilde{\nu}_0/kT$, $\delta = \alpha/B_0$, and $\gamma = (D/4B_e)^{1/2} = (B_e/\tilde{\nu}_e)$, then the additional correction to the partition functions can be expressed, after some simplification, by

$$\ln q_{\text{corr}} = 2xu\{\exp(u) - 1\}^{-2} + \delta\{\exp(u) - 1\}^{-1} + 8\gamma/u.$$

The resulting corrections to the thermodynamic properties then follow as functions of x, δ, γ, and u, all of which are derivable from spectroscopy. Tabulated values of the functions involving u are available,[3] which facilitate the calculation, an example of which is given in Table 2.

These first-order corrections need to be supplemented, for the most accurate calculation for a wide range of temperatures, by taking into account low-temperature effects of non-classical rotation and the uncoupling of the electronic spin from the internuclear axis.[20]

For diatomic molecules, accurate evaluation of the rotational contributions requires consideration of the ground electronic state. Where this is a $^1\Sigma$ state, equation (13) is satisfactory, but to be rigorous, molecules in Σ states of higher multiplicity should be treated somewhat differently [21] and

[18] W. H. Evans, R. Jacobson, T. R. Munson, and D. D. Wagman, *J. Res. Nat. Bur. Stand.*, 1955, **55**, 83.

[19] W. H. Evans, T. R. Munson, and D. D. Wagman, *J. Res. Nat. Bur. Stand.*, 1955, **55**, 147.

[20] L. Haar and A. S. Friedman, *J. Chem. Phys.*, 1955, **23**, 869.

[21] L. R. Sitney, 'Pubco-II, an IBM 704 code for computing the Ideal Thermodynamic Functions of a diatomic gas molecule', Los Alamos Scientific Laboratory, New Mexico, 1960.

Table 2 *Contributions*[a] *to thermodynamic functions of nitrogen at* $T =$ 2000 K

		$-\{G^\ominus(T) - H^\ominus(0)\}/T$ / $J\ K^{-1}\ mol^{-1}$	$C_p^\ominus(T)$ / $J\ K^{-1}\ mol^{-1}$
Translation	$M/g\ mol^{-1} = 28.016$	169.088	20.786
Rigid rotation	$y = 0.001439$	48.647	8.314
Harmonic oscillator	$u = 1.6766$	1.720	6.611
Vibrational anharmonicity	$x = 0.00613$	0.008	0.096
Rotation–vibration	$\delta = 0.00935$	0.017	0.088
Rotational stretching	$\gamma = 0.000852$	0.033	0.067
Total		219.513	35.962

[a] Molecular data used: $\tilde{\nu}_e = 2359.6\ cm^{-1}$, $x_e\tilde{\nu}_e = 14.46\ cm^{-1}$, $B_e = 2.010\ cm^{-1}$, and $\alpha = 0.0187\ cm^{-1}$; electronic state $^1\Sigma_g^+$.

the situation for $^2\Sigma$ and $^3\Sigma$ states has been discussed.[22] However, use of equation (13) gives only small errors at low temperatures and is correct at high temperatures.[21] The details for $^2\Pi$ molecules have been given [20] and extended [21] to $^2\Delta$ and $^2\Gamma$ molecules by using expressions [23] for the rotational energies of doublet molecules with Λ-type doubling. [In ref. (21) some corrections are made to the results of ref. (20).] The rotational partition function for triplet molecules with Λ-type doubling has been derived [21] using expressions [24] for the rotational energy levels for any degree of spin uncoupling for triplet states.

Polyatomic Molecules. General expressions for the effects of anharmonicity and rotational–vibrational interaction for polyatomic molecules have been derived,[25, 26] and have been used for several triatomic molecules.[27–30] Such calculations including the higher approximations are not necessarily free from errors. For example, for hydrogen cyanide, ref. 29 corrects small errors in ref. 28; and for carbon disulphide the calculations of ref. 29 give values for $C_p^\ominus$ which differ from experimental results [31] by as much as 0.5%. (The values for carbon disulphide are important in view of its use as a reference substance for vapour-flow calorimetry, see Chapter 6; the

[22] G. Herzberg, 'Molecular Spectra and Molecular Structure, I Spectra of Diatomic Molecules', Van Nostrand, New York, 1950.
[23] E. L. Hill and J. H. Van Vleck, *Phys. Rev.*, 1928, **32**, 250.
[24] A. Budo, *Z. Physik*, 1935, **96**, 219.
[25] R. E. Pennington and K. A. Kobe, *J. Chem. Phys.*, 1954, **22**, 1442.
[26] H. W. Woolley, *J. Res. Nat. Bur. Stand.*, 1956, **56**, 105.
[27] L. Haar, J. C. Bradley, and A. S. Friedman, *J. Res. Nat. Bur. Stand.*, 1955, **55**, 285.
[28] J. C. Bradley, L. Haar, and A. S. Friedman, *J. Res. Nat. Bur. Stand.*, 1956, **56**, 197.
[29] B. J. McBride and S. Gordon, *J. Chem. Phys.*, 1961, **35**, 2198.
[30] W. H. Evans and D. D. Wagman, *J. Res. Nat. Bur. Stand.*, 1952, **49**, 141.
[31] G. Waddington, J. C. Smith, K. D. Williamson, and D. W. Scott, *J. Phys. Chem.*, 1962, **66**, 1074.

most accurate calculations are those of ref. 31.) An error in equation (23) of ref. 25 has been noted.[31]

Additional features arise with degenerate vibrations,[32] and for linear molecules there is a further complication arising from the dependence of the rotational energy upon the quantum number for the degenerate vibration.[26]

For polyatomic molecules, the vibrational energy levels, written in terms of the observed fundamental wave numbers $\tilde{\nu}_i$ become, when anharmonicity is included:

$$(\varepsilon - \varepsilon_0)/hc = \Sigma_i \tilde{\nu}_i v_i - \Sigma_i X_{ii} v_i (v_i - 1) - \Sigma_{i<j} X_{ij} v_i v_j,$$

where the summation must be taken over all the fundamental wave-numbers.[32, 33] Here the X_{ii} and X_{ij} are anharmonicity coefficients expressed as wavenumbers; if they are small, then their contributions to the Boltzmann factor can be expanded. The resulting correction term to the harmonic partition function is then $(1 + \Sigma_{i \leqslant j} f_{ij})$ where

$$f_{ij} = d_i d_j X_{ij} hc[kT\{\exp(u_i) - 1)\}\{\exp(u_j) - 1\}]^{-1},$$

$$f_{ii} = d_i(d_i + 1) X_{ii} hc[kT\{\exp(u_i) - 1\}^2]^{-1}.$$

In these expressions d_i is the degeneracy of the ith fundamental and $u_i = hc\tilde{\nu}_i/kT$; by using them the corrections to the thermodynamic functions can be calculated[32, 34] by the equations of p. 271, provided the X_{ii} are known.

For some polyatomic molecules the X_{ii} are available from spectroscopy. An example is acetylene, for which values of the rotational stretching and rotational–vibrational interaction coefficients are also available, thus allowing rather complete calculations of the correction terms to be made.[33] For other molecules attempts have been made to deduce values of the anharmonicity coefficients from those for related molecules: for example, on the assumption[35] that the X_{ii} for a stretching vibration is approximately that for a diatomic molecule of the same wavenumber and bond energy, and also that $X_{ij} = a(\tilde{\nu}_i + \tilde{\nu}_j)$, where a is a constant. Such a procedure was adopted for sulphur dioxide[30] and sulphur trioxide,[32] but the calculations for the latter, and their repetition,[30] are no longer valid since the vibrational assignment on which they are based is now known to be in error.[36] Accurate experimental values of the vapour heat capacity of benzene revealed a contribution attributed to anharmonicity, which was treated[34] in a similar way. The constant a was adjusted to give agreement between observed and calculated values of $C_p^\ominus$; its magnitude corresponded to anharmonicity

[32] W. H. Stockmayer, G. M. Kavanagh, and H. S. Mickley, *J. Chem. Phys.*, 1944, **12**, 408.

[33] D. D. Wagman, J. E. Kilpatrick, K. S. Pitzer, and F. D. Rossini, *J. Res. Nat. Bur. Stand.*, 1945, **35**, 467.

[34] D. W. Scott, G. Waddington, J. C. Smith, and H. M. Hufmann, *J. Chem. Phys.*, 1947, **15**, 565.

[35] O. Redlich, *J. Chem. Phys.*, 1941, **9**, 298.

[36] R. W. Lovejoy, J. H. Colwell, D. F. Eggers, and G. D. Halsey, *J. Chem. Phys.*, 1962, **36**, 612; R. Bent and W. R. Ladner, *Spectrochim. Acta*, 1963, **19**, 931.

coefficients ranging from 0.7 cm^{-1} for the lowest fundamentals to 5.5 cm^{-1} for the carbon–hydrogen stretching modes, which are reasonable values for a molecule such as benzene.

A calculation[37] of the forms of the vibrational modes of cyclohexane suggested that two fundamentals are rather anharmonic. For one, at 382 cm^{-1}, a value for X of 3.0 cm^{-1} was inferred from the observed overtone; a value for X of 4.0 cm^{-1} was chosen for the other, at 231 cm^{-1}, in order to achieve satisfactory agreement with the observed vapour heat capacity.

In recent years, accurate measurements by vapour flow calorimetry of the vapour heat capacity of several molecules have shown that a small but experimentally significant contribution remains when the translational, rigid rotational, and harmonic oscillator contributions have been satisfactorily accounted for. The data[38] for thian (thiacyclohexane) in Table 3

Table 3 *Observed and calculated values of vapour heat capacity of thian (thiacyclohexane)*

T/K	399.20	423.20	453.20	483.20
$C_p^{\ominus}(T)$/J K^{-1} mol^{-1}, observed	149.08	158.49	170.29	181.21
$C_p^{\ominus}(T)$/J K^{-1} mol^{-1}, calculated [a]	147.33	156.29	167.15	177.42
$C_p^{\ominus}(T)$/J K^{-1} mol^{-1}, anharm. [b]	1.75	2.27	2.97	3.75
$C_p^{\ominus}(T)$/J K^{-1} mol^{-1}, total calc.	149.08	158.56	170.12	181.17

[a] Contributions of translation and rigid rotation and harmonic oscillator and chair-boat equilibrium; [b] calculated from equation (15c) using $Z = 17.09$, and $\tilde{\nu} = 1025$ cm^{-1}.

are typical. The calculations for this molecule involve a contribution from the tautomerism of chair and boat configurations, and quantitative discussion of the relative magnitudes of this contribution and that from anharmonicity is somewhat arbitrary. However, if anharmonicity is neglected no agreement within experimental error is possible between the measured and calculated values of the gaseous heat capacity.

For the treatment of the anharmonicity of such molecules, semi-empirical equations for the total anharmonicity contributions have been developed.[38] These involve two adjustable parameters: an arbitrary harmonic oscillator of wavenumber $\tilde{\nu}$ and a coefficient Z. Writing $u = hc\tilde{\nu}/kT$, these equations reduce to

$$[\{H^{\ominus}(T) - H^{\ominus}(0)\}/LkT]_{\text{anh}} = Zu\{2u\exp(u) - \exp(u) + 1\} \times\{\exp(u) - 1\}^{-3} = Z\,{}^{5}\phi/2, \quad (15a)$$

$$-[\{G^{\ominus}(T) - H^{\ominus}(0)\}/LkT]_{\text{anh}} = Zu\{\exp(u) - 1\}^{-2} = Z\,{}^{4}\phi/2, \quad (15b)$$

$$\{C_p^{\ominus}(T)/Lk\}_{\text{anh}} = 2Zu^2\exp(u) \times\{2u\exp(u) - 2\exp(u) + u + 2\} \times\{\exp(u) - 1\}^{-4} = Z\,{}^{6}\phi/2. \quad (15c)$$

[37] C. W. Beckett, K. S. Pitzer, and R. Spitzer, *J. Amer. Chem. Soc.*, 1947, **69**, 2488.
[38] J. P. McCullough, H. L. Finke, W. N. Hubbard, W. D. Good, R. E. Pennington, J. F. Messerly, and G. Waddington, *J. Amer. Chem. Soc.*, 1954, **76**, 2661.

Tabulated values of the functions ${}^4\phi$, ${}^5\phi$, and ${}^6\phi$ in terms of u are available,[25] which facilitate the use of this treatment. It should be noted that "this empirical function really serves not only as a representation of the anharmonicity contributions, but also as a sort of 'catch all' for all the other small effects neglected in the simplified treatment. It is an adequate representation over the range of the experimental heat-capacity values; how adequate a representation it remains at higher temperatures determines the accuracy of the calculated thermodynamic functions for those temperatures."[39]

3 Molecules with Internal Rotation

Although it was earlier believed that internal rotation about single bonds is unrestricted, Kemp and Pitzer in 1937 showed [40] that agreement between third law and calculated values of the entropy of ethane required the internal rotation about the C—C bond to be restricted by a potential barrier of about 13 kJ mol^{-1}. A large number of thermodynamic and spectroscopic studies have established the presence of barriers of this order of magnitude restricting internal rotation about single bonds. In only a very few cases can internal rotation be considered to be unrestricted.

Where the barrier to internal rotation is high such that there is negligible passage of rotating groups from one potential minimum to another, thermodynamic functions may be calculated by treating the internal rotation as a harmonic or anharmonic oscillation. In general, however, barriers to internal rotation are too low for internal rotation to be considered in this way. In these cases it is convenient to consider internal rotation in the total rotational partition function, leaving the translational partition function and, usually, the vibrational partition function to be calculated as described above, except that the number of vibrational degrees of freedom is reduced by one for each type of internal rotation.

It is convenient to divide types of internal rotation according to whether they originate from rotation of symmetric tops against a rigid frame, rotation of asymmetric tops against a rigid frame, or from compound rotation. Procedures for calculating thermodynamic properties of the first class are well established, whereas calculations for the latter two classes are generally more approximate.

Symmetric Tops attached to a Rigid Frame.—To a good approximation the rotational partition function q_R of a molecule having several symmetric tops attached to a rigid frame may be expressed by:[41, 42]

$$q_R = q_{OR} \times q_{IR}, \tag{16}$$

[39] D. W. Scott and G. A. Crowder, *J. Chem. Phys.*, 1967, **46**, 1054.
[40] J. D. Kemp and K. S. Pitzer, *J. Amer. Chem. Soc.*, 1937, **59**, 276.
[41] K. S. Pitzer and W. D. Gwinn, *J. Chem. Phys.*, 1942, **10**, 428.
[42] B. L. Crawford, *J. Chem. Phys.*, 1940, **8**, 273.

where $q_{\rm OR}$ and $q_{\rm IR}$ are the partition functions for overall and internal rotation, respectively. The partition function $q_{\rm OR}$ may be calculated as described above (p. 274), using the symmetry number obtained by considering the internal rotation of each top to be frozen in one configuration. The overall rotation symmetry number ($\sigma_{\rm OR}$) for ethane, for example, is 6, while for toluene $\sigma_{\rm OR}$ is 1.

Free Internal Rotation of a Single Symmetric Top. Although there are few examples of free internal rotation, it is instructive to consider this case first since restricted internal rotation has been generally considered as a perturbation of free internal rotation. The partition function for a single free internal rotation has been obtained by classical mechanics,[43] and may be conveniently expressed by:[41]

$$q_{\rm f}^{m} = (8\pi^3 I_m kT)^{1/2}(h\sigma_{{\rm IR},m})^{-1}. \qquad (17)$$

The practical equation is

$$q_{\rm f}^{m} = 2.7930\{(I_m \times 10^{38}/{\rm g\ cm^2})(T/K)\}^{1/2}(\sigma_{{\rm IR},m})^{-1},$$

where I_m is the reduced moment of inertia of the top and $\sigma_{{\rm IR},m}$ is the internal rotation symmetry number of the top.

For a single symmetric top the reduced moment I_m is given by

$$I_m = A_m\left(1 - \sum_{i=1}^{3} A_m\lambda_i/I_i\right), \qquad (18)$$

where A_m is the moment of inertia of the top about its symmetry axis, λ_i is the direction cosine between the axis of the top and the ith principal axis of the whole molecule, and I_i is the moment of inertia about the ith principal axis of the whole molecule.

The symmetry number $\sigma_{{\rm IR},m}$ is the number of identical configurations into which the molecule can be transformed on internal rotation. In toluene, for example, $\sigma_{{\rm IR},m}$ is 6, while in but-2-yne $\sigma_{{\rm IR},m}$ is 3. Contributions to thermodynamic functions from a single free internal rotation are readily calculated from equation (17), and the practical equations are summarized in Table 4. A breakdown of a typical calculation for a molecule with free internal rotation is illustrated by data for toluene in Table 5.

Table 4 *Contributions to the thermodynamic functions from a single free internal rotation*

$$[\{H^\ominus(T) - H^\ominus(0)\}/T]/{\rm J\ K^{-1}\ mol^{-1}} = 4.1572,$$

$$-[\{G^\ominus(T) - H^\ominus(0)\}/T]/{\rm J\ K^{-1}\ mol^{-1}} = 8.5398 + 19.1444\log_{10}\{(I_m \times 10^{38}/{\rm g\ cm^2})(T/{\rm K})(\sigma_{{\rm IR},m})^{-1}\},$$

$$S^\ominus(T)/{\rm J\ K^{-1}\ mol^{-1}} = 12.6970 + 19.1444\log_{10}\{(I_m \times 10^{38}/{\rm g\ cm^2})(T/{\rm K})(\sigma_{{\rm IR},m})^{-1}\},$$

$$C_p^\ominus(T)/{\rm J\ K^{-1}\ mol^{-1}} = 4.1572.$$

[43] M. L. Eidinoff and J. G. Aston, *J. Chem. Phys.*, 1935, **3**, 379.

le 5 *Calculated and observed thermodynamic properties of toluene* [46]

	$S^{\ominus}(298.15\ K)$ / $J\ K^{-1}\ mol^{-1}$	$S^{\ominus}(1000\ K)$ / $J\ K^{-1}\ mol^{-1}$	$C_p^{\ominus}(427.20\ K)$ / $J\ K^{-1}\ mol^{-1}$	$C_p^{\ominus}(1000\ K)$ / $J\ K^{-1}\ mol^{-1}$
ıslational contribution	165.16	190.31	20.79	20.79
rall rotational ntribution	107.08	122.18	12.47	12.47
internal rotational ntribution	14.80	19.83	4.16	4.16
ational contribution	33.62	209.04	111.15	223.50
armonicity contribution	0.01	1.89	0.60	4.04
l calculated	320.67	543.25	149.17	264.96
erved	321.21		149.16	

Free internal rotation has been assumed in thermodynamic calculations of nitromethane,[44, 45] toluene,[46] ααα-trifluorotoluene,[47] *p*-fluorotoluene,[48] 2-picoline,[49a] 3-picoline,[49b] and but-2-yne.[33] For all these molecules satisfactory agreement between calorimetric data and statistical calculations could be obtained only by assuming the internal rotation to be essentially unrestricted.

Independent evidence for free rotation in nitromethane has come from an analysis of its microwave spectrum, which has shown[50, 51] that the potential barrier restricting internal rotation is only 25.2 J mol^{-1}; for thermodynamic purposes this barrier is effectively zero. The low barrier in nitromethane is believed to be due to the possession of 6-fold symmetry about the internal rotation axis. A Fourier cosine series expansion of the restricting potential in the form:

$$V_m(\phi_m) = \sum_{K=1}^{\infty} (1/2) V_{m,K}\{1 - \cos(K\phi_m)\}, \tag{19}$$

therefore has as non-zero terms only $V_{m,6}, V_{m,12}, V_{m,18}, \ldots$. Microwave analysis of nitromethane has shown that $V_{m,6}$ is very small and that higher terms are probably negligible. Statistical thermodynamic calculations indicate that a similar situation applies to toluene,[46] ααα-trifluorotoluene,[47] and *p*-fluorotoluene,[48] which also have 6-fold internal rotation symmetry.

[44] K. S. Pitzer and W. D. Gwinn, *J. Amer. Chem. Soc.*, 1941, **63**, 3313.
[45] J. P. McCullough, D. W. Scott, R. E. Pennington, I. A. Hossenlopp, and G. Waddington, *J. Amer. Chem. Soc.*, 1954, **76**, 4791.
[46] D. W. Scott, G. B. Guthrie, J. F. Messerly, S. S. Todd, W. T. Berg, I. A. Hossenlopp, and J. P. McCullough, *J. Phys. Chem.*, 1962, **66**, 911.
[47] D. W. Scott, D. R. Douslin, J. F. Messerly, S. S. Todd, I. A. Hossenlopp, T. C. Kincheloe, and J. P. McCullough, *J. Amer. Chem. Soc.*, 1959, **81**, 1015.
[48] D. W. Scott, J. F. Messerly, S. S. Todd, I. A. Hossenlopp, D. R. Douslin, and J. P. McCullough, *J. Chem. Phys.*, 1962, **37**, 867.
[49a] D. W. Scott, W. N. Hubbard, J. F. Messerly, S. S. Todd, I. A. Hossenlopp, W. D. Good, D. R. Douslin, and J. P. McCullough, *J. Phys. Chem.*, 1963, **67**, 680.
[49b] D. W. Scott, W. D. Good, G. B. Guthrie, S. S. Todd, I. A. Hossenlopp, A. G. Osborn, and J. P. McCullough, *J. Phys. Chem.*, 1963, **67**, 685.
[50] E. Tannenbaum, R. D. Johnson, R. J. Myers, and W. D. Gwinn, *J. Chem. Phys.*, 1954, **22**, 949.
[51] E. Tannenbaum, R. J. Myers, and W. D. Gwinn, *J. Chem. Phys.*, 1956, **25**, 42.

Although 2-picoline and 3-picoline have 3-fold internal rotation symmetry, they can be considered to have local 6-fold symmetry. The essentially free rotation in but-2-yne determined from thermodynamic calculations has been confirmed by recent spectroscopic measurements.[52, 53] The small $V_{m,3}$ term is probably the result of the significant spectral separation between the methyl groups. Free rotation has also been postulated in dimethylcadmium.[54]

Free Internal Rotation of Several Symmetric Tops. Extension of statistical thermodynamic calculations from internal rotation of one top to several tops is straightforward. The rotational partition function may be expressed by

$$q_{\mathrm{R}} = q_{\mathrm{OR}} \prod_{m=1}^{N} q_{\mathrm{f}}^{m},$$

where N is the number of symmetric tops, and each q_{f} is given by equation (17).

The first approximation for the reduced moment of the mth top is given by equation (18). The next approximation is given by

$$I_m = I_m^{\circ} - (1/2) \sum_{m' \neq m} \Lambda_{mm'}^2 / I_{m'}^{\circ}, \tag{20}$$

where

$$\Lambda_{mm'} = A_m A_{m'} \sum_{i=1}^{3} \lambda_{mi} \lambda_{m'i} / I_i,$$

and I_m° is given by the value in equation (18).

Thermodynamic contributions from the internal rotation of several symmetric tops may be readily calculated by appropriate summation of terms in Table 4. Few reliable calculations, however, have been reported. Thermodynamic properties of propane and several methyl-substituted benzenes have been reported,[55] for example, but subsequent more accurate work has shown the necessity for considering that the internal rotation may be restricted.[56–58] Although the subsequent calculations for *m*-xylene and *p*-xylene used 6-fold internal rotation barriers of 2.1 to 3.1 kJ mol^{-1},[57, 58] more recent statistical calculations for toluene[46] employing the presence of free rotation suggest that internal rotation in the two xylenes may be effectively unrestricted.

Restricted Internal Rotation of a Single Symmetric Top. The potential restricting internal rotation may be expressed as a Fourier cosine and sine sum series, similar to equation (19). In practice, however, the restricting

52 P. R. Bunker and H. C. Longuet-Higgins, *Proc. Roy. Soc.*, 1964, **A280**, 340.
53 B. Kirtman, *J. Chem. Phys.*, 1964, **41**, 775.
54 J. C. M. Li, *J. Amer. Chem. Soc.*, 1956, **78**, 1081.
55 L. S. Kassel, *J. Chem. Phys.*, 1936, **4**, 276.
56 K. S. Pitzer, *J. Chem. Phys.*, 1937, **5**, 473.
57 K. S. Pitzer and D. W. Scott, *J. Amer. Chem. Soc.*, 1943, **65**, 803.
58 W. J. Taylor, D. D. Wagman, M. G. Williams, K. S. Pitzer, and F. D. Rossini, *J. Res. Nat. Bur. Stand.*, 1946, **37**, 95.

potential for symmetric top rotors is generally adequately expressed by the leading non-zero cosine term in the series. Thus the potential may be expressed by

$$V_m(\phi) = (1/2)V_{m,n_m}\{1 - \cos(n_m\phi)\}, \quad (21)$$

where n_m is the number of potential minima per revolution and is generally equal to $\sigma_{IR,m}$, the internal rotation symmetry number.

Pitzer,[59] and Pitzer and Gwinn[41] have reported extensive calculations of thermodynamic functions for the internal rotation of a symmetric top restricted by a simple cosine potential. They have published tables of thermodynamic functions for various combinations of V_m/RT and $1/q_f^m$, where q_f^m is given by equation (17). The tables were calculated from the energy levels of two coaxial rotors, one symmetrical and the other 'accidentally' symmetrical. It was subsequently shown, however, that the tables are also valid for a symmetric top attached to a rigid frame of any symmetry, provided that equation (21) is satisfied.[60] The appropriate energy levels are calculated as regions within which individual levels may be found, such that for a fixed combination of quantum states of the remaining degrees of freedom, one energy level is found within each region. Well below the potential barrier the levels are essentially vibrational in character, whereas above the barrier the regions are wide and the levels are more rotational in character. For a given potential barrier, symmetry number, moment of inertia, and temperature, there are two limiting values of the partition function and its derivatives. The calculated values in the tables, however, are restricted to those regions of the above variables such that the difference between the limiting values of the calculated thermodynamic functions is negligible.

The tables of thermodynamic functions have been extended by Li and Pitzer to include internal rotation of symmetric tops having very low moments of inertia.[61] These extended tables were calculated for two coaxial symmetric tops, though they are expected to apply to molecules containing symmetric tops attached to a rigid frame of any symmetry.

The validity of the simple cosine function in equation (21) as a description of the restricting potential of a symmetric top has been questioned by Blade and Kimball.[62] They examined the applicability of a two-parabola potential for the internal rotation in ethane, 1,1,1-trifluoroethane, and methanol, and concluded that the simple cosine function is not satisfactory. This conclusion for ethane has been criticized in detail by Pitzer, who has shown that the two-parabola potential does not provide a significantly better fit to the calorimetric data than does the simple cosine potential.[63] Experimental heat capacities for ethane over the range 90 to 305 K are

[59] K. S. Pitzer, *J. Chem. Phys.*, 1937, **5**, 469.
[60] K. S. Pitzer and W. D. Gwinn, *J. Chem. Phys.*, 1941, **9**, 485.
[61] J. C. M. Li and K. S. Pitzer, *J. Phys. Chem.*, 1956, **60**, 466.
[62] E. Blade and G. M. Kimball, *J. Chem. Phys.*, 1950, **18**, 630.
[63] K. S. Pitzer, *Discuss. Faraday Soc.*, 1951, No. 10, p. 66.

best fitted by a potential barrier that has a height 11.5 to 12.6 kJ mol^{-1} and that is very close to the simple cosine function.[63] Thermodynamic evidence for the simple cosine function has been supported by several microwave and far-i.r. studies of molecules containing single methyl groups, since analysis of the spectra in terms of a potential:

$$V_m(\phi) = (1/2)V_{m,3}(1 - \cos 3\phi) + (1/2)V_{m,6}(1 - \cos 6\phi),$$

has given values of $|(V_{m,6}/V_{m,3})|$ that are typically only 0.01 (Table 6).[64–67] The effect of small changes in the shape of the potential on thermodynamic properties has been estimated.[41] If $|(V_{m,6}/V_{m,3})|$ is of the order only of 0.01, and higher terms can be neglected, then the tables of thermodynamic functions for internal rotation of a symmetric top can be used with confidence.

Table 6 *Some values of $V_{m,3}$ and $V_{m,6}$ for internal rotation of symmetric tops*

Molecule	$V_{m,3}$ / kJ mol^{-1}	$V_{m,6}$ / kJ mol^{-1}	*Ref.*
CH_3CH_2Cl	15.44	0	64
$CH_3CF{=}CF_2$	9.79	0.05	64
$CH_3CH{=}CH_2$	8.51	− 0.19	64
CH_3OH	4.484	< 0.048	65
CH_3SH	5.310	< 0.048	66
$CH_3CH{-}CH_2$ (epoxide ring, O bridging CH and CH_2)	10.77	− 0.11	64
$CH_3N{-}CH_2$ (ring, CH_2 bridging N and CH_2)	14.98	− 0.31	67

Application of the tables[41, 61] to calculate thermodynamic contributions requires knowledge of molecular geometry to determine the appropriate reduced moment of inertia. In cases where the potential barrier height is known from spectroscopic measurements, the tables may be used directly to calculate thermodynamic properties. Examples of molecules for which this procedure has been used include methanol,[68, 69] methanethiol,[70] fluoroethane,[71] 1,1-difluoroethane,[71] 1,1,1-trifluoroethane,[72] pentafluoro-chloroethane,[73] chloroethane,[74] 1,1,1,2-tetrachloroethane,[75] pentachloro-

64 W. G. Fateley and F. A. Miller, *Spectrochim. Acta*, 1963, **19**, 611.
65 D. G. Burchard and D. M. Dennison, *J. Mol. Spectroscopy*, 1959, **3**, 299.
66 T. Kojima, *J. Phys. Soc. Japan*, 1960, **15**, 1284.
67 T. Ikeda, *J. Mol. Spectroscopy*, 1970, **36**, 268.
68 J. O. Halford, *J. Chem. Phys.*, 1950, **18**, 1051.
69 E. V. Ivash, J. C. M. Li, and K. S. Pitzer, *J. Chem. Phys.*, 1955, **23**, 1814.
70 D. W. Scott and J. P. McCullough, *U.S. Bur. Mines, Bull.*, 1961, 595.
71 D. C. Smith, R. A. Saunders, J. R. Nielsen, and E. E. Fergusson, *J. Chem. Phys.*, 1952, **20**, 847.
72 J. R. Nielsen, H. H. Claasen, and D. C. Smith, *J. Chem. Phys.*, 1950, **18**, 1471.
73 J. G. Aston, P. E. Wills, and T. P. Zolki, *J. Amer. Chem. Soc.*, 1955, **77**, 3939.
74 J. H. S. Green and D. J. Holden, *J. Chem. Soc.*, 1962, 1794.
75 J. R. Nielsen, C. Y. Liang, and L. W. Daasch, *J. Opt. Soc. Amer.*, 1953, **43**, 1071.

ethane,[75, 76] bromoethane,[74] iodoethane,[77] methyl isothiocyanate,[78] propylene oxide,[79] methylamine,[80] acetyl chloride,[81] and propionitrile.[82, 83] To illustrate typical magnitudes of internal rotation contributions, data for chloroethane are presented in Table 7.

e 7 *Calculated and observed thermodynamic properties of chloroethane* [74]

	$S^{\ominus}(285.37\ K)$ / $J\ K^{-1}\ mol^{-1}$	$S^{\ominus}(1000\ K)$ / $J\ K^{-1}\ mol^{-1}$	$C_p^{\ominus}(200\ K)$ / $J\ K^{-1}\ mol^{-1}$	$C_p^{\ominus}(1000\ K)$ / $J\ K^{-1}\ mol^{-1}$
slational contribution	159.80	185.87	20.79	20.79
all rotational contribution	97.61	113.25	12.47	12.47
ricted internal rotational ntribution	7.54	18.07	7.03	6.70
ational contribution	8.29	75.31	8.94	91.76
l calculated	273.24	392.50	49.23	131.72
erved	273.26		49.45	

The tables [41, 61] may be used to calculate barrier heights in cases where appropriate spectroscopic data are not available, but where experimental values of heat capacity or entropy are known at one or more temperatures. The calorimetrically determined value of the barrier height may then be used in conjunction with the tables [41, 61] to calculate internal rotation contributions to thermodynamic properties over an extended temperature range. Examples of this procedure include calculations for ethane,[40, 56, 62, 63] propene,[56, 84] acetaldehyde,[85] buta-1,2-diene,[86] acetic acid,[87] hexafluoroethane,[88] 3-methylthiophen,[89] and 2-methylthiophen.[90] Where spectroscopic values of the barrier height have subsequently been determined, satisfactory agreement has been obtained with the earlier calorimetric values. The agreement between calorimetric ($8.16\ kJ\ mol^{-1}$) [84] and subsequent microwave [$(8.28 \pm 0.07)\ kJ\ mol^{-1}$] [91] values of the barrier height in propene

[76] K. A. Kobe and R. H. Harrison, *Petrol Refiner*, 1957, **36**, 155.
[77] D. R. Stull, E. F. Westrum, jun., and G. C. Sinke, 'The Chemical Thermodynamics of Organic Compounds', Wiley, New York, 1969, p. 551.
[78] H. Mackle and P. A. G. O'Hare, *Trans. Faraday Soc.*, 1963, **59**, 309.
[79] J. H. S. Green, *Chem. and Ind.*, 1961, 369.
[80] K. A. Kobe and R. H. Harrison, *Petrol Refiner*, 1954, **33**, 161.
[81] J. Overend, R. A. Nyquist, J. C. Evans, and W. J. Potts, *Spectrochim. Acta*, 1961, **17**, 1205.
[82] N. E. Duncan and G. J. Janz, *J. Chem. Phys.*, 1955, **23**, 434.
[83] L. A. Weber and J. E. Kilpatrick, *J. Chem. Phys.*, 1962, **36**, 829.
[84] J. E. Kilpatrick and K. S. Pitzer, *J. Res. Nat. Bur. Stand.*, 1946, **37**, 163.
[85] K. S. Pitzer and W. Weltner, *J. Amer. Chem. Soc.*, 1949, **71**, 2842.
[86] J. E. Kilpatrick, C. W. Beckett, E. J. Prosen, K. S. Pitzer, and F. D. Rossini, *J. Res. Nat. Bur. Stand.*, 1949, **42**, 225.
[87] W. Weltner, *J. Amer. Chem. Soc.*, 1955, **77**, 3941.
[88] D. R. Stull, E. F. Westrum, jun., and G. C. Sinke, ref. 77, p. 499.
[89] J. P. McCullough, S. Sunner, H. L. Finke, W. N. Hubbard, M. E. Gross, R. E. Pennington, J. F. Messerly, W. D. Good, and G. Waddington, *J. Amer. Chem. Soc.*, 1953, **75**, 5075.
[90] R. E. Pennington, H. L. Finke, W. N. Hubbard, J. F. Messerly, F. R. Frow, I. A. Hossenlopp, and G. Waddington, *J. Amer. Chem. Soc.*, 1956, **78**, 2055.
[91] D. R. Lide and D. E. Mann, *J. Chem. Phys.*, 1957, **27**, 868.

is excellent. Fair agreement is obtained between the calorimetric (4.18 kJ mol^{-1})[85] and microwave [(4.86 ± 0.12) kJ mol^{-1}][92] values of the barrier height in acetaldehyde. In general, spectroscopic, and particularly microwave, values of barrier heights are more accurate than calorimetric values, and should be used when available. Where no appropriate calorimetric or spectroscopic data are available, such as for but-1-yne, an approximate value of the barrier height may be estimated by analogy with selected compounds.[93]

Restricted Internal Rotation of Several Symmetric Tops. The tables of thermodynamic functions for an internal rotation of a single symmetric top[41, 61] may be used for several symmetric tops [with moments of inertia calculated from equation (20)] provided both potential energy and kinetic energy cross-terms between the tops can be neglected. Both assumptions have been generally made in calculations for molecules with several tops. Where there are reliable calorimetric data at one or more temperatures, the tables have been used to calculate appropriate potential barriers. Using this procedure thermodynamic contributions have been calculated for propane,[56, 94] 2-methylpropane,[95, 96] 2,2-dimethylpropane,[56, 97] *cis*-but-2-ene,[54] *trans*-but-2-ene,[84] isobutene,[84] *o*-xylene,[57, 58] *m*-xylene,[57, 58] *p*-xylene,[57, 58] 1,2,3-trimethylbenzene,[58, 98, 99] 1,2,4-trimethylbenzene,[58, 98, 99] dimethyl sulphide,[100] 2-chloro-2-methylpropane,[101, 102] and dimethylamine.[103] In several cases thermodynamic contributions have been calculated using potential barriers estimated from those of related molecules. Examples of this procedure are found in calculations for 2-fluoro-2-methylpropane,[104] 2-chloropropane,[105] 2-bromopropane,[106] 2-iodopropane,[107] 2,2-dichloropropane,[108] 2-bromo-2-methylpropane,[109] 2-iodo-2-methylpropane,[110] 1,3,5-trimethylbenzene,[58] 1,2,3,5-tetramethylbenzene,[99] 1,2,4,5-tetramethylbenzene,[99] 1,2,3,4-tetramethylbenzene,[99] pentamethylbenzene,[99] and hexamethylbenzene.[99]

[92] R. W. Kilb, C. C. Lin, and E. B. Wilson, *J. Chem. Phys.*, 1957, **26**, 1695.
[93] D. D. Wagman, J. E. Kilpatrick, K. S. Pitzer, and F. D. Rossini, *J. Res. Nat. Bur. Stand.*, 1945, **35**, 467.
[94] K. S. Pitzer, *J. Chem. Phys.*, 1944, **12**, 310.
[95] J. G. Aston, R. M. Kennedy, and S. C. Schuman, *J. Amer. Chem. Soc.*, 1940, **62**, 2059.
[96] K. S. Pitzer, *Chem. Rev.*, 1940, **27**, 39.
[97] K. S. Pitzer and J. E. Kilpatrick, *Chem. Rev.*, 1946, **39**, 435.
[98] R. D. Taylor, B. H. Johnson, and J. E. Kilpatrick, *J. Chem. Phys.*, 1955, **23**, 1225.
[99] S. H. Hastings and D. E. Nicholson, *J. Phys. Chem.*, 1957, **61**, 730.
[100] J. P. McCullough, W. N. Hubbard, F. R. Frow, I. A. Hossenlopp, and G. Waddington, *J. Amer. Chem. Soc.*, 1957, **79**, 561.
[101] K. E. Howlett, *J. Chem. Soc.*, 1951, 1409.
[102] K. E. Howlett, *J. Chem. Soc.*, 1955, 1784.
[103] J. G. Aston, M. L. Eidinoff, and W. S. Forster, *J. Amer. Chem. Soc.*, 1939, **61**, 1539.
[104] P. P. Rodionov and A. A. Zenkin, *Russ. J. Phys. Chem.*, 1968, **42**, 1622.
[105] D. R. Stull, E. F. Westrum, jun., and G. C. Sinke, ref. 77, p. 519.
[106] D. R. Stull, E. F. Westrum, jun., and G. C. Sinke, ref. 77, p. 541.
[107] D. R. Stull, E. F. Westrum, jun., and G. C. Sinke, ref. 77, p. 553.
[108] D. R. Stull, E. F. Westrum, jun., and G. C. Sinke, ref. 77, p. 521.
[109] D. R. Stull, E. F. Westrum, jun., and G. C. Sinke, ref. 77, p. 543.
[110] D. R. Stull, E. F. Westrum, jun., and G. C. Sinke, ref. 77, p. 554.

Only a few cases have been reported where spectroscopically determined barriers have been used in conjunction with the tables[41, 61] to calculate thermodynamic properties. These include trimethylamine,[111] dimethyl ether,[112, 113] and acetone.[114]

The significance of kinetic and potential energy cross-terms between rotating tops has been discussed.[41] It has been suggested that since kinetic energy cross-terms are likely to result in both positive and negative perturbations of energy levels, the overall effect on thermodynamic functions is likely to be small.[41] Pitzer has attempted to estimate the magnitude of the potential energy cross-terms in dimethyl ether and propane and he has concluded that these terms are probably small, but not negligible.[112] This conclusion should be considered with caution, however, since an incorrect assignment was used for dimethyl ether, and incorrect reduced moments were used in the calculation for propane.

An interesting approximate procedure has been developed by Pitzer and Gwinn for calculating internal rotation contributions for molecules having significant potential energy cross-terms.[41] The internal rotation partition function q_{IR} for a molecule having N internal rotors attached to a rigid frame is given approximately by

$$q_{\text{IR}} = q^{\text{IR}}_{\text{class}}(q^{\text{IR}}_{\text{HO,quant}}/q^{\text{IR}}_{\text{HO,class}}), \tag{22}$$

where $q^{\text{IR}}_{\text{class}}$ is the classical partition function for internal rotation, and $q^{\text{IR}}_{\text{HO}}$ is the harmonic oscillator partition function of either quantum mechanical or classical form.

By developing the treatment described by Eidinoff and Aston for the classical partition function for internal rotation[43] as

$$q^{\text{IR}}_{\text{class}} = P_{\text{class}}(2\pi kT/h^2)^{N/2} \prod_{m=1}^{N} I_m^{1/2}, \tag{23}$$

where

$$P_{\text{class}} = \int_0^{2\pi/n_1} \ldots \int_0^{2\pi/n_N} e^{-V/kT}\, d\phi_1 \ldots d\phi_N,$$

simple approximate expressions for contributions to thermodynamic functions may be obtained. These expressions are satisfactory provided each torsional fundamental wavenumber $\tilde{\nu}_{\text{tors}}$ is sufficiently low that $hc\tilde{\nu}_{\text{tors}}/kT$ is not much greater than one.

Asymmetric Tops attached to a Rigid Frame.—The calculation of thermodynamic properties of molecules containing asymmetric tops attached to a rigid frame is in general complicated by the change of molecular moments of inertia on internal rotation. Furthermore, the restricting potential for an asymmetric top is likely to be more complex than for a symmetric top.

111 J. G. Aston, M. L. Sagenkahn, G. J. Szasz, G. W. Moessen, and H. F. Zuhr, *J. Amer. Chem. Soc.*, 1944, **66**, 1171.
112 K. S. Pitzer, *J. Chem. Phys.*, 1942, **10**, 605.
113 D. R. Stull, E. F. Westrum, jun., and G. C. Sinke, ref. 77, p. 412.
114 R. E. Pennington and K. A. Kobe, *J. Amer. Chem. Soc.*, 1957, **79**, 300.

In spite of these complexities, several approximate methods for calculating thermodynamic properties for these types of molecule have been developed.[39, 115–122]

Pitzer has divided asymmetric tops into the following three classes: those that are 'balanced', those that have a small 'off-balance factor', and those that have a large 'off-balance factor', where the term 'off-balance factor' U_m is defined below.[115] Examples of the first class are phenyl and nitro-groups. These tops are balanced, since for each the internal rotation axis passes through the centre of gravity of the top. The second class, having a small 'off-balance factor', includes hydroxy-, thiol, and amino-groups. Here the centre of gravity of the top is only slightly displaced from the internal rotation axis. The overall moments of inertia of molecules having balanced or slightly unbalanced tops are expected to change only slightly with internal rotation. Since calculated values of thermodynamic properties are not very sensitive to small changes in molecular moments of inertia, these changes may be neglected and contributions calculated for the molecular equilibrium configuration.

The third class having large 'off-balance factors', include $-CHO$, $-CH_2Cl$, $-CHBr_2$, $-COCl$, $-SCl$, and $-NO$. Here the moments of inertia change significantly with internal rotation, and are too large to dismiss. In the classical approximation, however, change in the overall rotational partition function due to change of molecular moments of inertia is cancelled by change in the vibrational partition function, provided that there are no potential energy cross-terms between internal rotation and vibrational motion.[115] Although molecules generally have several high-frequency vibrational modes, that are non-classical, it has been found that usually only the low-frequency fundamentals, associated with skeletal bending, change significantly with internal rotation. Thus for all three classes of asymmetric top, as a first approximation the rotational and vibrational partition functions may be calculated using moments of inertia and vibrational wavenumbers for the molecular equilibrium configuration. The rotational partition function can be factored into overall and internal rotation partition functions [equation (16)] using the reduced moment of inertia of the top I_m as defined below.

The reduced moment of inertia (I_m) of a single asymmetric top attached to a rigid frame is given by[115]

$$I_m = A_m - \Lambda_{mm} = A_m - \sum_{(i=a,b,c)} \{(\alpha_m^{iy} U_m)^2/M + (\beta_m^i)^2/I_i\}, \qquad (24)$$

[115] K. S. Pitzer, *J. Chem. Phys.*, 1946, **14**, 239.
[116] K. S. Pitzer, *J. Chem. Phys.*, 1940, **8**, 711.
[117] W. D. Gwinn and K. S. Pitzer, *J. Chem. Phys.*, 1948, **16**, 303.
[118] J. G. Aston, G. Szasz, H. W. Woolley, and F. G. Brickwedde, *J. Chem. Phys.*, 1946, **14**, 67.
[119] J. O. Halford, *J. Chem. Phys.*, 1948, **16**, 410.
[120] J. O. Halford, *J. Chem. Phys.*, 1948, **16**, 560.
[121] D. W. Scott and J. P. McCullough, *U.S. Bur. Mines, Report Invest.*, 1962, 5930.
[122] D. Smith, P. J. Devlin, and D. W. Scott, *J. Mol. Spectroscopy*, 1968, **25**, 174.

where M is the mass of the whole molecule, I_i is the ith principal moment of inertia of the whole molecule, and α_m^{iy} is the direction cosine between the ith principal axis of the whole molecule and the y axis of the top (m). The axes (x, y, z) of the top are defined such that the rotation axis is taken as z, the x axis passes through the centre of gravity of the top, and the y axis is perpendicular to x and z. Both the top axes (x, y, z) and the molecular axes ($i = a, b, c$) are uniformly right-handed or left-handed. Finally

$$\beta_m^i = \alpha_m^{iz}A_m - \alpha_m^{ix}B_m - \alpha_m^{iy}C_m + U_m(\alpha_m^{i-1,y}r_m^{i+1} - \alpha^{i+1,y}r_m^{i-1}),$$

where the superscripts $i - 1$ and $i + 1$ refer to cyclic shifts of the principal axes (a, b, c), *i.e.* for $i = a$, then $i - 1 = c$; and for $i = c$, then $i + 1 = a$; and $\mathbf{r}_m$ is the vector from the centre of gravity of the molecule to the centre of the co-ordinates of the top, having components r_m^a, r_m^b, and r_m^c on the principal axes.

The moments A_m, B_m, C_m, and U_m for the top (m) are for the top moment about z:

$$A_m = \Sigma_j m_j(x_j^2 + y_j^2),$$

for the xz product of inertia:

$$B_m = \Sigma_j m_j x_j z_j,$$

for the yz product of inertia:

$$C_m = \Sigma_j m_j y_j z_j,$$

and for the 'off-balance' factor:

$$U_m = \Sigma_j m_j x_j,$$

where m_j is the mass of the jth atom of the top.

The first approximation to the reduced moment for a molecule containing several tops attached to a rigid frame is given by equation (24).[115] The next approximation is given by equation (20) where

$$\Lambda_{mm'} = \sum_{i=a,b,c} [\{\alpha_m^{iy}\alpha_{m'}^{iy} U_m U_{m'}\}/m + \{\beta_m^i\beta_{m'}^i\}/I_i], \tag{25}$$

and I_m° is given as the value in equation (24).

The potential energy function for internal rotation of an asymmetric top is generally more complex than that for a symmetric top. In addition, for molecules having several tops the potential is further complicated by cross-terms between the tops. In practical applications, however, the cross-terms have been invariably neglected, and so thermodynamic properties have been calculated by considering contributions from the several tops separately.

Thermodynamic calculations have been reported for several molecules containing both symmetric and asymmetric tops attached to a rigid frame. With the exception of propane-2-thiol,[122] contributions from the internal rotation of symmetric tops have been calculated using procedures discussed on p. 288 and they are not further discussed here. Barrier heights for the

—SH and one —CH_3 internal rotation in propane-2-thiol were determined from i.r. data, but agreement between calculated and experimental values of entropy and heat capacity could not be obtained by fitting a 3-fold cosine potential to the second methyl torsion.[122] This was attributed to interaction between methyl groups. Satisfactory agreement between the calculated and experimental data was obtained by arbitrarily treating the second methyl torsion as an anharmonic oscillator having fundamental wavenumber $\tilde{\nu}$ = 257 cm^{-1} and anharmonicity coefficient $X = -6.3$ cm^{-1} (see p. 279).[122]

Free Internal Rotation of an Asymmetric Top. Contributions to thermodynamic functions from asymmetric tops having free internal rotation may be calculated by using reduced moments obtained from equations (24) or (20) and (25) in the practical equations in Table 4 (p. 282). Few examples of asymmetric tops with free rotation have been reported. Free, or nearly free, rotation in biphenyl has been postulated[123] on the basis of fair agreement between calculated and experimental values of its entropy at 528.39 K. The vibrational assignment of biphenyl, however, is not completely satisfactory. Free rotation has also been postulated for benzenethiol on the basis of agreement between experimental and calculated values of entropy and heat capacity,[124] though a more recent calculation[125] using a slightly vibrational assignment is consistent with a 2-fold cosine barrier height of 0.8 kJ mol^{-1}.

Internal Rotation of an Asymmetric Top restricted by an n-*Fold Potential Barrier.* A Fourier cosine series expansion of the internal rotation potential for a top having 2-fold symmetry about the axis of rotation may be expressed as:

$$V_m(\phi) = (1/2)V_{m,2}(1 - \cos 2\phi_m) + (1/2)V_{m,4}(1 - \cos 4\phi_m) + \ldots$$

It has been generally assumed in thermodynamic calculations that only the leading term is significant, and that other terms may be neglected. On this assumption the tables of thermodynamic functions for symmetric tops[41, 61] may be used. This procedure has been followed, using values of $V_{m,2}$ from spectroscopic measurements, in calculations for phenol,[126, 127] aniline,[128] hydrogen nitrate,[129] and methyl nitrate.[130] Calculations for styrene[131] and ethylbenzene[58] have been reported using values of $V_{m,2}$ determined by fitting calculated to calorimetric values of entropy.

[123] J. E. Katon and E. R. Lippincott, *Spectrochim. Acta*, 1959, **15**, 627.
[124] D. W. Scott, J. P. McCullough, W. N. Hubbard, J. F. Messerly, I. A. Hossenlopp, F. R. Frow, and G. Waddington, *J. Amer. Chem. Soc.*, 1956, **78**, 5463.
[125] M. Z. El-Sabban and D. W. Scott, *U.S. Bur. Mines, Bull.*, 1970, 654.
[126] J. C. Evans, *Spectrochim. Acta*, 1960, **16**, 1382.
[127] J. H. S. Green, *J. Chem. Soc.*, 1961, 2236.
[128] W. E. Hatton, D. L. Hildenbrand, G. C. Sinke, and D. R. Stull, *J. Chem. and Eng. Data*, 1962, **7**, 229.
[129] A. Palm and M. Kilpatrick, *J. Chem. Phys.*, 1955, **23**, 1562.
[130] D. R. Stull, E. F. Westrum, jun., and G. C. Sinke, ref. 77, p. 483.
[131] K. S. Pitzer, L. Guttman, and E. F. Westrum, jun., *J. Amer. Chem. Soc.*, 1946, **68**, 2209.

Some of the few $V_{m,2}$ values that have been reported are listed in Table 8. The large values of $V_{m,2}$ for a few of the molecules in Table 8 are of chemical

Table 8 *Some values of $V_{m,2}$ for internal rotation of 2-fold symmetric tops*

Molecule	$V_{m,2}$/kJ mol^{-1}	*Ref.*
$C_6H_5CH_2CH_3$	4.5	58
$C_6H_5CH{=}CH_2$	9.2	131
C_6H_5OH	14.1	126, 127
$C_6H_5NH_2$	14.4	128
C_6H_5CHO	19.5	132
CH_3ONO_2	38.1	130
$HONO_2$	40.8	129

interest since they indicate partial double-bond character between the top and the frame.

Molecules having an asymmetric top with no symmetry about the internal rotation axis generally consist of geometrically distinct conformers. If the conformers have very similar energies, however, the potential hindering internal rotation may be approximated to a simple n_m-fold cosine function [equation (21)], and again the tables for symmetric tops [41, 61] may be used to calculate thermodynamic properties. To determine $1/q_f^m$ for use in these tables, $\sigma_{IR,m}$ in equation (17) should be replaced by n_m so that the correct energy levels are used.[115] But since the rotating asymmetric top has symmetry number 1 and not n_m, the value of the partition function calculated using the tables is too small by a factor $(1/n_m)$. Thus $Lk\ln(n_m)$ should be added to values of $S^{\ominus}(T)$ and $\{G^{\ominus}(T) - H^{\ominus}(0)\}/T$ computed from the tables in order to obtain correct values for these functions.[115]

The potentials hindering internal rotation of —OH and —SH groups in aliphatic alcohols and thiols, respectively, have generally been assumed to have 3-fold symmetry. The appropriate barrier heights in ethanol,[133–135] isopropanol,[136] 2-methylpropan-2-ol,[137] propane-2-thiol,[122] and 2-methylpropane-2-thiol,[138] have been determined by fitting calculated and experimental values of entropy and heat capacity. In a calculation of 2-mercaptopropionic acid a 3-fold barrier height for the —SH rotation was assumed by analogy with related compounds.[79]

Internal rotation of hydroxy-group against an aromatic frame has been assumed to be restricted by a potential with 2-fold symmetry. The —OH internal rotation barriers in three cresols have been assumed [139] to be the

[132] W. G. Fateley, R. K. Harris, F. A. Miller, and R. E. Witkowski, *Spectrochim. Acta*, 1965, **21**, 231.
[133] J. O. Halford, *J. Chem. Phys.*, 1949, **17**, 111.
[134] G. M. Barrow, *J. Chem. Phys.*, 1952, **20**, 1739.
[135] J. H. S. Green, *Trans. Faraday Soc.*, 1961, **57**, 2132.
[136] J. H. S. Green, *Trans. Faraday Soc.*, 1963, **59**, 1559.
[137] E. T. Beynon and J. J. McKetta, *J. Phys. Chem.*, 1963, **67**, 2761.
[138] J. P. McCullough, D. W. Scott, H. L. Finke, W. N. Hubbard, M. E. Gross, C. Katz, R. E. Pennington, J. F. Messerly, and G. Waddington, *J. Amer. Chem. Soc.*, 1953, **75**, 1818.
[139] J. H. S. Green, *Chem. and Ind.*, 1962, 1575.

same as in phenol.[140] 2-Fold barriers have also been assumed in calculations for sulphur monochloride [141, 142] and methyl nitrite.[143]

Internal Rotation of an Asymmetric Top Restricted by a Complex Potential Barrier. Simple general procedures for calculating thermodynamic contributions from internal rotation of asymmetric tops restricted by a potential more complex than a simple cosine function have not been developed. Nevertheless, several calculations using appropriate approximations have been reported, and are discussed below.

For light rotors it may be possible to observe sufficient torsional transitions for the internal rotation partition function to be determined by direction summation (p. 271). If only a few low-lying torsional levels are observed, they may be fitted to a suitable potential from which higher torsional energy levels can be computed. Thermodynamic functions of hydrogen peroxide calculated by use of this procedure give a value of $S^{\ominus}(298.15\text{ K})$ in excellent agreement with the calorimetric value.[144]

For heavy asymmetric rotors the torsional levels may be sufficiently low for thermodynamic properties to be estimated by using the classical approximation. Gwinn and Pitzer have expressed the classical approximation to the complete internal partition function as [117]

$$q^{\text{internal}} = \int_0^{2\pi} \{q_{\text{F}}^{\text{internal}}(\phi)/2\pi\}\exp\{-V(\phi)/LkT\}\,\text{d}\phi, \qquad (26)$$

where $q_{\text{F}}^{\text{internal}}(\phi)$ is identical in form to the internal partition function for the molecule with free internal rotation. Thus

$$q_{\text{F}}^{\text{internal}}(\phi) = q_{\text{OR}}(\phi) \times q_{\text{f}}(\phi) \times q_{\text{vib}}(\phi). \qquad (27)$$

Although $q_{\text{F}}^{\text{internal}}(\phi)$ is a function of ϕ, within the classical approximation change in $q_{\text{F}}^{\text{internal}}(\phi)$ due to change in the molecule moments of inertia on internal rotation is expected to be cancelled by the change in $q_{\text{F}}^{\text{internal}}(\phi)$ due to change in the vibrational wavenumber (see p. 290).[115] Thus, to a first approximation, $q_{\text{F}}^{\text{internal}}(\phi)$ is independent of ϕ, and so the internal partition function may be expressed as [117]

$$q^{\text{internal}} = \{q_{\text{F}}^{\text{internal}}(\phi)/2\pi\}\int_0^{2\pi} \exp\{-V(\phi)/LkT\}\,\text{d}\phi. \qquad (28)$$

Equation (28) has been used to calculate thermodynamic functions for 1,2-dichloroethane,[117] 1,1,2-trichloroethane,[145] and but-1-ene.[84] For each molecule a potential function was selected to give satisfactory agreement between calculated and calorimetric values of entropy and heat capacity.

[140] T. Kojima, *J. Phys. Soc. Japan*, 1960, **15**, 284; *cf.* T. Pedersen, N. W. Larsen, and L. Nygard, *J. Mol. Struct.*, 1969, **4**, 59.
[141] N. W. Luft and K. H. Todhunter, *J. Chem. Phys.*, 1953, **21**, 2225.
[142] K. K. Kelley and E. G. King, *U.S. Bur. Mines, Bull.*, 1961, 592.
[143] P. Gray and M. W. T. Pratt, *J. Chem. Soc.*, 1958, 3403.
[144] H. J. Spangenberg and L. Zülicke, *Z. phys. Chem.* (*Frankfurt*), 1967, **52**, 336.
[145] R. H. Harrison and K. A. Kobe, *J. Chem. Phys.*, 1957, **26**, 1411.

A breakdown of part of the calculation for 1,2-dichloroethane is given in Table 9. The potential selected to give best fit to calorimetric and dipole moment data is [117]

$$V/\text{kJ mol}^{-1} = 3.83(1 - \cos\phi) + 4.81(1 - \cos 3\phi), \quad (0 \leqslant \phi \leqslant 130°),$$

$$V = \infty, \quad (180° \geqslant \phi > 130°),$$

where $\phi = 0$ for the *trans* conformation.

able 9 *Calculated and observed thermodynamic properties of* 1,2-*dichloroethane* [117]

	$S^{\ominus}(298.15\text{ K})/\text{J K}^{-1}\text{ mol}^{-1}$	$S^{\ominus}(1000\text{ K})/\text{J K}^{-1}\text{ mol}^{-1}$	$C_p^{\ominus}(428\text{ K})/\text{J K}^{-1}\text{ mol}^{-1}$	$C_p^{\ominus}(1000\text{ K})/\text{J K}^{-1}\text{ mol}^{-1}$
ranslational contribution	166.05	191.20	20.79	20.79
verall rotational contribution	103.10	118.19	12.47	12.47
estricted internal rotational contribution	20.95	32.44	11.30	5.75
ibrational contribution	18.76	97.20	50.21	99.19
otal calculated	308.86	439.03	94.77	138.20
bserved	308.19		94.93	

To simplify the problem of determining approximate potential functions for the internal rotation of asymmetric tops, Scott and McCullough have tabulated contributions to thermodynamic properties calculated within the classical approximation for potential functions consisting of segments of $(1 - \cos 3\phi)$ curves.[121] In certain favourable cases these tables can be used to derive potential functions that are consistent with calorimetric data.

Pitzer and Gwinn have considered calculations for molecules having potential energy cross-terms between vibrational and internal rotational motion.[117] Here $q_{\text{F}}^{\text{internal}}(\phi)$ is not, even to a first approximation, independent of ϕ. However, $q_{\text{F}}^{\text{internal}}(\phi)$ may be approximated by a step function with different values for the different observable molecular conformations.[117] The internal rotation partition function for 1,2-dibromoethane, for example, which has been observed in *trans* and *gauche* conformations, may be expressed as [117]

$$q^{\text{internal}} = \{q_{\text{F}}^{\text{internal}}(trans)/\pi\}\left[\int_0^{\pi/3} \exp\{-V(\phi)/LkT\}\,\mathrm{d}\phi + \omega\int_{\pi/3}^{\pi} \exp\{-V(\phi)/LkT\}\,\mathrm{d}\phi\right],$$

where

$$\omega = \{q_{\text{F}}^{\text{internal}}(gauche)\}/\{q_{\text{F}}^{\text{internal}}(trans)\}.$$

Where the classical approximation is invalid, and where insufficient energy levels are known for thermodynamic properties to be computed by direct summation, alternative computational procedures must be adopted. These useful approximations are discussed below.

Halford has developed a method for using the Wilson and Sommerfeld quantization rule to determine energy levels for the calculation of thermodynamic functions.[146, 147] Although this procedure does not give the correct energy levels, by adopting additional conditions values of thermodynamic properties may be calculated with an accuracy comparable to that obtained using the symmetric top tables.[41, 61] This treatment has been extended to two types of asymmetric potential: one is a double minimum potential having maxima of equal energy but unequal minima;[120] the other potential has two equal minima but unequal maxima.[119] For both potentials, tables of thermodynamic functions have been computed as functions of $1/q_f$, V_1/LkT, and V_2/LkT, where V_1 and V_2 are the barrier heights.[119, 120]

A procedure has been proposed by Pitzer,[56] and developed by Aston and co-workers,[118] where thermodynamic properties of a molecule with an asymmetric top are calculated by considering the molecule to consist of an equilibrium mixture of conformers. If the molecule can be considered to consist of two conformers A and B, with B having the lower energy, then the molecular thermodynamic functions can be expressed as contributions from A, B, and their mixing, giving [118]

$$\begin{aligned}\{H^{\ominus}(T) - H^{\ominus}(0)\}/T = {} & \chi_A[\{H^{\ominus}_A(T) - H^{\ominus}_A(0)\}/T + \{H^{\ominus}_A(0) - H^{\ominus}_B(0)\}/T] \\ & + \chi_B\{H^{\ominus}_B(T) - H^{\ominus}_B(0)\}/T, \qquad (29)\end{aligned}$$

$$\begin{aligned}\{G^{\ominus}(T) - H^{\ominus}(0)\}/T = {} & \chi_A[\{G^{\ominus}_A(T) - H^{\ominus}_A(0)\}/T + \{H^{\ominus}_A(0) - H^{\ominus}_B(0)\}/T] \\ & + \chi_B\{G^{\ominus}_B(T) - H^{\ominus}_B(0)\}/T \\ & + Lk\{\chi_A \ln \chi_A + \chi_B \ln \chi_B\}, \qquad (30)\end{aligned}$$

$$\begin{aligned}S^{\ominus}(T) = {} & \chi_A S^{\ominus}_A(T) + \chi_B S^{\ominus}_B(T) \\ & - Lk\{\chi_A \ln \chi_A + \chi_B \ln \chi_B\}, \qquad (31)\end{aligned}$$

$$\begin{aligned}C^{\ominus}_p(T) = {} & \chi_A C^{\ominus}_{p,A}(T) + \chi_B C^{\ominus}_{p,B}(T) \\ & + (\chi_A\chi_B/Lk)[\{H^{\ominus}_A(T) - H^{\ominus}_A(0)\}/T \\ & + \{H^{\ominus}_A(0) - H^{\ominus}_B(0)\}/T \\ & - \{H^{\ominus}_B(T) - H^{\ominus}_B(0)\}/T]^2, \qquad (32)\end{aligned}$$

where χ_A and χ_B are the mole fractions of A and B, respectively, and

$$Lk \ln(\chi_A/\chi_B) = - [\{G^{\ominus}_A(T) - H^{\ominus}_B(0)\}/T - \{G^{\ominus}_B(T) - H^{\ominus}_B(0)\}/T].$$

Application of this procedure requires knowledge of the partition functions for each molecular conformer, and the energy differences between the conformers. Calculations have been reported for hydrazine,[148] 1-chloropropane,[74] 1-bromopropane,[74] and buta-1,3-diene.[118] For each of these molecules, the calculation was simplified by assuming that the

[146] J. O. Halford, *J. Chem. Phys.*, 1947, **15**, 645.
[147] J. O. Halford, *J. Chem. Phys.*, 1950, **18**, 444.
[148] D. W. Scott, G. D. Oliver, M. E. Gross, W. N. Hubbard, and H. M. Huffman, *J. Amer. Chem. Soc.*, 1949, **71**, 2293.

internal partition function $q_{F}^{internal}(\phi)$ is the same for both conformers. The energy difference between the conformers in the halogenopropanes was determined spectroscopically.[74] Only one (*trans*) conformer has been observed for buta-1,3-diene. Satisfactory agreement between experimental and calculated values of entropy and heat capacity could be obtained, however, only if a second (*cis*) conformer is postulated to have an energy about 9.6 kJ mol^{-1} higher than that of the *trans*-conformer.[118]

A third procedure for calculating approximate thermodynamic properties for internal rotation of an asymmetric top has been proposed.[122] This consists of approximating the torsional energy levels of the asymmetric rotor to the levels of a slightly perturbed symmetric top. It is expected to be accurate only for molecules that have small differences between the potential minima of the different conformers. This procedure has been used to calculate thermodynamic properties of ethanethiol, which is believed to exist in two energetically equivalent C_1 conformers and a slightly higher energy C_s conformer.[122] The first approximation to the energy levels of the thiol torsion is obtained using a 3-fold cosine potential with V_3 = 5.94 kJ mol^{-1}, as determined from the thiol torsional fundamental 191 cm^{-1}. A small perturbation of the A levels by an amount ΔE, the energy difference between the conformers, allows for the higher energy of the C_s conformer. Calculation of thermodynamic properties by direct summation gives[122] agreement with calorimetric values of entropy and heat capacity on the assumption that $\Delta E = 1.2$ kJ mol^{-1}.

Compound Rotation.—Procedures for calculating appropriate moments of inertia for molecules having rotating groups attached to a rotating group, thus having compound rotation, are more complicated than the methods discussed above for rotating tops attached to a rigid frame. A general formalized procedure for calculating elements of the kinetic energy matrix for compound rotation has been described by Kilpatrick and Pitzer,[149] however, and it has been used extensively. The reduced moments of inertia for rotating tops attached to a rigid frame, given by equations (18), (20), and (24), are also obtained from the elements of this rotational matrix.

The form of the rotational kinetic energy matrix (S) for a molecule having N internal rotations is $(3 + N) \times (3 + N)$, originating from the 3 overall rotational co-ordinates and the N internal rotational co-ordinates. In general, some or all of the elements of the matrix (S) may vary with internal rotation, but to a first approximation for thermodynamic purposes the matrix may be computed for the molecular equilibrium configuration. Within this approximation, appropriate diagonalization about the elements for overall rotation leads to the $(N \times N)$ internal rotational matrix (D).[149]

If off-diagonal elements in (D) are small compared with the diagonal elements, such that the square root of the determinant of (D) is not changed significantly by omission of the off-diagonal elements, the diagonal elements

[149] J. E. Kilpatrick and K. S. Pitzer, *J. Chem. Phys.*, 1949, **17**, 1064.

may be taken as the appropriate reduced moments of inertia.[149] In the calculations discussed below this approximation has been generally adopted. If the off-diagonal elements in (D) are not small compared with the diagonal elements, it has been suggested that the approximate treatment described by equations (22) and (23) may be followed, using the determinant of the matrix (D) in place of I_m.

Many applications of Kilpatrick and Pitzer's procedure for calculating thermodynamic properties of molecules with compound rotation have been reported. In all cases possible potential energy cross-terms between rotating tops have been neglected. Contributions from internal rotation of symmetric tops have been calculated using the appropriate tables.[41, 61] These tables have also been used in calculations for the internal rotation of asymmetric tops hindered by a simple n-fold cosine potential. 3-Fold potential barriers have been assumed in calculations for the —OH rotations in propanol[150] and 1-methylpropanol,[151] the —SH rotations in propane-1-thiol,[125, 152] butane-2-thiol,[125, 153] 2-methylpropane-1-thiol,[125, 154] and 2-methylbutane-2-thiol,[125, 155] the C—S skeletal rotations in ethyl methyl sulphide,[125, 156] diethyl sulphide,[125, 157] isopropyl methyl sulphide,[125, 158] and t-butyl methyl sulphide,[125, 159] and the C—C skeletal rotations in 2,3-dimethylbutane,[160] and 2-methylpropane-1-thiol.[154] 2-Fold cosine potential barriers have been assumed in calculations in the S—S skeletal rotations in dimethyl disulphide[125, 161] and diethyl disulphide.[125, 161]

The classical approximation described on p. 294 by equations (26), (27), and (28) has been used for calculating thermodynamic contributions for the C—C skeletal rotation in 2-methylbutane,[160] and for the C—S skeletal rotation in diethyl disulphide,[161] by using assumed restricting potentials. Contributions from the skeletal torsion in 2-methylbutane-2-thiol[155] and

150 J. F. Matthews and J. J. McKetta, *J. Phys. Chem.*, 1961, **65**, 758.
151 N. S. Berman and J. J. McKetta, *J. Phys. Chem.*, 1962, **66**, 1444.
152 R. E. Pennington, D. W. Scott, H. L. Finke, J. P. McCullough, J. F. Messerly, I. A. Hossenlopp, and G. Waddington, *J. Amer. Chem. Soc.*, 1956, **78**, 3266.
153 J. P. McCullough, H. L. Finke, D. W. Scott, R. E. Pennington, M. E. Gross, J. F. Messerly, and G. Waddington, *J. Amer. Chem. Soc.*, 1958, **80**, 4786.
154 D. W. Scott, J. P. McCullough, J. F. Messerly, R. E. Pennington, I. A. Hossenlopp, H. L. Finke, and G. Waddington, *J. Amer. Chem. Soc.*, 1958, **80**, 55.
155 D. W. Scott, D. R. Douslin, H. L. Finke, W. N. Hubbard, J. F. Messerly, I. A. Hossenlopp, and J. P. McCullough, *J. Phys. Chem.*, 1962, **66**, 1334.
156 D. W. Scott, H. L. Finke, J. P. McCullough, M. E. Gross, K. D. Williamson, G. Waddington, and H. M. Huffmann, *J. Amer. Chem. Soc.*, 1951, **73**, 261.
157 D. W. Scott, H. L. Finke, W. N. Hubbard, J. P. McCullough, G. D. Oliver, M. E. Gross, C. Katz, K. D. Williamson, G. Waddington, and H. M. Huffman, *J. Amer. Chem. Soc.*, 1952, **74**, 4656.
158 J. P. McCullough, H. L. Finke, J. F. Messerly, R. E. Pennington, I. A. Hossenlopp, and G. Waddington, *J. Amer. Chem. Soc.*, 1955, **77**, 6119.
159 D. W. Scott, W. D. Good, S. S. Todd, J. F. Messerly, W. T. Berg, I. A. Hossenlopp, J. L. Lacina, A. Osborn, and J. P. McCullough, *J. Chem. Phys.*, 1962, **36**, 406.
160 D. W. Scott, J. P. McCullough, K. D. Williamson, and G. Waddington, *J. Amer. Chem. Soc.*, 1951, **73**, 1707.
161 W. N. Hubbard, D. R. Douslin, J. P. McCullough, D. W. Scott, S. S. Todd, J. F. Messerly, I. A. Hossenlopp, A. George, and G. Waddington, *J. Amer. Chem. Soc.*, 1958, **80**, 3547.

butan-2-one[162] have been calculated using the tables of Scott and McCullough.

The approximation of internal rotation as an equilibrium of rotational conformers (p. 296) described by equations (29)—(32) has been followed in calculations of contributions from the C—C skeletal torsion in n-propanol,[150] s-butanol,[151] propane-1-thiol,[152] and butane-2-thiol.[153] Contributions for each of the conformers were calculated assuming 3-fold cosine potentials and using the tables of thermodynamic functions for symmetric tops.[41, 61]

The accurate calculation of thermodynamic properties of molecules having more than three internal rotational degrees of freedom is generally restricted by lack of data describing the restricting potential function. To facilitate calculations of internal rotation contributions for more complex molecules, Pitzer has extended the equilibrium of conformers procedure, described on p. 296, to molecules containing several conformers.[116] The restricting potential for a conformer over an internal rotation angle $2\pi/3$ is approximated by

$$\Delta E_i + (V_i/2)(1 - \cos 3\phi)$$

where ΔE_i is the difference in energy between the *i*th conformer and the most stable conformer and V_i is the appropriate sinusoidal barrier height. Integration of this potential in the classical approximation gives a partition function that can be factored into a part that is a function of V_i and ϕ_i (and that may be calculated as discussed above), and a part that is a function of ΔE_i, called the steric partition function. It is given by[116]

$$q_{\text{steric}} = \Sigma \exp(-\Delta E_i/kT),$$

where the sum is over all conformers.

Contributions to thermodynamic properties using this steric partition function have been estimated for paraffins using a simple model of additive steric repulsions.[96, 116, 163, 164] The *trans,trans* conformation of a straight-chain paraffin, for example, is assumed to have the lowest energy, while other conformations are assigned a steric energy na, where n is the number of appropriate steric interactions and a is an energy factor. Similar procedures have been used in the calculation of thermodynamic properties of five- and six-membered ring compounds exhibiting ring tautomerism.[37, 39, 125, 165] Calculations using this procedure have also recently been reported for several organosulphur compounds, using values of ΔE_i that were determined by spectroscopic methods, by fitting calculated values of thermodynamic properties to calorimetric data, or in some cases by comparison with related compounds.[125]

Calculations for complex hydrocarbons, using the above procedures, are restricted by lack of data for calculating not only the internal rotational

[162] G. C. Sinke and F. L. Oetting, *J. Phys. Chem.*, 1964, **68**, 1354.
[163] K. S. Pitzer, *J. Amer. Chem. Soc.*, 1941, **63**, 2413.
[164] K. S. Pitzer and D. W Scott, *J. Amer. Chem. Soc.*, 1941, **63**, 2419.
[165] C. W. Beckett, N. K. Freeman, and K. S. Pitzer, *J. Amer. Chem. Soc.*, 1948, **70**, 422[illegible].

partition function but also the overall rotational and vibrational partition functions. To overcome these difficulties Pitzer has developed an alternative statistical treatment based on a small perturbation from the classical approximation for the partition function.[116] To a first approximation a paraffin is considered to consist of a skeleton having CH_3, CH_2, and CH groups as single units. Contributions from protons are considered as a perturbation. The potential energy of the skeleton includes terms for bond bending and stretching, internal rotation, and steric interactions. The general expression for a thermodynamic function F has the form:

$$F = F_0(T) + N_1(\text{C—C stretch}) + N_2(\text{C—C bend}) + N_3(\text{internal rotation}) + F(\text{steric}) + F(\sigma) + N_4(\text{CH}_3) + N_5(\text{CH}_2) + N_6(\text{CH}),$$

where $F_0(T)$ is a function of T and does not depend on the molecule, (C—C stretch) is a function of the C—C stretching force constant, (C—C bend) is a function of the C—C bending force constant, (internal rotation) is a function of a potential barrier V_0, F(steric) is a function of the energy of various conformers, $F(\sigma)$ is a function of the symmetry number σ, N_1, N_2, and N_3 are the numbers of C—C stretching, C—C bending, and internal rotation degrees of freedom, respectively, and N_4, N_5, and N_6 are the number of CH_3, CH_2, and CH groups, respectively.

The evaluation of these terms has been described by Pitzer[116] and has been reviewed.[96, 166] The resulting practical formulae have been revised using more recent experimental data.[167] This procedure has been used to calculate the thermodynamic properties of many linear and branched paraffins. It forms the basis of calculations using incremental methods, which are not discussed here.

4 Molecules exhibiting Inversion

Molecular inversion involves tunnelling between two equivalent configurations that are related by a planar reflection of one or more atoms. Inversion occurs for all non-planar molecules, but it is of thermodynamic significance only for the few molecules, such as ammonia and its derivatives, that have low inversion barriers.

The internal partition function for molecules having inversion may be factored, to a good approximation, into overall rotational and vibrational partition functions. Although inversion tunnelling results in a splitting of rotational energy levels, the statistical weights are such that the classical formulae for rotational contributions to thermodynamic functions may be used. The appropriate symmetry number depends on the procedure used to calculate the vibrational partition function.

Vibrational energy levels are split into doublets by each type of molecular inversion. In ammonia and its derivatives, however, the vibrational energy

[166] G. J. Janz, 'Estimation of Thermodynamic Properties of Organic Compounds', Academic Press, New York, 1958, p. 35.

[167] W. B. Person and G. C. Pimentel, *J. Amer. Chem. Soc.*, 1953, **75**, 532.

levels of only one of the normal modes are significantly perturbed by inversion at nitrogen. Thermodynamic contributions from the remaining fundamentals may be calculated using the standard procedures described on p. 275.

The calculation of contributions to thermodynamic functions from a double minimum vibrational degree of freedom have been discussed in general terms by Pitzer.[168] On the assumption that the inversion potential is a slightly perturbed parabolic function for low-barrier cases, and that for high-barrier cases it can be expressed by [169]

$$V = (1/2)f(|x| - l)^2,$$

where f is a force constant, x is a spatial co-ordinate, and $2l$ is the separation between the minima along this co-ordinate, then, if the inversion barrier is near or above the vibrational level $v = 2$, the heat capacity at the practical temperatures is very similar to that for an harmonic oscillator having the same 0—1 vibrational transition wavenumber.[168] The entropy is $Lk \ln 2$ larger than that for the analogous harmonic oscillator. Compensation for this factor may be obtained by using the appropriate symmetry number in calculating the rotational contribution to the entropy. Thus, provided the inversion barrier is sufficiently high, thermodynamic contributions from a double minimum vibration may be calculated from the analogous harmonic oscillator provided that the symmetry number is determined by neglecting the possibility of inversion.

If the inversion barrier is too low for the harmonic oscillator approximation to be valid, or if the potential differs significantly from that described above,[168] thermodynamic contributions should be estimated by direct summation of observed and calculated vibrational energy levels (p. 271).

The entropy and Gibbs free energy of ammonia have been determined both by direct summation of observed and calculated vibrational levels, using a symmetry number of 6, and by the harmonic oscillator approximation, using a symmetry number of 3.[170] Good agreement was obtained between values calculated by the two methods. The harmonic oscillator approximation has also been used in calculation of the thermodynamic contributions from double minimum vibrations in methylamine,[80] dimethylamine,[103] trimethylamine,[111] aziridine,[171] hydrazine,[148] and aniline.[128] There is good agreement between the calculated and calorimetric values of entropy and heat capacity of the three methylamines.[80, 103, 111] A recent analysis of the electronic spectrum of aniline,[172] however, suggests that here the harmonic oscillator approximation is incorrect, and so the calculated functions for aniline [128] may be in error.

168 K. S. Pitzer, *J. Chem. Phys.*, 1939, **7**, 251.
169 F. T. Wall and G. Glockler, *J. Chem. Phys.*, 1937, **5**, 314.
170 C. C. Stephenson and H. O. McMahon, *J. Amer. Chem. Soc.*, 1939, **61**, 437.
171 D. R. Stull, E. F. Westrum, jun., and G. C. Sinke, ref. 77, p. 468.
172 J. C. D. Brand, D. R. Williams, and T. J. Cook, *J. Mol. Spectroscopy*, 1966, **20**, 359.

The harmonic oscillator approximation has generally been used in calculations of thermodynamic contributions from inversion at tetrahedral carbon. Inversion in a molecule having four different groups bound to a carbon atom results in the presence of two optical isomers. To calculated values of entropy and Gibbs free energy a contribution $Lk \ln 2$ should be added for each pair of optical isomers.

5 Molecules with Ring Motions

Some special features arise with molecules containing flexible rings, and thermodynamics has made important contributions to the molecular energetics of such systems. Statistical calculations for alicyclic six-membered rings – cyclohexane and its derivatives – involve consideration of the conformational equilibrium involved and, although the number of possible conformers can become large, no new principles are involved in the treatment. With five-membered rings the wholly new concept of pseudo-rotation arises from consideration of the form of the ring-bending motions. For four-membered rings the ring puckering motions involve a potential with both quartic and harmonic terms, possibly with a barrier; provided the levels are known, the calculation is straightforward.

Six-membered Rings.—The calculations for six-membered rings have been undertaken using the methods of conformational equilibria described on p. 296.

It is well established that the stable configuration of cyclohexane is the chair of symmetry D_{3d}. Other, higher energy, forms are present only in quite small amounts at room temperature. However, their energies are low enough to contribute to the thermodynamic properties; the proportion of the boat form increases rapidly with temperature and since the conversion of chair to boat is strongly endothermic, this conversion contributes appreciably to the heat capacity.

In the original calculations [37] the contribution of chair–boat equilibrium was evaluated by assuming an energy difference of 23.4 kJ K^{-1} mol^{-1}, *i.e.* twice the internal rotation barrier in ethane since the change from chair to boat twists two C—C bonds from their potential minima to their maxima. The entropy change was taken as $Lk \ln 3$, since the symmetry number of the chair form is 6 and that of the boat is 2. At 500 K the contributions to the heat capacity are: translational and rotational, 33.26; harmonic vibrational, 151.74; anharmonicity, 2.42; and chair–boat equilibrium, 2.83 J K^{-1} mol^{-1}, which total 190.25 J K^{-1} mol^{-1}; the experimental value is (191 $\pm$ 2.0) J K^{-1} mol^{-1}. A similar good agreement exists over the range 370 to 544 K, and between the measured and calculated entropy at 298 K.

The value used for the energy difference between chair and (skew) boat forms has received subsequent support: several experimental estimates [173]

[173] E. L. Eliel, N. L. Allinger, S. J. Angyal, and G. A. Morrison, 'Conformational Analysis', Interscience, New York, 1965, p. 38.

range from 20 to 25 kJ K^{-1} mol^{-1} and the average is 23.0 kJ K^{-1} mol^{-1}. The vibrational spectra of cyclohexane have been investigated further[174] and a revision of the calculations is possible; nevertheless, the most recently tabulated values are the original ones.

Thian. The treatment[38] for this molecular paralleled that for cyclohexane; the numerical values were subsequently changed very slightly.[125] The appreciable contribution of anharmonicity, and its treatment, has been mentioned (p. 280). One fundamental was unobserved: its value, and that of the energy difference between the boat (or skew boat) and chair conformations of the ring were then chosen to obtain agreement between the observed and calculated thermodynamic data.[125]

Cyclohexanethiol. In substituted cyclohexanes the number of possible conformations increases considerably, since the equatorial–axial equilibrium has to be considered in addition to that of the chair–skew-boat equilibrium. For thermodynamic calculations the contributions of these two equilibria have to be treated separately.[39]

Of the 24 possible conformations for cyclohexanethiol, the spectroscopic evidence is that only those with the chair form of the ring are present in high enough concentration to be observed at room temperature. The torsion of the —SH group was treated separately as a 3-fold cosine barrier. There are three conformations with respect to that torsion which are likely to be of high energy and they were excluded. The partition function then becomes

$$q = (1/3)\{3 + 2\exp(-\Delta E_1/LkT) + 9\exp(-\Delta E_2/LkT) + 7\exp\{-(\Delta E_1 + \Delta E_2)/LkT\},$$

where the four terms correspond successively to the equatorial–chair, axial–chair, equatorial–skew-boat, and axial–skew-boat equilibria, and ΔE_1 is the energy difference between the axial and equatorial conformations, and ΔE_2 that between the skew-boat and chair conformation. With ΔE_2 taken as 23.0 kJ K^{-1} mol^{-1}, as for cyclohexane, the value of ΔE_1 required to give agreement between observed and calculated thermodynamic properties was 4.0 kJ K^{-1} mol^{-1}. Investigations of the equatorial–axial equilibrium for the liquid by n.m.r. spectroscopy[175] gave a value which corresponds to $\Delta E_1 = 2.9$ kJ K^{-1} mol^{-1}, an agreement with the value from thermodynamics which is very satisfactory.

Pseudo-rotation.—The concept of pseudo-rotation was introduced by Kilpatrick, Pitzer, and Spitzer[176] to account for the thermodynamic and spectroscopic properties of cyclopentane. Although the basis of the

[174] F. A. Miller and H. R. Golob, *Spectrochim. Acta*, 1964, **20**, 1517, and references therein; H. Takahashi, T. Shimanouchi, K. Fukushima, and T. Miyazawa, *J. Mol. Spectroscopy*, 1964, **13**, 43.

[175] E. L. Eliel and B. P. Thill, *Chem. and Ind.*, 1963, 88.

[176] (*a*) J. E. Kilpatrick, K. S. Pitzer, and R. Spitzer, *J. Amer. Chem. Soc.*, 1947, **69**, 2483; (*b*) K. S. Pitzer, *J. Amer. Chem. Soc.*, 1958, **80**, 6697.

concept has been challenged, there is now much evidence for its validity for several molecules, both from spectroscopic observations and from comparisons between calculated and accurately measured thermodynamic properties.

Analysis of the configurations of cyclopentane showed that in this five-membered ring the angular-strain forces arising from the non-tetrahedral CCC angles are nearly cancelled by the torsional forces from the hydrogen–hydrogen repulsions. As a result the ring is slightly puckered, but the puckering is of an indefinite type because the potential energy is essentially the same for all types of puckering. The doubly degenerate out-of-plane ring-bending motion can be expressed in terms of the amplitude q of ring puckering, and the phase angle ϕ of puckering around the ring. The wave equation in these two co-ordinates can be approximately separated. To solve for the amplitude q, a harmonic oscillator potential about an equilibrium position q_0 is assumed and the energy is then specified by a vibrational quantum number v and a vibrational frequency. The solution for the angular part ϕ gives the energy levels, which are those for a one-dimensional rotor:

$$E_{\phi} = n^2h^2/8\pi^2mq_0^2, \qquad (33)$$

where n is a quantum number and m is the reduced mass appropriate to the problem, so that mq_0^2 is essentially a moment of inertia, the effective reduced moment of inertia I_r for pseudo-rotation.

The two degrees of freedom associated with the ring puckering are, therefore, an ordinary vibration and a type of one-dimensional rotation in which the phase of the puckering moves around the ring; the latter is not, however, a true rotation since there in no angular momentum about the axis of rotation, and so is described as a 'pseudo-rotation'. This separation of the wave equation is not exact, but it has been stated[177] that exact separation is possible and, on the assumptions of harmonic oscillations and small amplitudes of vibrations, leads to the same results as those given.

The contributions of pseudo-rotation to $C_p^{\ominus}/Lk$ is then 0.5 and to $S^{\ominus}/Lk$ is $[\frac{1}{2} + \frac{1}{2}\{\ln(8\pi^3I_rkT/h^2\sigma^2)\}]$ where σ is the symmetry number and has the value 10 for cyclopentane. (The value 5 is in error;[176a] a further factor of two for the identity of the two sides of the molecule is required.[176b]) In the original calculations, the puckering vibration was assigned to a Raman line at 288 cm^{-1} (with an anharmonicity coefficient of 0.03) and a value for q_0 of 23.6 pm (corresponding to $I_r = 10.468 \times 10^{-40}$ g cm^2) was chosen to give agreement with the measured entropy and heat capacity.

Later workers contended that pseudo-rotation was an unnecessary concept and made vibrational assignments for cyclopentane in which the ring-puckering motion was treated as two genuine degrees of freedom.[178]

[177] J. P. McCullough, R. E. Pennington, J. C. Smith, I. A. Hossenlopp, and G. Waddington, *J. Amer. Chem. Soc.*, 1959, **81**, 5880.

[178] F. A. Miller and R. G. Inskeep, *J. Chem. Phys.*, 1950, **18**, 1519; B. Curnutte and W. H. Shaffer, *J. Mol. Spectroscopy*, 1957, **1**, 239.

Such assignments are in error, however, even though one of them is based on a normal-co-ordinate calculation, and are quite unable, with or without a contribution of pseudo-rotation, to reproduce the measured heat capacity. The most accurate values of the measured thermodynamic properties require, for agreement within 0.1%, that $I_r = 10.59 \times 10^{-40}$ g cm^2; a small contribution of anharmonicity was treated by equations (15). A subsequent molecular vibrational analysis led to some revision of the vibrational assignment, but excellent agreement with the measured heat capacity values is still obtained from it – provided the contribution of pseudo-rotation is included.[179] The details of the selection rules for cyclopentane have been discussed.[180]

A high-resolution Raman investigation [181] of cyclopentane provided no positive evidence for pseudo-rotation, but a later i.r. investigation [182] provided such evidence in the form of a series of regular and well-pronounced Q branches in the methylene deformation vibration at 1460 cm^{-1}. From equation (33), with the selection rule $\Delta n = \pm 1$, the pseudo-rotation levels will give rise to absorptions at wavenumbers

$$\tilde{\nu} = h(2n + 1)/8\pi^2 cmq_0^2, \tag{34}$$

on either side of the fundamental, and the spacing between the bands will be $h/4\pi^2 cmq_0^2$. The twenty observed Q branches could be well-fitted to the value $I_r = mq_0^2 = (11.0 \pm 0.2) \times 10^{-40}$ g cm^2. This excellent agreement between the values for I_r derived from independent spectroscopic and thermodynamic data leaves little doubt that the cyclopentane molecule is undergoing pseudo-rotation.

With thiacyclopentane [183] an additional feature arises from the replacement of a CH_2 group in cyclopentane by a sulphur atom. Obviously the symmetry is thus removed and the minimum potential energy is now unlikely to be independent of the phase of puckering; that is, potential barriers must be surmounted as the phase of puckering moves around the ring. This situation has been described as restricted pseudo-rotation and the contributions to the thermodynamic functions have been calculated as those of an ordinary restricted internal rotation assumed to be of the simple cosine type (p. 285). The other puckering degree of freedom is treated, as for cyclopentane, as an ordinary vibration. In the calculations for thiacyclopentane the latter was taken as a Raman line at 297 cm^{-1} and the parameters V, the height of the restricting potential barrier, and (I_r/n^2), where I_r is the effective reduced moment of inertia and n is the number of potential minima in one complete pseudo-rotation, were adjusted to give agreement

179 F. H. Kruse and D. W. Scott, *J. Mol. Spectroscopy*, 1966, **20**, 276.
180 I. M. Mills, *Mol. Phys.*, 1971, **20**, 127.
181 K. Tanner and A. Weber, *J. Mol. Spectroscopy*, 1963, **10**, 381.
182 J. R. Durig and D. W. Wertz, *J. Chem. Phys.*, 1968, **49**, 2118.
183 W. N. Hubbard, H. L. Finke, D. W. Scott, J. P. McCullough, C. Katz, M. E. Gross, J. F. Messerly, R. E. Pennington, and G. Waddington, *J. Amer. Chem. Soc.*, 1952, **74**, 6025.

with the measured entropy and heat capacity. (The value of n is not required but is likely to be 2 since indistinguishable configurations occur twice in each pseudo-rotation.) The values $V = 11.7$ kJ mol^{-1} and $I_r/n^2 = 2.665 \times 10^{-40}$ g cm^2 were required; the latter was subsequently[125] changed to 2.925×10^{-40} g cm^2. On this restricted pseudo-rotation model the difference between the ground and first-excited state is calculated to be 99 cm^{-1}, which is in agreement with observed[184] i.r. absorption at 100 cm^{-1}. However, a difficulty exists: the predicted spacing of the upper-stage band is 2.6 cm^{-1} which would give a broad 'smeared out' contour, whereas the observed band has a well-defined contour indicating upper-stage bands nearly coincident with the fundamental. Microwave spectroscopy confirms the value of the fundamental and shows that the first five levels are a nearly harmonic series. Apparently the lower part of the potential functions for hindered rotation is more nearly parabolic than a simple cosine function.

A similar treatment was necessary for pyrrolidine;[185] again the pseudo-rotation was found to be slightly restricted, if a satisfactory fit to the calorimetric data was to be obtained; if pseudo-rotation was excluded no such fit was possible.

The results for tetrahydrofuran are important since there is substantial evidence for pseudo-rotation from both far-i.r. and microwave spectroscopy. A series of absorptions between 20 and 100 cm^{-1} can be fitted[186] to equation (34) with $mq_0^2 = (8.56 \pm 0.13) \times 10^{-40}$ and $(8.48 \pm 0.15) \times 10^{-40}$ g cm^2 for the vibrational states $v = 0$ and $v = 1$, respectively. For a fit to the measured thermodynamic properties a value for mq_0^2 of 8×10^{-40} g cm^2 had previously[187] been required, so the agreement is extremely satisfactory. Microwave spectroscopy has since[188] afforded values for the first 15 energy levels for pseudo-rotation and shown that the barrier is very low. Additional levels have been estimated from these, their contributions to the thermodynamic properties calculated, and the total values of the latter obtained and shown to be in excellent agreement with the observed values.[189] A calculation of the barrier to pseudo-rotation gave[190] a value of 10.5 kJ mol^{-1}. But it has been shown[191] that if, for this calculation, the barrier height is transferred from dimethyl ether[192] (11.4 kJ mol^{-1}) rather than from methanol[193] (4.48 kJ mol^{-1}), then the

184 G. A. Crowder and D. W. Scott, *J. Mol. Spectroscopy*, 1965, **16**, 122.
185 J. P. McCullough, D. R. Douslin, W. N. Hubbard, S. S. Todd, J. F. Messerly, I. A. Hossenlopp, F. R. Frow, J. P. Dawson, and G. Waddington, *J. Amer. Chem. Soc.*, 1959, **81**, 5884.
186 W. J. Lafferty, D. W. Robinson, R. V. St. Louis, J. W. Russell, and H. L. Strauss, *J. Chem. Phys.*, 1965, **42**, 2915.
187 D. W. Scott, presented at the Symposium on Molecular Structure and Spectroscopy, Columbus, Ohio, 1962.
188 G. G. Engerholm, A. C. Luntz, W. D. Gwinn, and D. O. Harris, *J. Chem. Phys.*, 1969, **50**, 2446.
189 D. W. Scott, *J. Chem. Thermodynamics*, 1970, **2**, 833.
190 K. S. Pitzer and W. E. Donath, *J. Amer. Chem. Soc.*, 1959, **81**, 3213.
191 J. R. Durig and D. W. Wertz, *J. Chem. Phys.*, 1968, **49**, 675.
192 P. H. Kasai and R. J. Myers, *J. Chem. Phys.*, 1959, **30**, 1096.
193 E. V. Ivash and D. M. Dennison, *J. Chem. Phys.*, 1953, **21**, 1804.

barrier to pseudo-rotation is only 0.71 kJ mol^{-1}, which is effectively free rotation.

Recent spectroscopic work has established the presence of pseudo-rotation in other molecules. For 1,3-dioxolan a series of absorptions [191] in the region 33 to 120 cm^{-1} has been assigned to hot-band transitions arising from excited states of the pseudo-rotational levels, and fitted to equation (34) with $I_r/n^2 = 7.4 \times 10^{-40}$ g cm^2. The barrier to pseudo-rotation is calculated to be 0.50 kJ mol^{-1}, which is essentially free-pseudo-rotation.

The far-i.r. spectrum of cyclopentanone similarly reveals [194] a series of Q branches which can be accounted for by pseudo-rotation with $I_r/n^2 = 14.5 \times 10^{-40}$ g cm^2 and restricted by a barrier of (11.3 ± 3.3) kJ mol^{-1}. From the microwave spectrum the lower bound to the barrier was set at about 4.8 kJ mol^{-1}, and it was inferred that the nature of the ring-puckering potential is such that the ring is permanently twisted and is undergoing bending and twisting vibrations quite independently.[195]

The question arises whether these series of Q branches could be interpreted on a slightly anharmonic potential function rather than on a cosine function type of pseudo-rotation. For a low barrier ($\approx$ 7.5 kJ mol^{-1}) the contribution to the entropy at 298 K of a hindered pseudo-rotator differs from that of a harmonic oscillator of the same frequency by 6 J K^{-1} mol^{-1}, which is certainly measurable. But for a high barrier ($\approx$ 14.6 kJ mol^{-1}) the difference between the entropy contributions is 0.6 J K^{-1} mol^{-1} at 298 K and only 1.2 J K^{-1} mol^{-1} at 398 K. Therefore only for a low barrier can the two possibilities be distinguished by thermodynamic measurements. For γ-butyrolactone and ethylene carbonate the barriers are calculated [194] to be $\approx$ 46 kJ mol^{-1} and 75 kJ mol^{-1}, respectively, which would negate the effect of pseudo-rotation at reasonable temperatures. Thus, in molecules with a sufficiently high barrier the motion can be treated as an ordinary vibration in which the puckering oscillates about a stable configuration.

Pseudo-rotation has been investigated in chlorocyclopentane: the barrier has been calculated [196] as 4.61 kJ mol^{-1} and evidence for pseudo-rotation argued.[197] However, it seems more likely that the barrier is much greater, perhaps $\approx$ 75 kJ mol^{-1}, and that the chlorocyclopentane ring is not undergoing pseudo-rotation.[196] Evidence for other forms of potential function governing the out-of-plane ring bending modes of five-membered rings has been gained. For example, cyclopentene exhibits a double minimum function which is largely quadratic;[198] whilst 2,5-dihydrofuran also shows a largely quartic function but with a single minimum.[199]

[194] J. R. Durig, G. L. Coulter, and D. W. Wertz, *J. Mol. Spectroscopy*, 1968, **27**, 285; L. Cawiera and R. C. Lord, quoted therein as a personal communication.
[195] H. Kim and W. D. Gwinn, *J. Chem. Phys.*, 1969, **51**, 1815.
[196] J. R. Durig, J. M. Karricker, and D. W. Wertz, *J. Mol. Spectroscopy*, 1969, **31**, 237.
[197] J. Reisse, L. Nagels, and G. Chiurdoglu, *Bull. Soc. chim. belges*, 1965, **74**, 162.
[198] J. Laane and R. C. Lord, *J. Chem. Phys.*, 1967, **47**, 4941.
[199] T. Ueda and T. Shimanouchi, *J. Chem. Phys.*, 1967, **47**, 4042.

Four-membered Rings.—Early calculations [200] of the entropy of cyclobutane, assuming the ring to be planar and treating the ring-puckering fundamental as a quartic oscillator, were appreciably lower than that measured.[201] Agreement between calculated and experimental values of the entropy was subsequently obtained [202] by treating the ring-puckering motion as a perturbed harmonic oscillator potential with a double minimum, the barrier being $\approx$ 4 kJ mol^{-1}. The energy levels were estimated, and their contribution to the calculated thermodynamic functions evaluated by summation (*cf.* p. 271). Confirmation of the double minimum ring-puckering potential has been recently obtained by observation of several of the pure puckering transitions in the gas-phase Raman spectrum,[203] and by observation of several combination tones involving the puckering mode in the i.r. spectrum.[203, 204] Fitting the puckering transitions to a suitable quadratic potential gives a barrier height of (6.20 $\pm$ 0.06) kJ mol^{-1} and the equilibrium dihedral angle of the ring as 35°.

The suggestion [205] that the out-of-plane motion in four-membered rings might be quartic in nature has been further confirmed by microwave investigation. For oxetan [206] the ring-puckering motion has been found to contain both quartic and harmonic terms; the barrier in the double minimum function is sufficiently small that the molecule is essentially planar. For thietan [207] the function is similar, but the barrier is high enough to lead to a bent configuration in the lower vibrational states. Values for the first 15 energy levels have been obtained by microwave spectroscopy: from these a quadratic extrapolation formula was used to estimate the higher, and thermodynamically less-important, levels. The thermodynamic properties thus calculated,[125] by using a revised vibrational assignment, are in good agreement with measured values.[208]

6 Individual Molecules

Calculations by the preceding methods have now been made for several hundred molecules and the results form a substantial part of tabulated chemical thermodynamic data. Some values for simpler molecules are to be found in ref. 3, where a bibliography of tabulated data is also given.

200 T. P. Wilson, *J. Chem. Phys.*, 1943, **11**, 369; T. L. Cottrell, *Trans. Faraday Soc.*, 1948, **44**, 716.
201 G. W. Rathjens and W. D. Gwinn, *J. Amer. Chem. Soc.*, 1953, **75**, 5629.
202 G. W. Rathjens, N. K. Freeman, W. D. Gwinn, and K. S. Pitzer, *J. Amer. Chem. Soc.*, 1953, **75**, 5634.
203 F. A. Miller and R. J. Capwell, *Spectrochim. Acta*, 1971, **27A**, 947.
204 T. Ueda and T. Shimanouchi, *J. Chem. Phys.*, 1967, **47**, 5018; J. M. R. Stone and I. M. Mills, *Mol. Phys.*, 1970, **18**, 631.
205 R. P. Bell, *Proc. Roy. Soc.*, 1945, **A183**, 328.
206 S. I. Chan, J. Zinn, and W. D. Gwinn, *J. Chem. Phys.*, 1961, **34**, 1319.
207 D. O. Harris, H. W. Harrington, A. C. Luntz, and W. D. Gwinn, *J. Chem. Phys.*, 1966, **44**, 3467.
208 D. W. Scott, H. L. Finke, W. N. Hubbard, J. P. McCullough, C. Katz, M. E. Gross, J. F. Messerly, R. E. Pennington, and G. Waddington, *J. Amer. Chem. Soc.*, 1953, **75**, 2795.

The most recent critical survey [77] of predominantly organic molecules has the merit, not always found in tabulated data, of revealing explicitly for most of the compounds the data on which the calculations are based. In this section the treatments of some individual molecules are outlined. The examples are illustrative of the ways in which the calculational methods have been used in conjunction with spectroscopic data and considerations of molecular conformations and energetics. The status of the present data for the molecules is indicated.

Water.—A calculation by the methods of pp. 273–275 follows. Using the values [7] $I_A = 0.966$, $I_B = 1.908$, and $I_C = 2.981 \times 10^{-40}$ g cm^2, $\tilde{\nu}_1 = 3652$, $\tilde{\nu}_2 = 1595$, and $\tilde{\nu}_3 = 3756$ cm^{-1}, and $\sigma = 2$, the translational, rotational, and vibrational contributions to $S^{\ominus}$(g, 298.15 K) are 144.84, 43.82, and 0.00 J K^{-1} mol^{-1}, which total 188.7 J K^{-1} mol^{-1}. The results of a much more complete calculation [209] incorporating the corrections of p. 278 give $S^{\ominus}$(g, 298.15 K) = 188.73 J K^{-1} mol^{-1}. The size of the vibrational contribution exemplifies the relative unimportance of the higher vibrational fundamentals for the entropy at room temperature.

Carbon Tetrachloride.—The calculations for carbon tetrachloride are summarized in view of the assertion [7] that a substantial discrepancy exists between $S^{\ominus}$(g, 298.15 K) calculated (310.95 J K^{-1} mol^{-1}) and observed (297.7 J K^{-1} mol^{-1}). In fact, the most reliable experimental value [210] is (309.2 ± 0.8) J K^{-1} mol^{-1} (*cf.* ref. 211). Using a C—Cl bond length of 176.8 pm and wavenumbers of 458, 218 (2), 776 (3), and 314 (3) cm^{-1} (degeneracies in parentheses), the contributions to $S^{\ominus}$(g, 298.15 K) are translational, 171.20 J K^{-1} mol^{-1}; rotational, 98.89 J K^{-1} mol^{-1}; and vibrational, 39.69 J K^{-1} mol^{-1}, which total 309.78 J K^{-1} mol^{-1}. A similar calculation [212] gave 309.37 J K^{-1} mol^{-1}; the most reliable value is probably (310.12 ± 0.04) J K^{-1} mol^{-1} from work [213] in which the anharmonicity was estimated for all the chlorofluoromethanes by using experimental values [214] of the vapour heat capacity for CCl_2F_2. Therefore no discrepancy exists between measured and calculated values, and the latter are probably slightly the more accurate.

Carbon Suboxide.—A calculation for this molecule illustrates the evaluation of an unobserved vibrational fundamental by choosing a value to give equality of the observed and calculated thermodynamic properties. This is, of course, the traditional procedure for determining barrier heights to internal rotation. For carbon suboxide (C_3O_2) a linear structure was indicated by the vibrational spectra, but the lowest (π_u) bending

[209] A. S. Friedman and L. Haar, *J. Chem. Phys.*, 1954, **22**, 2051.
[210] D. L. Hildenbrand and R. A. MacDonald, *J. Phys. Chem.*, 1959, **63**, 1521.
[211] J. F. G. Hicks, J. G. Hooley, and C. C. Stephenson, *J. Amer. Chem. Soc.*, 1944, **66**, 1064; R. C. Lord and E. R. Blanchard, *J. Chem. Phys.*, 1936, **4**, 707.
[212] E. Gelles and K. S. Pitzer, *J. Amer. Chem. Soc.*, 1953, **75**, 5259.
[213] L. F. Albright, W. C. Galegar, and K. K. Innes, *J. Amer. Chem. Soc.*, 1954, **76**, 6017.
[214] J. F. Masi, *J. Amer. Chem. Soc.*, 1952, **74**, 4738; 1953, **75**, 2276.

fundamental had not been observed. From the measured entropy a value of (61.6 ± 2.6) cm^{-1} was deduced [215] for the fundamental. Subsequent observation [216] of the far-i.r. spectrum revealed weak absorption at 63 cm^{-1}.

Methane.—The most accurate values of the thermodynamic functions for methane are those derived solely from spectroscopic measurements. The values for the range 60 to 5000 K have been calculated,[217] taking into account low-temperature quantum effects, anharmonicity, centrifugal distortion, and rotational–vibrational interaction. Excluding the contribution of nuclear spin (p. 273) $S^\ominus$(g, 298.15 K) is found to be 186.27 J K^{-1} mol^{-1}. The value from calorimetric measurements [218] is (188.66 ± 0.40) J K^{-1} mol^{-1}, and to account for the discrepancy it is necessary to take into account the existence of different nuclear spin species (as with *ortho*- and *para*-hydrogen).

Fluorobenzene.—The calculations for fluorobenzene illustrate the accuracy possible for a large molecule, provided that there are reliable spectroscopic data. Using the first complete vibrational assignment $S^\ominus$(g, 298.15 K) was calculated [219] to be 302.6 J K^{-1} mol^{-1}, the experimental value [220] being 291.8 J K^{-1} mol^{-1}. But more accurate measurements of the entropy, combined with values of the vapour heat capacity from which the anharmonicity contribution could be evaluated (S_{anh} = 0.04 J K^{-1} mol^{-1} at 298.15 K), gave the value [221] for $S^\ominus$(g, 298.15 K) of 302.6 J K^{-1} mol^{-1}. For the other halogenobenzenes the calculated thermodynamic functions are tabulated [222] in preference to the older calorimetric values.

Ethane.—The tabulated thermodynamic functions [223] for ethane are based on a value of (12.03 ± 0.52) kJ mol^{-1} for the 3-fold barrier to internal rotation derived from entropy measurements.[63] This value also gave a good overall fit to the measured vapour heat capacity. A relatively early microwave value [224] for the barrier height is (12.7 ± 1.3) kJ mol^{-1}.

Propane.—The most recently calculated values for propane [223] can be traced back and found to be identical with those calculated in 1944, the values for the barrier heights being derived from measurements of the entropy on the assumption of independent methyl rotation.[94] The b_2 torsional frequency could not be detected in the i.r. spectrum and there is no microwave determination.

[215] L. A. McDougall and J. E. Kilpatrick, *J. Chem. Phys.*, 1965, **42**, 2311.
[216] F. A. Miller, D. H. Lemmon, and R. E. Witkowski, *Spectrochim. Acta*, 1965, **21**, 1709.
[217] R. S. McDowell and F. H. Kruse, *J. Chem. and Eng. Data*, 1963, **8**, 547.
[218] J. H. Colwell, E. K. Gill, and J. A. Morrison, *J. Chem. Phys.*, 1963, **39**, 635.
[219] D. H. Whiffen, *J. Chem. Soc.*, 1956, 1350.
[220] D. R. Stull, *J. Amer. Chem. Soc.*, 1937, **59**, 2726.
[221] D. W. Scott, J. P. McCullough, W. D. Good, J. F. Messerly, R. E. Pennington, T. C. Kincheloe, I. A. Hossenlopp, D. R. Douslin, and G. Waddington, *J. Amer. Chem. Soc.*, 1956, **78**, 5457.
[222] D. R. Stull, E. F. Westrum, jun., and G. C. Sinke, ref. 77, pp. 532, 548, 556.
[223] D. R. Stull, E. F. Westrum, jun., and G. C. Sinke, ref. 77, p. 244.
[224] D. R. Lide, *J. Chem. Phys.*, 1958, **29**, 1426.

Ethylene.—The tabulated data for ethylene[225] are the unrevised values from calculations[84] using an early vibrational assignment.[226] The a_{1u} fundamental, which is the twisting of one CH_2 group relative to the other, is inactive in optical spectra. An early value for this was 825 cm^{-1}, deduced from its overtone at 1654 cm^{-1}, and a more recent assignment[227] gives 1027 cm^{-1}, the use of which will lower the calculated heat capacity by $\approx$ 0.8 J K^{-1} mol^{-1} at 500 K and gives a rather better fit to the observed values.[14]

2-Methylbut-1-ene.—Thermodynamic functions for all the pentenes were originally calculated by an incremental method[228] from data for the lower mono-olefins. Later, experimental values of the entropy and heat capacity were used in a statistical calculation[229] which exemplifies the treatment necessary for several molecules.

Rotation of the ethyl group can give rise to distinguishable conformers, and the most stable is taken as that in which the plane of the skeleton of the ethyl group is rotated 60° out of the plane of the rest of the molecule. The barrier to this rotation was described by a height V_3, and an energy difference ΔE between the low- and high-energy conformations. The contributions to the thermodynamic functions were evaluated (*cf.* p. 296) as those for a simple 3-fold cosine type barrier of height V_3, plus those arising from the conversion of molecules from the low- to the high-energy conformation. The latter were calculated on the assumption that the entropy change for the conversion is $- Lk \ln 2$, since there are two conformations of low energy and one of high energy. The methyl rotations were treated as 3-fold cosine-type barriers of height 10.0 and 14.2 kJ mol^{-1} by comparison with those in isobutene and but-1-ene. Agreement within 0.15% of the measured thermodynamic properties was then possible with the values $V_3 = 4.81$ kJ mol^{-1} and $\Delta E = 14.4$ kJ mol^{-1}.

Styrene.—The tabulated values for styrene[230] are identical with those[131] given in 1946. They are based on a measured entropy at 298 K, and a calculation which gave the value 9.2 kJ mol^{-1} for the (2-fold) barrier to internal rotation of the vinyl group (*cf.* p. 272). Agreement with the one available value for the vapour heat capacity was satisfactory. A recalculation[231] of all the data, the full details of which are given, yielded 11.7 kJ mol^{-1} for the barrier height. Some details of the vibrational assignment remain uncertain,[232] and investigation of the i.r. spectrum to 80 cm^{-1} unfortunately failed to reveal the torsional fundamental.[132]

[225] D. R. Stull, E. F. Westrum, jun., and G. C. Sinke, ref. 77, p. 312.
[226] W. S. Gallaway and E. F. Barker, *J. Chem. Phys.*, 1942, **10**, 88.
[227] R. L. Arnett and B. L. Crawford, *J. Chem. Phys.*, 1950, **18**, 118.
[228] J. E. Kilpatrick, E. J. Prosen, K. S. Pitzer, and F. D. Rossini, *J. Res. Nat. Bur. Stand.*, 1946, **36**, 559.
[229] J. P. McCullough and D. W. Scott, *J. Amer. Chem. Soc.*, 1959, **81**, 1331.
[230] D. R. Stull, E. F. Westrum, jun., and G. C. Sinke, ref. 77, p. 386.
[231] A. Pacault and P. Bothorel, *Bull. Soc. chim. France*, 1956, 217.
[232] W. G. Fateley, G. L. Carlson, and F. E. Dickson, *Appl. Spectroscopy*, 1968, **22**, 650.

Acetylene.—The most recent calculations [233] for acetylene are the most accurate, and incorporate spectroscopic data for rotational stretching, overtone and combination anharmonicities, and rotational–vibrational stretching. These values replace those calculated previously [33] and which are those tabulated.[234]

Dimethylacetylene.—A discrepancy of 1.7 J K^{-1} mol^{-1} between the calculated and calorimetric entropy for dimethylacetylene was revealed long ago.[33] It cannot arise from hindered internal rotation since this would increase the discrepancy. The vibrational assignment is well established but there is a shift [46] of the lowest fundamental from 213 cm^{-1} for the liquid to 194 cm^{-1} for the vapour state, and use of the latter wavenumber gives excellent agreement between calculated and measured values.

This example, and that of toluene,[46] illustrate the importance, for accurate work, of using vapour-state frequencies. The tabulated data [235] are those [33] of 1946, and need revising.

Methanol.—The problem of internal rotation in methanol is complex and the calculation of the thermodynamic functions proved troublesome because at low temperatures the molecule falls outside the limits of validity of the original treatment.[41] A special method was developed [68] making possible the separation of the partition functions for internal rotation from that for overall rotation. This method is correct for all temperatures of practical importance, but a more general formulation has been given,[61, 69] by which the tabulated thermodynamic functions have been calculated.

The calculations yield $S^{\ominus}$(g, 298.15 K) = 239.7 J K^{-1} mol^{-1}. With the most recent value [236] for $S^{\ominus}$(l, 298.15 K) = (127.2 ± 0.13) J K^{-1} mol^{-1} replacing the previous value, (126.8 ± 0.8) J K^{-1} mol^{-1}, the experimental value for $S^{\ominus}$(g, 298.15 K) is then (239.7 ± 0.4) J K^{-1} mol^{-1}. The calculations used gave a value of 4.484 kJ mol^{-1} for the 3-fold barrier; more recent work [237] gives 4.493 kJ mol^{-1}; the best values for the fundamental frequencies of monomeric methanol are probably those now available [238] for an argon matrix at 20 K.

Methanethiol.—By contrast, the calculations for methanethiol are straightforward.[70] A satisfactory vibrational assignment can be made and the barrier to internal rotation is known from microwave spectroscopy.[66, 239] The value (5.310 ± 0.012) kJ mol^{-1} gives for the calculated entropy at 279.12 K the value [70] 251.7 J K^{-1} mol^{-1}; the experimental value [240] is (251.7 ± 0.42) J K^{-1} mol^{-1}.

[233] J. S. Gordon, *J. Chem. and Eng. Data*, 1963, **8**, 294.
[234] D. R. Stull, E. F. Westrum, jun., and G. C. Sinke, ref. 77, p. 334.
[235] D. R. Stull, E. F. Westrum, jun., and G. C. Sinke, ref. 77, p. 336.
[236] H. G. Carlson and E. F. Westrum, jun., *J. Chem. Phys.*, 1971, **54**, 1464.
[237] R. M. Lees and J. G. Baker, *J. Chem. Phys.*, 1968, **48**, 5299.
[238] A. J. Barnes and H. E. Hallam, *Trans. Faraday Soc.*, 1970, **66**, 1920.
[239] T. Kojima and T. Nishikawa, *J. Phys. Soc. Japan*, 1957, **12**, 680.
[240] H. Russell, D. W. Osborne, and D. M. Yost, *J. Amer. Chem. Soc.*, 1942, **64**, 165.

Ethanethiol.—A detailed treatment of the calorimetric data [241] for ethanethiol has been made [122] which utilizes the results of spectroscopic investigations, including a normal-co-ordinate analysis.[242] A contribution to the vapour heat capacity from conformational equilibrium arising from rotation of the —SH group was thus indicated. The simple 3-fold cosine-type barrier corresponding to zero energy difference between these conformers was replaced by a barrier of the form $(V_1/2)(1 + \cos\theta) + V_3/2(1 - \cos 3\theta)$, where $\theta = 0$ is the C_s configuration and $\theta \approx 2\pi/3$ and $4\pi/3$ are the C_1 configurations which, the spectroscopic evidence indicates, are the more stable. The average distribution of levels was then approximated by taking the Mathieu levels for a 5.94 kJ mol^{-1} 3-fold cosine-type barrier – the value from interpreting an i.r. absorption at 191 cm^{-1} as the $0 \rightarrow 1$ (E) transition – and raising the A levels by an amount ΔE, the energy difference between the conformations, the E levels being left unchanged (*cf.* p. 297). The location of individual levels may thus be in error, but only the average distribution is required for the partition function. To obtain agreement with the measured heat capacity and entropy, a value of 1.2 kJ mol^{-1} for ΔE was required, together with that of 15.7 kJ mol^{-1} for the barrier of the methyl rotation.

Propane-1-thiol.—In the original calculations [152] for propane-1-thiol, a treatment similar to that outlined above for 2-methylbut-1-ene was followed. The energy difference between the *trans*- and *gauche*-rotamers arising by rotation about the central C—C bond was chosen, together with the height of a 3-fold cosine-type barrier to this rotation, to give agreement with the measured values of heat capacity and entropy.

In recent work [125] the additional conformers arising from the —SH rotation were taken into account and the treatment of the —SH torsion was transferred from ethanethiol. That is, the energy of the *trans* orientation of the C—C—SH chain was taken as 1.3 kJ mol^{-1} higher than that of the *gauche* orientation. The rotation of the CH_2—CH_2 frame is now known from microwave spectroscopy to be complex in n-propyl fluoride and n-propyl chloride, so the treatment as a simple 3-fold cosine barrier was replaced by one as an anharmonic oscillator with a single average wavenumber $\tilde{\nu}$ and average anharmonicity coefficient X (*cf.* p. 292).

A similar treatment of the torsion of other groups (*e.g.* of one methyl group in butane-2-thiol, both methyl groups in 2-methylpropane-1-thiol, and of the ethyl group in 2-methylbutane-2-thiol) as anharmonic oscillators has been adopted in recent work.[125]

Trimethylborane.—The calculation of the entropy of trimethylborane is of interest in view of the large symmetry number involved, and the question as to whether rotation of the methyl groups is free [243] or hindered.[244]

[241] J. P. McCullough, D. W. Scott, H. L. Finke, M. E. Gross, K. D. Williamson, R. E. Pennington, G. Waddington, and H. M. Hufmann, *J. Amer. Chem. Soc.*, 1952, **74**, 2801.
[242] D. W. Scott and M. Z. El-Sabban, *J. Mol. Spectroscopy*, 1969, **30**, 317.
[243] L. A. Woodward, J. R. Hall, R. N. Dixon, and N. Sheppard, *Spectrochim. Acta*, 1959, **15**, 249.
[244] J. E. Stewart, *J. Res. Nat. Bur. Stand.*, 1956, **56**, 337.

The experimental value[245] for $S^\ominus$(g, 199 K) is 285.7 J K^{-1} mol^{-1}. For free rotation, the total symmetry number is the product of 6 for the BC_3 skeleton and (3 × 3 × 3) for the three methyl groups, that is, 162. This contributes − 42.30 J K^{-1} mol^{-1} to the entropy, the total contributions of overall and free rotation being 123.68 J K^{-1} mol^{-1}. The translational and vibrational contributions are 150.58 J K^{-1} mol^{-1} and 9.75 J K^{-1} mol^{-1}, respectively, giving a total calculated entropy at 199 K of 284.0 J K^{-1} mol^{-1}. Any hindering of internal rotation will lower this value and further increase the difference from the measured value, a difference which is probably due to experimental errors (the limits for the measured value are not given) and uncertainties in the vibrational frequencies and assignments. On the other hand, erroneous use of symmetry numbers of 81 (point group C_{3h}) or 27 (C_s) affords calculated entropies higher than observed and requiring restriction of the rotation by 3.14 or 6.86 kJ mol^{-1}, respectively.

Tetramethylsilane.—The torsional frequencies for tetrahedral $(CH_3)_4X$ molecules are forbidden in the i.r. spectra and the only source of information on the barrier heights is a comparison of measured and calculated thermodynamic functions, on the assumption that the internal rotations are independent. For tetramethylsilane such a comparison[246] has been made for the entropy at 227 and 298.16 K, and affords barrier height values of 6.91 and 6.69 kJ mol^{-1}, respectively. These are to be compared with values from microwave spectroscopy of (7.11 ± 0.42) kJ mol^{-1} for methylsilane[247] and of (6.97 ± 0.04) kJ mol^{-1} for dimethylsilane.[248]

Tetramethyl-lead.—Calorimetric measurements[249] give $S^\ominus$(g, 298.15 K) = (420.4 ± 0.84) J K^{-1} mol^{-1}, whereas calculations assuming free rotation give[250] a value 417·8 J K^{-1} mol^{-1}; this discrepancy would be made greater by any hindrance to rotation. These findings provide another example of the need to use vibrational frequencies from the vapour spectrum: the lowest f_2 fundamental is at 130 cm^{-1} for the liquid, but shifts to 120 cm^{-1} for the vapour.[251] Some additional problems arise with other details of the assignment, which can be solved with the aid of a normal-co-ordinate calculation. The calculated entropy at 298.15 K is then found to be 420.5 J K^{-1} mol^{-1} on the basis of free rotation, although a slight hindrance to rotation, with a corresponding lower wavenumber for one (e) fundamental is possible.[251]

Methylbenzenes.—The barriers to internal rotation in some methylbenzenes are considered as the final examples, since in recent years results from a

[245] G. T. Furukawa and R. P. Park, quoted in ref. 244 as a personal communication.
[246] D. W. Scott, J. F. Messerly, S. S. Todd, G. B. Guthrie, I. A. Hossenlopp, R. T. Moore, A. Osborn, W. T. Berg, and J. P. McCullough, *J. Phys. Chem.*, 1961, **65**, 1320.
[247] R. W. Kilb and L. Pierce, *J. Chem. Phys.*, 1957, **27**, 108.
[248] L. Pierce, *J. Chem. Phys.*, 1959, **31**, 547.
[249] W. D. Good, D. W. Scott, J. L. Lacina, and J. P. McCullough, *J. Phys. Chem.*, 1959, **63**, 1139.
[250] E. R. Lippincott and M. C. Tobin, *J. Amer. Chem. Soc.*, 1953, **75**, 4141.
[251] G. A. Crowder, G. Gorin, F. H. Kruse, and D. W. Scott, *J. Mol. Spectroscopy*, 1965, **16**, 115.

variety of spectroscopic techniques have confirmed and extended those previously available only from statistical calculations and thermodynamic properties.

For toluene,[46] accurate experimental data, and calculations using the vapour phase value for the lowest fundamental, showed that rotation of the methyl group is essentially free. Previous results[57] had indicated an upper bound of 6.7 kJ mol^{-1} for this barrier and similarly of 4.2 and 3.8 kJ mol^{-1} for the barriers in *m*- and *p*-xylene, respectively. Recent determinations by microwave spectroscopy have given for the 6-fold barrier for toluene[252] the value (58.32 ± 0.42) J mol^{-1}, with almost identical values for *p*-fluoro-[253] and *p*-chloro-toluene.[254] For *m*-fluorotoluene[255] the same method gives a value of $\approx$ 0.55 kJ mol^{-1} for the 3-fold barrier, with a comparable but equally small 6-fold component.

The situation with *o*-xylene is more complex, since it is difficult to express a satisfactory potential function for the two interacting, 'meshing,' methyl groups. On the assumption of cosine-type potential functions for each group, separately, the barrier height of each was found from the entropy measurements, giving a value in satisfactory agreement with the heat capacity data.[57] A repetition of the calculations[58] gave the value 8.79 kJ mol^{-1} and another, with a different vibrational assignment,[99] gave 7.53 kJ mol^{-1}. From a treatment of the heat capacity data for solid *o*-xylene a barrier height of 7.74 kJ mol^{-1} was deduced.[256] Recently, the technique of neutron scattering afforded[257] the value (8.20 ± 0.59) kJ mol^{-1}, from an interpretation of observed peak energies at (169 ± 6) cm^{-1} as the $1 \rightarrow 0$ torsional transitions, again on the assumption of simple 3-fold cosine potentials. Observed 'hot' bands in the u.v. absorption spectrum of *o*-xylene have been attributed to excited levels of torsional vibrations of the methyl groups and a barrier height of 7.47 kJ mol^{-1} deduced.[258] The Raman line[259] and weak i.r. absorption[260] at $\approx$ 180 cm^{-1} have been attributed[257, 258] to the torsional vibration and would then indicate a barrier of $\approx$ 8.97 kJ mol^{-1}. However, it is not certain that these features arise from the torsional vibration since another a$''$ fundamental is to be expected in this region.[261]

[252] H. D. Rudolph, H. Dreizler, A. Jaeschke, and P. Wendling, *Z. Naturforsch.*, 1967, **22a**, 940.
[253] H. D. Rudolph and H. Seiler, *Z. Naturforsch.*, 1965, **20a**, 1682.
[254] G. E. Herberich, *Z. Naturforsch.*, 1967, **22a**, 761.
[255] H. D. Rudolph and A. Trimkaus, *Z. Naturforsch.*, 1968, **23a**, 68.
[256] C. A. Wulff, *J. Chem. Phys.*, 1963, **39**, 1227.
[257] J. J. Rush, *J. Chem. Phys.*, 1967, **47**, 3936.
[258] K. C. Ingham and S. J. Strickler, *J. Chem. Phys.*, 1970, **53**, 4313.
[259] E. Herz, *Monatsh.*, 1947, **76**, 1.
[260] J. H. S. Green, W. Kynaston, and H. A. Gebbie, *Spectrochim. Acta*, 1962, **19**, 807; G. W. F. Pardoe, S. J. Larson, H. A. Gebbie, S. J. Strickler, K. C. Ingham, and D. G. Johnson, *J. Chem. Phys.*, 1970, **52**, 6426; W. G. Fateley, F. A. Miller, and R. E. Witkowski, Technical Documentary Report ML-TDR-64-158, Office of Technical Services, U.S. Dept. of Commerce, Washington, 1964.
[261] J. H. S. Green, *Spectrochim. Acta*, 1970, **26A**, 1913.

It is relevant that the microwave spectrum of *o*-fluorotoluene[262] leads to a value of 2.72 kJ mol^{-1} for the 3-fold barrier. Another recent determination is that[263] by n.m.r. spectroscopy of a barrier of $\approx$ 5 kJ mol^{-1} for methyl rotation in *o*-chlorotoluene dissolved in the nematic phase of 4-amino-4-methoxybenzylidene-α-methylcinnamic acid n-propyl ester. The position of minimal energy was inferred to be that in which the chlorine is staggered to the methyl group. This conclusion has also been reached[264] from semi-empirical MO calculations by the extended Hückel theory, which give a barrier height of 7.41 kJ mol^{-1}.

Entropy measurements have been made[98] for 1,2,4- and 1,2,3-trimethylbenzene, and from these barrier heights of 5.86 and 13.39 kJ mol^{-1} for the *ortho* and central methyl groups were deduced,[99] again on the assumption that their rotations are independent. On the same assumption, a neutron-scattering peak for 1,2,3-trimethylbenzene at 180 cm^{-1} corresponds to a barrier of (9.33 ± 0.71) kJ mol^{-1} for the *ortho*-methyl groups; and an additional broad band with a maximum near 100 cm^{-1} was attributed to a second set of torsional levels arising from the central methyl group.[257]

The entropy of hexamethylbenzene has been used to investigate the barriers to internal rotation in this compound. Vapour-pressure measurements by the effusion technique[265] gave erroneous values which were subsequently replaced by those obtained by the transpiration method[266] and combined with heat capacity measurements[267] for the solid, to obtain $S^{\ominus}$(g) over the range 303 to 343 K. There are some uncertainties about the vibrational assignment,[99, 268] but the barrier to fit the observed data was calculated for ring symmetry numbers of 3 (for a non-planar S_6 configuration) and 12 (for a planar D_{6h} configuration); the average values were 14.53 and 10.05 kJ mol^{-1}, respectively. However, both neutron scattering[257] and n.m.r. spectra[269] have afforded lower values, in the range 4.5 to 7.9 kJ mol^{-1}. For hexamethylbenzene any derivations of barrier heights from torsional frequencies can at present yield only some average values and are seriously limited by the treatment of the rotations as independent cosine functions.

[262] J. Susskind, *J. Chem. Phys.*, 1970, **53**, 2492.
[263] P. Diehl, P. M. Henricks, and W. Niederberger, *Mol. Phys.*, 1971, **20**, 139.
[264] J. F. Yan, F. A. Momany, and H. A. Scheraga, *J. Amer. Chem. Soc.*, 1970, **92**, 1109.
[265] S. Seki and H. Chihara, *Sci. Papers Fac. Sci. Osaka Univ.*, 1949, **1**, 1.
[266] J. E. Overberger, W. A. Steele, and J. G. Aston, *J. Chem. Thermodynamics*, 1969, **1**, 535.
[267] M. Frankosky and J. G. Aston, *J. Phys. Chem.*, 1965, **69**, 3126.
[268] M. A. Kovner and A. M. Bogomolov, *Fiz. Sbornik L'vovsk. Gos. Univ., Ser. Khim.*, 1950, **11**, 1.
[269] P. S. Allen, E. R. Andrew, and A. Cowking, Proc. XIV Colloque Ampère, Ljubljana, Yugoslavia, 1966.

9

Metallurgical Thermochemistry at High Temperatures

BY O. KUBASCHEWSKI, P. J. SPENCER, AND W. A. DENCH

1 Introduction

Thermodynamic properties of metal compounds and alloys are being measured in quite a large number of laboratories all over the world. There are two main reasons for this amount of work. Systematic measurements of enthalpies and entropies of formation provide the data which are needed to test the various models designed to explore the causes of chemical stability. Secondly, the data converted into Gibbs energies are needed to assess the feasibility of industrial processes and to deal with other practical problems. These two major aims are, of course, common to chemical thermodynamics in general, but metallurgical thermochemistry presents special problems which will be briefly indicated.

The molecular structure of gaseous species including organic molecules is now fairly well understood. Metallurgy is mainly concerned with condensed metallic phases, the structures of which are much more complex and still far from being understood quantitatively. The chemical stability of such phases is believed to be the result of an overlap of metallic, covalent, and polar bonding but none of these bond types is clearly defined. Theoretical physicists are trying to derive internal energies and entropies of chemical substances from fundamental quantum theory, but whether this approach will reach a stage where practical use can be made of it for the calculation of accurate thermochemical properties is still somewhat in doubt. Thermochemists are content to take the thermodynamic properties of the elements for granted and to consider only the changes in enthalpy and entropy that take place on combination. Even so, the simultaneous presence of the various bonding mechanisms makes analyses very difficult. However, some success has already been achieved.

The concept of electronegativity, based on considerations of the Born cycle and Coulomb's law, is now well established and as a result it is possible to estimate the enthalpies of formation of strongly polar, *i.e.* ionic, compounds [1] with a fair degree of reliability, the accuracy depending on the extent to which covalent bonding occurs. On the other hand, the enthalpies of formation of predominantly metallic compounds in which

[1] O. Kubaschewski and W. Slough, *Progr. Mater. Sci.*, 1969, **14**, No. 1.

polar and covalent bonding is largely absent can be estimated from the change in co-ordination, making the assumption that the bond energies of pairs of unlike atoms are additive relative to the bond energies of pairs of like atoms.[2, 3]

Whereas it is possible, for materials that are predominantly ionic or metallic, to estimate fairly good enthalpies of formation on the basis of sensible models, the covalent bond in metallic phases has so far defied quantitative description. The situation is even more complex in phases which exhibit a proportion of all three bond types, as occurs for example in most concentrated metallic solutions. Other attempts to predict unknown thermodynamic values have therefore been based on entirely empirical rules. There seems to exist, for instance, a 'linear' relation between the enthalpies and excess entropies of formation of concentrated solutions independent of bond mechanism.[1, 3] Other investigators are at present engaged in testing various empirical equations for the estimation of the thermodynamic properties of multicomponent solutions from those of the binaries.[4]

Apart from the fundamental interest in the nature of chemical bonding and stability, these efforts have a very practical aim, namely the provision of further thermodynamic data without recourse to experimental measurements. Such data are needed for the second major aim of chemical thermodynamics mentioned above, namely their application to industrial problems. This type of application may be demonstrated by means of a few examples.

The thermodynamic quantity usually required for this purpose is the standard Gibbs energy of reaction (ΔG°_r) whereas the fundamental considerations are based on standard enthalpies (ΔH°_f) and standard entropies (ΔS°_f) of formation. The most important relation may be restated, namely

$$\Delta G^\circ_r = \Delta H^\circ_r - T\Delta S^\circ_r = -RT \ln K_p, \qquad (1)$$

where K_p, the equilibrium constant, is the quantity usually sought. Enthalpies and entropies of formation may be obtained from appropriate tabulations, where necessary with the application of Hess' and Kirchhoff's laws.

A relatively simple example of the practical application of equation (1) is the silicothermic production of magnesium from calcined dolomite:*

$$2\langle CaO{\cdot}MgO\rangle + \langle Si\rangle = \langle Ca_2SiO_4\rangle + 2(Mg).$$

Here $K_p = \{p(Mg)\}^2$. Since all the relevant enthalpies and entropies of formation of the substances involved are well known, the magnesium

[2] O. Kubaschewski, 'Physical Chemistry of Metallic Solutions and Intermetallic Compounds', Proceedings N.P.L. Symposium No. 9, 1958, HMSO, 1959, Paper 3C.

[3] O. Kubaschewski, 'Phase Stability in Metals and Alloys', ed. P. S. Rudman, J. Stringer, and R. I. Jaffee, Batt. Mem. Inst. Coll., 1967, p. 63.

[4] P. J. Spencer, F. H. Hayes, and L. Elford, 'Applications in Ferrous Metallurgy', International Symposium Metall. Chem., Sheffield, 1197; P. J. Spencer, F. H. Hayes, and O. Kubaschewski, *Rev. Chim. Minérale*, 1972, **9**, 13.

* Throughout this Chapter, the state of aggregation of a substance is indicated as follows: ⟨solid⟩, {liquid}, (gaseous), [solute]$_{solvent}$.

pressures can be calculated and a convenient temperature for the practical performance selected.

A reaction in which two gaseous species participate is the refining of aluminium *via* its gaseous subchloride:

$$2\{Al\} + (AlCl_3) = 3(AlCl).$$

Here $K_p = \{p(AlCl)\}^3/p(AlCl_3)$. Again, all the relevant thermodynamic quantities are available and K_p can be calculated. Two temperatures must be selected for the refining process, one for the forward reaction for which K_p should preferably be about 1 atm, and another for its reversal for which $p(AlCl)$ should be negligibly small.

Many reactions occurring in practice for which equilibrium constants are required involve the formation of metallic solutions. Examples are the deoxidation (or desulphurization) of molten metals by suitable additions. Liquid iron is usually deoxidized by the addition of small amounts of ferroaluminium or ferrosilicon. In the latter case, the pertinent reaction may be written* as follows:

$$\langle SiO_2\rangle = [Si]_{\{Fe\}} + 2[O]_{\{Fe\}}.$$

The deoxidation constant, $K_p = x(Si)\{x(O)\}^2$ where x denotes mole fraction, can again be calculated from available data and is found to agree with practical experience. In the corresponding case for aluminium:

$$\langle Al_2O_3\rangle = 2[Al]_{\{Fe\}} + 3[O]_{\{Fe\}},$$

a discrepancy between the calculated and observed deoxidation constants has existed for many years, but this has recently been resolved in favour of the thermochemist with some implications for the practical performance.

Another example involving dilute solutions is the deoxidation of titanium by calcium:

$$\langle CaO\rangle = \{Ca\} + [O]_{Ti}.$$

The Gibbs energy of solution of oxygen in titanium was not known when the process was first considered, and had to be determined. In view of the high affinity of titanium for oxygen, a new experimental method had to be devised to study the system and details will be described later.

The number of chemical reactions of practical significance worthy of consideration by the thermochemist could be extended *ad infinitum*, but only one more example will be mentioned here. The wide use that industry is making of metallic materials requires in the first place the availability of good equilibrium diagrams for alloy systems. Such diagrams have been studied on a large scale by conventional methods. Since the equilibrium diagram of a system forms part of the Gibbs energy diagram in the three co-ordinates $\Delta G°$, temperature, and composition, thermodynamic data have been systematically applied to the evaluation of metallurgical equilibrium

* As these equations are written, the equilibria refer to oxidation rather than deoxidation, but this reversal has been metallurgical usage for decades.

diagrams in the authors' laboratory.[5] One important observation that has been made from these studies is that certain types of phase boundary, such as solidus curves and, more important, solid solubilities, are frequently unreliable when determined by conventional methods. The crucial difference between the conventional and the thermochemical approaches is that the latter is more or less independent of the kinetic restrictions on phase boundary determination. With thermochemical measurements, the temperature can be selected such that the rate of equilibration is sufficiently fast, whereas the conventional methods are hampered by diffusional delays and the resulting slow approach to equilibrium.

2 Thermochemical Quantities to be Measured

It is apparent from the preceding section that the thermochemical quantities required for metallurgical substances are the enthalpy, entropy, and Gibbs energy of formation. Since the three functions are inter-related, only two of them need be determined experimentally for any one substance. Since the enthalpy and entropy of formation are not independent of temperature, heat capacities should also be known, although these can often be estimated for condensed phases with some reliability. It may be recalled that the heat capacity enters the relevant calculations in the following form:

$$\Delta H^\circ(T) = \Delta H^\circ(298.15\ \mathrm{K}) + \int_{298.15\ \mathrm{K}}^{T} \Delta C_p\,\mathrm{d}T, \qquad (2)$$

$$\Delta S^\circ(T) = \Delta S^\circ(298.15\ \mathrm{K}) + \int_{298.15\ \mathrm{K}}^{T} (\Delta C_p/T)\,\mathrm{d}T.$$

If any solid–solid, solid–liquid, solid–gas, or liquid–gas transformation enters the calculation of the Gibbs energy of a particular reaction, the enthalpies and entropies of transformation must also be known. The following quantities are then tabulated to describe the thermodynamics of a particular chemical compound in full:

$\Delta H^\circ(298.15\ \mathrm{K})$, the standard enthalpy of formation at 298.15 K

$S^\circ(298.15\ \mathrm{K})$, the standard entropy at 298.15 K

C_p, the heat capacity usually given as an expression with two, three, or four terms of an expansion in powers of T

ΔH_{m}, ΔS_{m}, the enthalpy and entropy of fusion

ΔH_{tr}, ΔS_{tr}, the enthalpy and entropy of transition

ΔH_{v}, ΔS_{v} or ΔH_{s}, ΔS_{s}, the enthalpy and entropy of evaporation or sublimation.

[5] O. Kubaschewski, 'The Study of Equilibrium Diagrams by Thermochemical Methods', The Carter Memorial Lecture, *Metallurgical Journal*, University of Strathclyde, 1971, No. 21, p. 9.

Instead of tabulating standard entropies, it might well be preferable to tabulate entropies of formation at 298.15 K. For many substances of metallurgical interest, the applicability of the third law is in doubt; in other words, the zero-temperature entropy $S(T \rightarrow 0)$ in the equation:

$$S^\circ(298.15\ \mathrm{K}) = \int_0^{298.15\ \mathrm{K}} (\Delta C_p/T)\,\mathrm{d}T + S(T \rightarrow 0), \qquad (3)$$

may be significant owing to frozen-in disorder which may be difficult to remove experimentally. In metallurgical thermochemistry it is often advisable to obtain entropies of formation from measurements of Gibbs energies and enthalpies of formation made at fairly high temperatures.

Similarly, ΔC_p, 'the heat capacity of formation', might be tabulated instead of C_p, but the reason for this is different. The amount of heat capacity data for metallurgical substances is still rather restricted and, since the data are mostly obtained by drop calorimetry (see below), often not very reliable. Since it is much easier to estimate changes in, rather than actual, heat capacities, the former might be tabulated.

The metallurgist is very interested in non-stoicheiometric phases and solid and liquid solutions, and thus another variable, namely composition, must be taken into account. The position may be illustrated by just one example out of very many. Figure 1 shows the equilibrium diagram of the system titanium + oxygen.[6] There is an extensive solubility of oxygen in the solid metal and the phases low in oxygen show substantial deviations from stoicheiometry. Even titanium dioxide seems to be stable only with a deficit in oxygen in its lattice, TiO_{2-x}, the extent of the deficit depending on the chemical potential of oxygen in the surroundings.

Consideration of the composition dependence requires the introduction of a further set of thermodynamic quantities. So far it has been assumed that the enthalpies, entropies, and Gibbs energies of formation are the molar quantities for the formation of the compound from the component elements. These are termed the integral quantities. However, for experimental as well as practical reasons, the corresponding partial quantities must be defined. The partial molar quantities, say ΔZ_B, are related to the integral quantity by the formula:

$$\Delta Z = x_A \Delta Z_A + x_B \Delta Z_B + \ldots, \qquad (4)$$

where $Z = H, S, G$, *etc.*, and $x_A, x_B, \ldots$, denote the mole fractions of the constituents A, B, ...

In multicomponent systems, the partial molar Gibbs energy (or chemical potential) of only one component can, as a rule, be determined accurately. In this case, a Duhem–Margules integration must be invoked for a full evaluation. For a binary system, this may be written as follows:

$$\Delta Z = x_B \int_0^{x_A} (\Delta Z_A / x_B^2)\,\mathrm{d}x_A, \quad (Z = H, S, G;\ x_A + x_B = 1). \qquad (5)$$

[6] P. G. Wahlbeck and P. W. Gilles, *J. Amer. Ceram. Soc.*, 1966, **49**, 180.

The partial molar quantities for component B can then be obtained by means of equation (4).

Since operations with both partial and integral quantities are important for the metallurgical thermochemist, the connection may further be

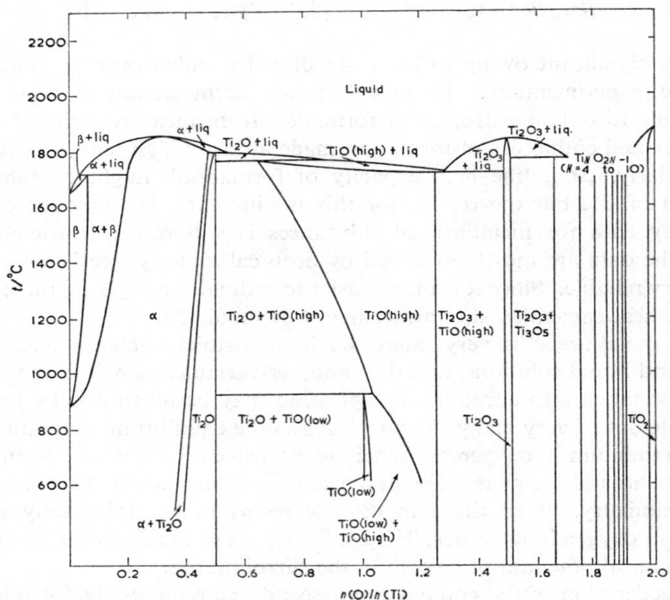

Figure 1 *Equilibrium diagram of the system titanium + oxygen. The Celsius temperature t is plotted against the mole ratio n*(O)/*n*(Ti)
Reproduced by permission from *J. Amer. Ceram. Soc.*, 1966, **49**, 180)

illustrated graphically. The continuous lines in Figure 2a show the integral molar Gibbs energies of formation of a hypothetical binary system, exhibiting one compound and a large solubility range for B in A. The tangent to the full line at any point cuts the $x_A = 0$ and $x_A = 1$ ordinates at the values of the partial molar Gibbs energies of solution of B and A respectively. Whereas these vary continuously through the solid solution phase, they remain constant in heterogeneous ranges, as may be seen from Figure 2(b).

Gibbs energy measurements are, by nature, determinations of partial quantities whereas calorimetric measurements mostly produce integral quantities. However, sensitive calorimetric methods are now being devised which allow the determination of the enthalpy of assimilation of the second component by a solution without significant change in composition. A direct method for the determination of partial entropies of solution cannot yet be envisaged. There is also some interest in values for partial changes in heat capacity, but the only ways to obtain such values at present are by differentiating accurate curves of integral heat capacity against mole fraction or by measuring good partial enthalpies of solution at various temperatures.

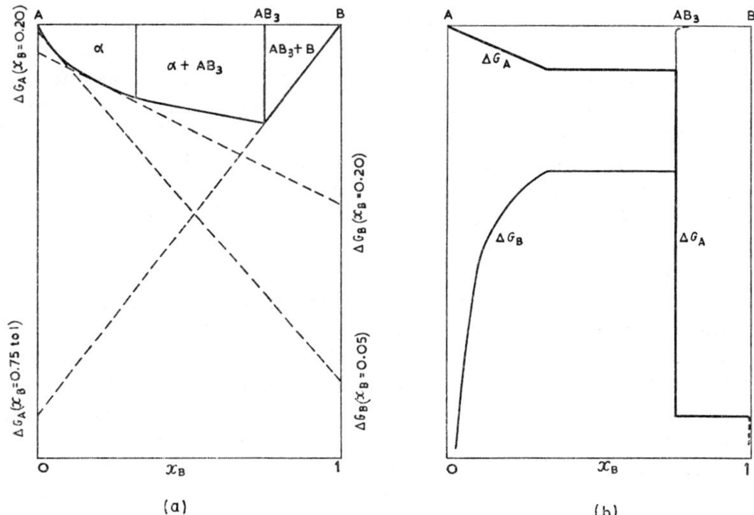

Figure 2 *Relation between partial and integral molar Gibbs energies for a hypothetical system.* (a) *Full lines represent integral molar Gibbs energies of formation.* (b) *Corresponding partial molar Gibbs energies of solution*

It is thus seen that not all the thermodynamic quantities required for the complete description of a particular system can be obtained directly by experiment, but, since the values are interconnected, a complete description can still be achieved.

3 Experimental Methods

For the determination of standard Gibbs energies of reaction, a wide variety of experimental methods have been devised. These may be subdivided into e.m.f. measurements, equilibria with a gaseous phase, and distribution equilibria. From the temperature coefficients of the Gibbs energies, enthalpies and entropies of reaction can be deduced, but experience has shown that these cannot be relied upon when one or more solid phase takes part in the reaction, and errors are very difficult to assess. In such cases, it is recommended that the enthalpies of reaction are measured calorimetrically and combined with the standard Gibbs energies to yield standard entropies of reaction. Calorimetric methods are also used to determine heat capacities, enthalpies of transformation, and enthalpies of fusion. Only for the determination of enthalpies of evaporation may second-law evaluations be preferable to direct calorimetric methods. Of course, when entropies are available from spectroscopic data or low-temperature heat capacities, third-law evaluations give the most reliable results.

When experimental methods are considered, the first question that arises concerns the accuracy at which one should aim. This question is not as easily answered as one may think. The higher the required accuracy, the more sophisticated and the more cumbersome the apparatus must be. Thus, the answer very much depends on the problem under consideration. In the search for empirical relations involving metallurgical thermodynamic quantities, the investigator may be content with a relatively low accuracy, say, $\pm$ 2 $kcal_{th}$ mol^{-1}.* In the natural course of things, he will then see his satisfaction dwindle as soon as more accurate values become available and are used to test his relation.

When the accuracy of measurements for practical applications is considered, again no general ruling is possible. In the first dealings with a metallurgical process or an engineering problem, quite rough figures may be sufficient to determine whether a particular line ought to be pursued. Standard Gibbs energies with a reliability of $\pm$ 10 $kcal_{th}$ mol^{-1} may already provide valuable information. However, if it is then discovered that the problem in hand deserves further study by the thermochemist, a much higher accuracy for his relevant quantities is likely to be found desirable and the demands may become rather high. The present authors have encountered problems where the error in the final answer should have been less than $\pm$ 100 cal_{th} mol^{-1} to be of real value.

The optimum accuracy in the determination of enthalpies and standard Gibbs energies one might strive for is, say, $\pm$ 20 cal_{th} mol^{-1}. At this level, other factors such as strain energies and surface and interface energies become effective and cannot normally be assessed. Chemical factors, such as impurity effects and reactions between the substances investigated and their surroundings, may be more serious. It is essential to define the physical state and chemical composition of samples, and to consider carefully all possible interfering reactions if good results are to be obtained. In short, the thermochemist must in the first place also be a good analyst – a demand the late Professor W. A. Roth repeatedly made.

4 Calorimetric Methods

Calorimeters vary so widely in their design that it is difficult to derive for them a simple, all-embracing classification system. In discussing design, the best that can be done is to select from possible constructional and operational features those that are present in each particular design. A calorimeter may include (i) a fixed mass of material comprising the specimen, (ii) a block of material of high thermal conductivity or a vessel containing a stirred liquid,

* cal_{th} = 4.184 J.

often called the calorimeter, separated from (iii) an outer enclosure having uniform temperature distribution. The purpose of (ii) is to provide (*a*) a reference-mass of known heat capacity and uniform temperature distribution to which a temperature-measuring element may meaningfully be applied, (*b*) a mass with which the specimen is free to come into thermal equilibrium but which serves to isolate it from its surroundings, and (*c*) a mass which presents to the surroundings a surface layer, with a well-defined outer boundary, at uniform temperature and amenable to temperature measurement, enabling its heat-exchange with the surroundings to be measured or controlled.

However, not all of these components are necessarily included in every calorimeter design, and their arrangement may be altered. As examples, the separate *calorimeter* may be dispensed with, the specimen in effect constituting the *calorimeter*; the specimen may be replaced by a continuous flow of fluid reactants; an insulating material may occupy the space between (i) and (ii).

Names have been given to calorimeters (usually having these three components) that are operated in certain modes. In the *isothermal* calorimeter the temperature T_c of the *calorimeter* is kept equal to the temperature T_s of the surroundings and both are held constant. In the *adiabatic* calorimeter T_s is kept equal to T_c although both may change. In the *isoperibol* calorimeter T_s is kept constant and T_c, usually initially near T_s, undergoes an excursion. In the *constant heat-flow* calorimeter $(T_s - T_c)$ is kept constant.

These calorimeters and some others will be discussed in turn with reference to metallurgical reactions.

Isothermal Calorimeters.—The ice calorimeter or Bunsen-type calorimeter is the best known representative of the truly isothermal calorimeters. Proposals have been made to replace the water by an organic liquid. Of the substances investigated, diphenyl ether (melting temperature: 300.0 K) has become quite popular. Attempts to replace the water by molten metals have so far failed. Since thermochemists are familiar with the principle of the ice calorimeter and since this chapter is concerned only with high-temperature methods, no further description will be given.

The *drop* or *mixing method* is one of the oldest tools of the thermochemist. As no fundamental change has been made in recent years a detailed description is unnecessary. However, very high temperatures have now been studied by the application of modern methods of temperature measurement and heating. For example, Fredrickson *et al.*[7a] have used

[7] (*a*) D. R. Fredrickson, R. Kleb, E. L. Nuttall, and W. N. Hubbard, *Rev. Sci. Instr.*, 1969, **40**, 1022; (*b*) L. S. Levinson, *Rev. Sci. Instr.*, 1962, **33**, 639; (*c*) E. D. West and S. Ishihara, 'Advances in Thermophysical Properties at Extreme Temperatures and Pressures', A.S.M.E., 1965, pp. 146–151; (*d*) M. Hoch and H. L. Johnston, *J. Phys. Chem.*, 1961, **65**, 855; (*e*) L. W. Mishler, L. Leibowitz, and M. G. Chasanov, *Trans. Metallurg. Soc. A.I.M.E.*, 1969, **245**(5), 981; (*f*) W. Oelsen, O. Oelsen, and D. Thiel, *Z. Metallkunde*, 1955, **46**, 555.

electron beam heating up to 1900 K, Levinson[7b] has used a graphite resistance element up to 2775 K, and electromagnetic induction heating has been used up to 2600 K by West and Ishihara,[7c] up to 3000 K by Hoch and Johnston,[7d] and up to 3600 K by Leibowitz *et al.*[7e] In a modification of the drop method devised by Oelsen,[7f] the result of each drop is a complete C_p *versus* T curve instead of the usual single enthalpy value. The hot specimen is placed on pin-points inside the calorimeter, which should preferably be an adiabatic one. The method constitutes a comparison between two calorimeters: the specimen itself with the unknown energy equivalent ε_c, and the observed variable temperature T_c, and the adiabatic calorimeter having a known energy equivalent ε_a, and a variable temperature T_a. Assuming quantitative heat transfer from the specimen to the calorimeter as well as adiabatic conditions, the unknown energy equivalent is obtained from the equation:

$$\varepsilon_c \, \delta T_c = \varepsilon_a \, \delta T_a.$$

The accuracies obtained (better than ± 1%) exceed those of many other calorimetric methods and are definitely superior to the ordinary drop calorimeter for the determination of heat capacities and enthalpies of transformation and fusion.

The ordinary drop calorimeter is widely in use – considering the limited number of substances to which it can be applied, rather too widely. Its application should really be limited to metals and stoichiometric compounds. If the substance undergoes a phase transformation in the temperature range studied or another solid-state phase change, the rapid cooling may produce undefined end states in the specimen. However, the resulting enthalpies may still be useful for approximate conversions of enthalpies of reaction from one temperature to another but should not be differentiated to obtain true heat capacities.

The furnaces of many existing drop calorimeters reach a maximum temperature of 1800 K. For further applications of the drop method, this temperature should be extended to 2000 K and higher. An improvement would be brought about if the specimen, instead of being dropped into a room-temperature calorimeter, were instead dropped into a high-temperature calorimeter (say 1000 K) to reduce the effect of rapid quenching on the final physical state of the specimen.

Adiabatic Calorimeters.—The adiabatic principle has also been applied in calorimetry for a long time and adiabatic calorimeters can be used for the determination of heat capacities and enthalpies of transformation and fusion, as well as enthalpies of reaction. The first adiabatic calorimeters were designed for room temperature. Low-temperature heat capacities for the determination of standard entropies, *etc.*, are also measured by adiabatic methods. However, low-temperature calorimetry is the subject of Chapter 4.

All recent developments in adiabatic calorimetry at high temperatures seem to go back to a design by Moser.[8] Figure 3 shows a calorimeter with

[8] H. Moser, *Phys. Z.*, 1936, **37**, 737.

working range 775 to 1675 K built by Dench.[9] The calorimeter consists of a small cylindrical box constructed from tantalum radiation shielding which forms the adiabatic enclosure and which contains the specimen. The box is within a tantalum-wound furnace, which provides uniform heating. Three differential thermocouples, connected in series to a galvanometer, are

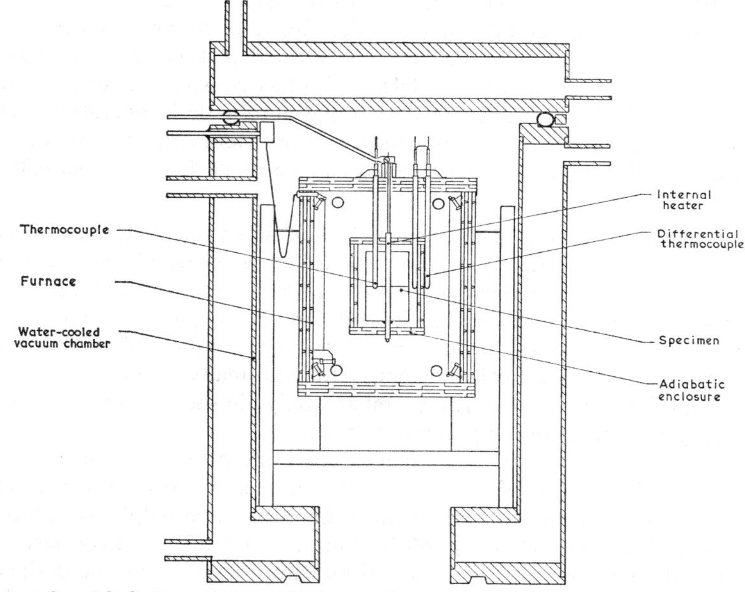

Figure 3 *Adiabatic reaction calorimeter for use in the temperature range* 775 *to* 1675 K
(Reproduced by permission from *Trans. Faraday Soc.*, 1963, **59**, 1279)

positioned around the adiabatic enclosure with their junctions against the inner and outer surfaces. An e.m.f. produced by the thermocouples indicates a temperature difference between the outside and inside of the enclosure. During experimental measurements, the specimen is heated through a selected temperature range by means of a small heater passing through its centre. The electrical energy to the furnace is continually adjusted so that the temperature difference across the walls of the box remains as small as possible. For determinations of enthalpies of formation of alloys, a specimen consisting of a compacted mixture of the powdered component metals is heated from an initial temperature where the rate of alloying is negligibly slow to a final temperature where alloying is complete in a short period of time. The energy input is recorded by a Watt-hour meter. A second specimen, consisting of the same quantities of the *unmixed* pure components, is heated through the same temperature range and the

[9] W. A. Dench, *Trans. Faraday Soc.*, 1963, **59**, 1279.

difference in energy input between this run and the first, after corrections for heat losses, provides the enthalpy of formation of the alloy. The calorimeter can also be used for average heat capacity determinations over 20 to 30 K intervals and for measurements of enthalpies of transformation.

Predel and Mohs[10] have recently described an adiabatic calorimeter capable of operating at temperatures up to 1875 K and have used it to measure the enthalpies of fusion of Fe, Co, and Ni and to determine enthalpies of mixing of liquid Co + Ni alloys. A further recent modification has been introduced by Sale[11] who has constructed an adiabatic calorimeter which is spherical in design and automatic in operation. The shape of the apparatus leads to a more uniform temperature distribution within the calorimeter. Measurements can be made in the range 650 to 1750 K.

Isoperibol Calorimeters.—The isoperibol calorimeter, operated in a thermoconstant jacket and therefore formerly called 'isothermal', is the most widely used type of calorimeter at room temperature. Enthalpies of solution in aqueous and organic solvents and enthalpies (or energies) of combustion in oxygen, fluorine, or chlorine are more often than not determined in isoperibol calorimeters. Although many of these results are pertinent to metallurgical thermochemistry, only high-temperature methods are being considered in the present chapter.

The pioneering though rather inaccurate work of Kawakami[12] on the enthalpies of mixing of liquid alloys may be mentioned in passing. The much improved calorimeter for the determination of enthalpies of solution in liquid tin constructed by McKisson and Bromley[13] may also be mentioned. The most sensitive isoperibol calorimeter is the one designed by Calvet.[14] In this design, two radial thermopiles enclosing reaction vessels are situated in wells in a massive thermoconstant block. The two thermopiles are connected differentially and the reaction is made to occur in only one reaction vessel. It is possible with this type of calorimeter to measure the enthalpy of solution of, say, a small additional amount of gas in a solution of the gas in a metal without changing the concentration significantly, that is, to measure quasi-partial enthalpies of solution (see p. 322). An example is the study of the partial enthalpy of solution of oxygen in the homogeneity range of the phase UO_{2+x} at 1073 K by Gerdanian and Dodé.[15] Calvet calorimeters for measurements at temperatures up to 1275 K are available commercially.

[10] B. Predel and R. Mohs, *Arch. Eisenhüttenw.*, 1970, **41**, 143.
[11] F. R. Sale, *J. Sci. Instr.*, 1970, **3**, 653.
[12] M. Kawakami, *Sci. Reports Tôhoku Imp. Univ.*, 1927, **16**, 915; 1930, **19**, 521.
[13] R. L. McKisson and L. A. Bromley, *J. Amer. Chem. Soc.*, 1951, **73**, 314.
[14] E. Calvet and H. Prat, 'Microcalorimétrie: Applications Physico-Chimiques et Biologiques', Masson, Paris, 1956; 'Recent Progress in Microcalorimetry', ed. H. A. Skinner, Pergamon, London, 1963.
[15] P. Gerdanian and M. Dodé, 'Thermodynamics of Nuclear Materials', Proceedings I.A.E.A. Symposium, Vienna, 1968, p. 41.

Accurate work at temperatures above about 575 K employing isoperibol calorimeters has also been done by Wittig[16] and Kleppa.[17] The former measured mostly enthalpies of mixing in liquid alloys, whereas the latter studied in addition enthalpies of mixing in liquid halide and nitrate systems as well as enthalpies of solution of oxides or double oxides in liquid oxides. Kleppa's calorimetric designs have therefore found fairly wide application. As in the previous section, only one representative calorimeter of a particular type has been selected for a more detailed description. In this case, the adaptation of Kleppa's design by Pool[18] as used for the determination[19] of the enthalpies of formation of copper + manganese alloys has been chosen. The apparatus is illustrated in Figure 4.

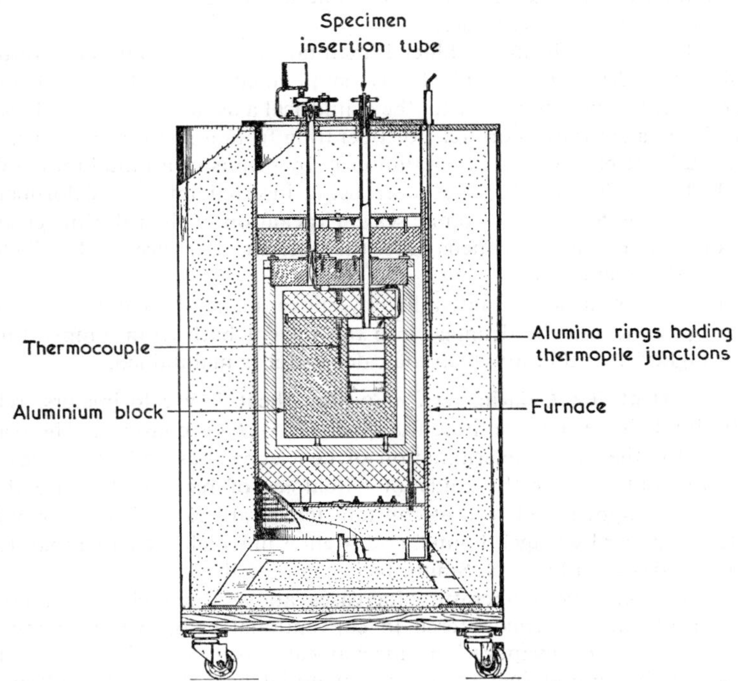

Figure 4 *Isoperibol liquid-metal solution calorimeter of Pool*

This calorimeter also operates on a differential twin-well basis in that, during a set of measurements, two of three identical calorimeter wells are used together, one well being 'active', the other 'passive'. The wells contain

[16] F. E. Wittig, W. Schilling, and G. Keil, *Z. Metallkunde*, 1959, **10**, 610; 1963, **54**, 576.
[17] O. J. Kleppa, *J. Phys. Chem.*, 1960, **64**, 1937; T. Østvold and O. J. Kleppa, *Inorg. Chem.*, 1969, **8**, 78.
[18] M. J. Pool, *Trans. Metallurg. Soc. A.I.M.E.*, 1965, **233**, 1711.
[19] P. J. Spencer, Final Tech. Report, NSF Grant No. GK-2752, 1969.

a liquid-metal solvent bath and are situated in a massive aluminium block which provides a constant-temperature surrounding. Forty-eight junctions of a thermopile are distributed over the surface of each well immediately adjacent to the solvent bath and the other forty-eight junctions are next to the aluminium calorimeter block. During a series of measurements the thermopile of the 'active' well is connected to that of the 'passive' well to record differences in e.m.f. Any e.m.f. resulting from a temperature change which occurs at the outer junctions, and which will be the same for the two wells because of the high conductivity of aluminium, will thus cause no shift of the zero differential reading. The electrical output is thus indicative of the true temperature difference between the 'active' and 'passive' wells. The e.m.f. from the differential thermopile is fed to a d.c. amplifier and thence to a strip chart recorder.

The aluminium calorimeter block is maintained at a constant temperature within a cylindrical furnace which has main and control heaters parallel to, and end heaters perpendicular to, the cylindrical axis of the furnace. Power input is *via* a saturable core reactor operating from a resistance bridge.

The solvent bath is stirred continuously by means of a tantalum paddle attached to a Pyrex tube, the latter being inserted into the calorimeter through a double O-ring dynamic seal. Evacuation of the calorimeter and introduction of the specimens into the solvent bath are both effected through the Pyrex tube.

A preheating furnace situated directly above the 'active' calorimeter well allows specimens to be dropped into the solvent bath from temperatures much higher than that at which the calorimeter is maintained.

Constant Heat-flow Calorimeters.—Constant heat-flow calorimeters, relatively simple in design, are those for which $(T_s - T_c)$ is constant. These are suitable for the direct determination of heat capacities and enthalpies of transition but not for the study of enthalpies of reaction because they cannot be 'stopped' and held at constant temperature. Reference may be made to a paper by Hagel, Pound, and Mehl [20] in which such an apparatus, originally designed by C. S. Smith, has been described.

As the operating temperature increases above the region of 1500 K severe problems are encountered in conventional calorimeters owing to increased rates of evaporation, mechanical creep, diffusion into, and reaction with, the container and support material, heat transfer by radiation, *etc*. The effects of the increased rates can be reduced by shortening the duration of the experiment, and this has led to the development of *pulse* or *high-speed* calorimeters for the measurement of heat capacities and enthalpies of transformation, using heating durations usually shorter than 10 s. They have been applied so far mainly to electrical conductors and semiconductors, and could be better classified in a non-quantitative manner as calorimeters making use of direct electrical resistive heating for supplying energy to a specimen in the form of a wire, rod, or tube.

[20] W. C. Hagel, G. M. Pound, and R. F. Mehl, *Acta Metallurgica*, 1956, **4**, 37.

The energy may be supplied from a battery or a regulated power supply, the power input being determined by measurements of current and potential difference, or it may be supplied as a discrete pulse by discharging a calibrated capacitor-bank through the specimen, the energy being determined by the capacitance and the initial potential.

The specimen temperature may be measured by means of a thermocouple attached to the specimen, or it may be derived from the electrical resistance (this requiring calibration in separate steady-state experiments), or it may be measured by automatic optical methods. The potential differences produced by the transducers used in the measurements are recorded either by using an oscillograph, or by photographing oscilloscope traces, or by using high-speed digital data-acquisition systems.

Several different modes of operation may be distinguished, among them the following.

(*i*) The specimen is initially in thermal equilibrium with the surroundings, which are kept at room temperature. A continuous current is then supplied to the specimen, and measurements of potential difference, current, and temperature are continuously recorded *versus* time during the heating, giving a heat capacity *versus* temperature curve for the whole of the range covered. For example, Cezairliyan *et al.*[21a] measured the heat capacity of molybdenum between 1900 and 2800 K, as well as its electrical resistivity and thermal radiation properties.

(*ii*) The specimen is initially in thermal equilibrium at an elevated temperature in a furnace. An increment in temperature of a few kelvins is produced by delivering a pulse of electrical energy to the specimen, one heat capacity value being obtained for each pulse. This procedure is then repeated for different starting temperatures. For example, Parker[21b] measured the enthalpy of transformation α- to β-titanium at 1155.6 K.

(*iii*) The surroundings are kept at room temperature and the specimen is held initially in thermal equilibrium at an elevated temperature by passing a steady current through it. The input current is then made to undergo a step-change, and measurements are made as in (*i*) during the subsequent heating. For example, Kollie[21c] measured the heat capacity of iron between 300 and 1175 K.

In (*ii*) the specimen is held for a relatively long time at elevated temperatures and some of the advantages of the pulse method are lost. However, it does allow steady-state measurements of starting-temperatures to be made and requires a less elaborate data-recording system than (*i*). Method (*iii*) would seem to have little advantage over (*i*) and has not been much used.

Cezairliyan has reviewed high-speed methods in two recent publications.[22]

[21] (*a*) A. Cezairliyan, M. S. Morse, H. A. Berman, and C. W. Beckett, *J. Res. Nat. Bur. Stand.*, 1970, **74A**, 65; (*b*) R. Parker, *Trans. Metallurg. Soc. A.I.M.E.*, 1965, **233**, 1545; (*c*) T. G. Kollie, *Rev. Sci. Instr.*, 1967, **38**, 1452.

[22] A. Cezairliyan, *High Temp.-High Press.*, 1969, **1**, 517; *Rev. Internat. Hautes Tempér. et Réfract.*, 1970, **7**, 215.

5 Electromotive Force Measurements

Many chemical reactions can be made to proceed in a galvanic cell and if the e.m.f., E, is measured by a compensation method, the Gibbs energy of the chemical reaction, ΔG, can be calculated by means of the equation:

$$\Delta G = - zFE,$$

where z is the stoichiometric number of electrons transferred in the equation for the cell reaction and F is Faraday's constant (23 066 cal_{th}/V). The best-known examples are probably the Daniell cell and the formation of water from the constituent gases, *e.g.*

$$Pt(H_2) \,|\, \{H_2O\} \,|\, (O_2)\, Pt.$$

If both hydrogen and oxygen are at atmospheric pressure, the e.m.f. of the cell at zero current-flow gives the standard Gibbs energy ($z = 2$) of the reaction:

$$(H_2) + \tfrac{1}{2}(O_2) = \{H_2O\}.$$

Every galvanic cell consists of an electrolyte which must be predominantly an ionic conductor and two electrodes which must also conduct electronically. Although the e.m.f. is determined when no current flows, in the actual measurement the transportation of some ions to the electrodes cannot be avoided. Unless diffusion rates are high, transport of materials may lead to changes in concentration in the electrodes and the electrolyte and thus falsify the measurements. This applies particularly when any part of the cell is a solid, whereas diffusion rates in liquids often suffice to secure equilibration. Hence amalgam electrodes have found a wide application at room temperature and have also been used to measure activities of metals dissolved in mercury.

Generally, however, it is necessary to raise the temperature substantially to increase the diffusion and equilibration rates and thus under these conditions the application of e.m.f. measurements to the determination of Gibbs energies is truly a high-temperature method.

A large variety of cells have been employed to obtain thermochemical data of metallurgical interest. The Gibbs energies of formation of metal chlorides have frequently been determined, starting with the early work of Lorenz, in cells of the type:

$$\langle C\rangle\{M\} \,|\, \{MCl_x\} \,|\, (Cl_2)\langle C\rangle.$$

According to Lorenz no significant junction potentials arise between the chlorides of different metals so that e.m.f.'s of cells of the type:

$$\{Zn\} \,|\, \{ZnCl_2\} \,|\, \{PbCl_2\} \,|\, \{Pb\},$$

produce good Gibbs energies, in this case for the reaction:

$$\{Zn\} + \{PbCl_2\} = \{ZnCl_2\} + \{Pb\}.$$

Molten halides, in particular those of the alkali metals, have always been popular solvents for the electrolyte because of their strongly ionic nature and have been applied on a large scale to the investigation of alloys. The determination of the partial molar Gibbs energy of solution of zinc in brasses of varying composition may serve as one example of many. A typical cell is:

$$\{Zn\} \mid [ZnCl_2]_{\{KCl + LiCl\}} \mid [Zn]_{brass}$$

Several per cent of zinc chloride is dissolved in some low-melting eutectic of alkali chlorides and, when the cell produces current, zinc ions move to the alloy where they are dissolved. To maintain equilibrium the precipitated zinc must dissolve quickly in the brass electrode so that no change in concentration occurs at its surface. This demonstrates the necessity of working at sufficiently high temperatures. Another important condition is that the two components of the alloy must differ substantially in electronegativity (say, by more than 0.5 V), otherwise concentration cells may be set up which would affect the results of the cell proper. This problem has been discussed in some detail by Wagner.[23]

The use of iodide instead of chloride electrolytes may sometimes offer a small advantage, namely when the difference in affinity of the component metals for iodine is more pronounced than that for chlorine. This is probably why Vecher and Plazun [24] used iodides for the investigation of Bi–Sb–Ag alloys in the cell:

$$\langle Ag \rangle \mid [AgI]_{\{KI + NaI\}} \mid [Ag]_{\{Ag + Bi + Sb\}}.$$

Of the various experimental problems connected with the use of molten chloride electrolytes, incomplete dehydration forms a major source of errors. Because of such difficulties and the need to work at higher temperatures than say 975 K, the search for other suitable electrolytes has been continuously maintained. The advantage of using solid electrolytes was gradually recognized. Sodium glasses are predominantly sodium ion conductors, and Hauffe [25] made use of this knowledge for the determination of the Gibbs energy of solution of sodium in liquid alloys, setting up and measuring the e.m.f. of cells of the type:

$$\{Na\} \mid \text{Glass} \mid [Na]_{\{Na + Cd\}}.$$

Kubaschewski and Huchler [26] pursued this approach. They first proved that, *e.g.*, silver dissolves in glass in the form of ions, and then used a glass saturated with silver as the electrolyte for the investigations (575 to 875 K) of the thermochemical properties of silver + gold alloys in the cell:

$$\langle Ag \rangle \mid [Ag^+]_{glass} \mid [Ag]_{\langle Ag + Au \rangle}.$$

Further progress was achieved by Kiukkola and Wagner [27] who introduced, as electrolytes, stable oxides of high melting temperature that are

[23] C. Wagner and C. Werner, *J. Electrochem. Soc.*, 1963, **110**, 326.
[24] R. A. Vecher and V. K. Plazun, *Zhur. fiz. Khim.*, 1968, **42**, 667.
[25] K. Hauffe, *Z. Elektrochem.*, 1940, **46**, 348.
[26] O. Kubaschewski and O. Huchler, *Z. Elektrochem.*, 1948, **52**, 170.
[27] K. Kiukkola and C. Wagner, *J. Electrochem. Soc.*, 1957, **104**, 308, 379.

oxygen-ion conductors even above 1275 K. The electrolyte used was a solution of about 18 moles per cent of CaO in ZrO_2. This electrolyte conducts electrical charges predominantly by oxygen ions at oxygen pressures as low as 10^{-17} atm* (1275 K) and has since become the most popular electrolyte for thermochemical investigations at high temperatures. The cell:

$$\langle Cu, Cu_2O\rangle \,|\, \langle ZrO_2 + CaO\rangle \,|\, \langle NiO, Ni\rangle,$$

was first studied by Kiukkola and Wagner and has since frequently been employed for calibration purposes.

The experimental procedure may be illustrated by a discussion of the cell constructed by Dench[28] and used for the investigation of cobalt + copper alloys by the cell

$$\langle Co, CoO\rangle \,|\, \langle Zr_{0.85}Ca_{0.15}O_{1.85}\rangle \,|\, \langle [Co]_{\langle Co + Cu\rangle}, CoO\rangle.$$

This particular cell has produced the most reproducible e.m.f.'s on alloys so far and shows some features which may be of general interest.

The electrolyte was used in the form of a stabilized zirconia tube 450 mm long, 12.7 mm o.d., 9 mm i.d., closed at one end. The electrodes, made by compacting the mixed metal and oxide powders in a cylindrical die, were pressed against the closed end of the zirconia tube. The cell was heated in an alumina work-tube in a vertical platinum alloy-wound resistance furnace. With the cell in the uniform temperature zone of the furnace the open end of the zirconia tube extended into the cold zone, and in initial experiments one electrode was placed inside and the other outside. The reason for using this configuration was to prevent interaction between the electrodes through the gas phase.

The cell was supported on an alumina platform suspended from a hoist fitted with a rotary shaft-seal. By means of the hoist, the cell, without being opened to air, could be lifted out of the work-tube into a chamber sealed to the top of the work-tube. This enabled the work-tube to be outgassed by heating *in vacuo* before the cell was lowered into the hot-zone, and also enabled the cell to be cooled down and the electrodes replaced, without cooling the furnace to room temperature. After outgassing the system was filled with purified argon.

The apparatus was first tested with the cell

$$\langle Cu, Cu_2O\rangle \,|\, \langle ZrO_2 + CaO\rangle \,|\, \langle Ni, NiO\rangle,$$

and the results showed agreement with previous measurements with this cell, the e.m.f. being about 270 mV at 1123 K. However, on testing the arrangement with the symmetrical cell

$$\langle Co, CoO\rangle \,|\, \langle ZrO_2 + CaO\rangle \,|\, \langle Co, CoO\rangle,$$

deviations from zero e.m.f. of about 1 mV were found. To determine whether these deviations were due to the geometrical asymmetry of the

[28] W. A. Dench and O. Kubaschewski, *High Temp.-High Press.*, 1969, **1**, 357.

* atm = 1.013 25 × 10^5 Pa.

cell, two electrodes were placed side-by-side as shown in Figure 5, the alumina platform being divided into two electrically separate parts. With all electrodes of Co, CoO the e.m.f. produced between the two outer electrodes was only about 0.07 mV, while about 1 mV was still produced between either outer electrode and the inner electrode.

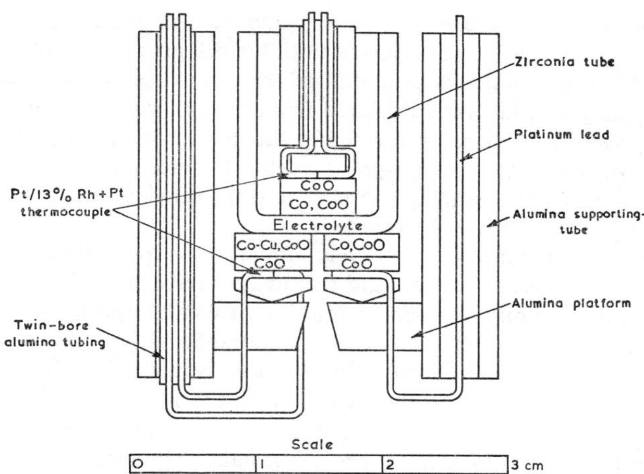

Figure 5 *Part of an e.m.f. cell employing a solid electrolyte*
(Reproduced by permission from *High Temp.-High Press.*, 1969, **1**, 357)

This modified arrangement was used for measuring the e.m.f.'s (between 0 and 6 mV) produced by two-phase Co + Cu alloys between 1025 and 1550 K, the electrode inside the tube now being, of course, redundant. The purpose of the CoO layers shown in Figure 5 was to prevent alloying between the platinum leads and the cobalt in the electrodes.

An example of the use of the zirconia + calcia electrolyte for the investigation of liquid alloys is provided by Fruehan [29] who determined the activity of chromium in molten chromium + nickel alloys at 1875 K by measuring the e.m.f. of the galvanic cell:

$$\langle \mathrm{Cr}, \mathrm{Cr_2O_3} \rangle \,|\, \langle \mathrm{ZrO_2} + \mathrm{CaO} \rangle \,|\, \{\mathrm{Ni} + \mathrm{Cr}\}, \langle \mathrm{Cr_2O_3} \rangle.$$

At 1875 K, the dissociation pressure of chromic oxide is about 10^{-12} atm, whereas Fruehan demonstrated that zirconia + calcia exhibits insignificant electronic conductivity down to about 3×10^{-13} atm at that temperature.

It has been mentioned above that zirconia + calcia solutions exhibit predominantly ionic conductivity at oxygen pressures of 10^{-17} atm. At lower oxygen pressures, however, electronic conductivity begins to limit the usefulness of the electrolyte. In other words, the electrolyte cannot be employed reliably in the usual cell arrangement with metals the affinity of

[29] R. J. Fruehan, *Trans. Metallurg. Soc. A.I.M.E.*, 1968, **242**, 2007.

which for oxygen is higher than that of, say, molybdenum and tungsten. Briggs, Dench, and Slough[30] tried to overcome this limitation by the following subterfuge. For the investigation of the thermochemical properties of chromium alloys, they set up the cell:

$$\text{Pt}, \langle \text{Cr alloy}, Cr_2O_3 \rangle \,|\, \langle Zr_{0.85}Ca_{0.15}O_{1.85} \rangle \,|\, (H_2O, H_2), \text{Pt}.$$

The composition of the hydrogen + water vapour mixture, which had a total pressure of 1 atm, was adjusted until the cell produced zero e.m.f., that is, until the same chemical potential of oxygen existed at both electrodes. This oxygen potential was calculated from the known hydrogen and water vapour pressures. Thus, by arranging the conditions in such a way that the method becomes a null-point method, interference by electronic conduction is made as small as possible and the range of utility of stabilized zirconia as an electrolyte is extended.

In essence, the null-point method is similar to the method of equilibration with $H_2O + H_2$ gas mixtures, and this defines the limits of its application. There is a certain minimum mole fraction of water vapour in the mixture, say 10^{-5}, which can be accurately maintained. Below this level, water vapour present as an impurity from various sources begins to interfere. For the investigation of reactions involving the more stable metal oxides it is therefore preferable to find oxides that retain predominantly oxygen-ion conduction at very low oxygen potentials. The electrolytes that have so far proved to be most efficient in this respect are based on thoria, suitable solutes being yttria and lanthana. Using this electrolyte, Zador[31] has, for instance, determined the partial molar Gibbs energies of solution of oxygen in the non-stoicheiometric range of the titanium dioxide phase.

The cell may be schematically represented as follows:

$$\langle \text{Fe}, \text{‘FeO’} \rangle \,|\, \langle Th_{0.9}Y_{0.1}O_{1.95} \rangle \,|\, \langle TiO_{2-x} \rangle,$$

and the cell reaction as

$$\tfrac{1}{2}(O_2) = [O]_{TiO_{2-x}},$$

where x indicates the deviation from stoichiometry $[2 - x = n(\text{O})/n(\text{Ti})]$. Non-stoicheiometric niobium dioxide, NbO_{2-x}, has also been investigated by Zador by this method, and the Gibbs energy of formation of niobium monoxide has been investigated by Worrell.[32a]

As in the case of liquid electrolytes, the mobile species can be transferred quantitatively from one electrode to the other by passing measured quantities of electricity through the cell. This process, solid-state coulometric titration, has been applied, for example, by Alcock and Belford[32b] to vary the oxygen content in the metal in the course of measuring the

[30] A. A. Briggs, W. A. Dench, and W. Slough, *J. Chem. Thermodynamics*, 1971, **3**, 43.

[31] S. Zador, ‘Electromotive Force Measurements in High-Temperature Systems’, ed. C. B. Alcock, Inst. Min. Metall., London, 1968, p. 145.

[32] (*a*) W. L. Worrell, ‘Thermodynamics’, Proceedings I.A.E.A. Symposium, 1965, Vienna, 1966, Vol. **1**, p. 131; (*b*) C. B. Alcock and T. N. Belford, *Trans. Faraday, Soc.*, 1964, **60**, 822; 1965, **61**, 443.

standard Gibbs energies of solution and solubilities of oxygen in lead and tin.

Another solid electrolyte that has recently found fairly wide application in thermochemical e.m.f. work is calcium fluoride which is reversible to fluoride ions. Galvanic cells employing this electrolyte have been used to obtain thermochemical data for the Gibbs energy of formation of metal fluorides, carbides, borides, and phosphides at temperatures of 875 to 1120 K. Typical examples are the cells:

$$\langle Ni, NiF_2\rangle \,|\, \langle CaF_2\rangle \,|\, \langle AlF_3, Al\rangle,$$

and

$$\langle Th, ThF_4\rangle \,|\, \langle CaF_2\rangle \,|\, \langle ThF_4, ThC_2, C\rangle.$$

Work in this field up to 1966 has been surveyed by Markin [33] in the 'Proceedings' of a symposium held by the Nuffield Research Group, Imperial College, London in 1967. This volume represents an excellent assessment of e.m.f. measurements in high-temperature systems and is recommended to any thermochemist working in this field or intending to enter it.

Whereas galvanic cells employing solid oxide or solid fluoride electrolytes seem to produce reliable results for the formation of metal oxides and metal fluorides, difficulties continue to be encountered when they are applied to homogeneous alloys and other non-stoichiometric phases. The main difficulty is that of maintaining exactly the right oxygen or fluorine pressure in the neighbourhood of the alloy electrode. Even small deviations from this pressure lead to preferential oxidation or reduction at the electrode-to-electrode interfaces resulting in drifts of the e.m.f.s which cannot easily be explained but which produce unreliable results. In addition, any oxidation or reduction changes the composition of the alloy and this cannot be verified by subsequent analysis. The best countermeasure is to select a standard electrode with an oxygen potential as near to that of the alloy electrode as possible and to use a 'getter' in the neighbourhood of the electrodes with an excess of the same mixture as that used to form the standard electrode.

In conclusion, the commercial use that is now widely made of standard electrodes encased in solid-oxide electrolytes should be mentioned. Such electrodes are used to measure the concentration of oxygen in steel, for instance during deoxidation, and also in liquid copper and silver. Reference may be made to articles by Fitterer.[34]

6 Equilibria with a Gaseous Phase

Partial molar Gibbs energies (chemical potentials) can be obtained by measurement of the equilibrium pressure of a gas phase above a condensed phase, for example above a compound or a solution of given composition.

[33] T. L. Markin, ref. 31, p. 91.

[34] G. R. Fitterer, *J. Metals*, 1966, Aug., p. 1; 1967, Sep., p. 92.

The gas phase may consist simply of one constituent of the substance, or a gas mixture that provides a quantitatively defined lower chemical potential of that constituent, such as mixtures of known composition of $H_2O + H_2$ or $CO_2 + CO$ for oxygen, of $H_2S + H_2$ for sulphur, $CH_4 + H_2$ for carbon, and $NH_3 + H_2$ for nitrogen. A few typical examples may be given for illustration.

In the system Ni + NiO, solid solubilities are small and the Gibbs energy of formation of the oxide can be obtained by measurement of the mole ratio $n(CO_2)/n(CO)$ in the reaction:

$$\langle NiO \rangle + (CO) = \langle Ni \rangle + (CO_2).$$

The titanium–oxygen system (Figure 1, p. 322) exhibits significant homogeneity ranges and the measured Gibbs energies of say oxygen, depend on composition within these phases (Figure 2, p. 323). The oxygen-rich phases may be studied by equilibration with $H_2O + H_2$ mixtures, *i.e.* by the reaction:

$$[O]_{TiO_{2-x}} + (H_2) = (H_2O).$$

For titanium oxides low in oxygen, more powerful methods must be employed, as will be seen below.

Gibbs energies can be obtained by measuring the dissociation pressure directly for alloy systems with one or more 'volatile' components. Taking again the example of brass: by the determination of the vapour pressure of zinc above the alloy as well as above, say, liquid zinc, one obtains the partial molar Gibbs energy of solution, ΔG_{Zn}, by means of the scheme:

$$[Zn]_{brass} = (Zn); \quad p'_{Zn},$$

$$\{Zn\} = (Zn); \quad p^\circ_{Zn}.$$

Hence, for $\{Zn\} = [Zn]_{brass}$

$$\Delta G_{Zn} = RT\ln(p'_{Zn}/p^\circ_{Zn}).$$

Similarly, the Gibbs energies of solution in salt melts can be determined by measuring the vapour pressures above the melts, for instance the vapour pressures of KCl and NaCl over KCl + NaCl melts.

The equilibrium of nitrides with ammonia + hydrogen mixtures is not very efficient because of the relatively low stability of ammonia gas. Direct dissociation of the nitride will often produce satisfactory results provided that a sufficiently high temperature of measurement can be attained experimentally. Thus, Pemsler [35] was able to measure the dissociation pressures of solid solutions of, for example, nitrogen in niobium:

$$[N]_{Nb} = \tfrac{1}{2}(N_2),$$

by using very sensitive methods of detection.

Methane also is not very stable and equilibration of carbides with $CH_4 + H_2$ mixtures finds very limited application. The thermodynamic

[35] J. P. Pemsler, *J. Electrochem. Soc.*, 1961, **108**, 744.

investigation of the very stable metal carbides is a great problem; the e.m.f. method mentioned in the previous section, however, seems to offer some real hope of progress.

A great variety of methods for the measurement of vapour pressures and similar equilibria involving gases have been devised. These will be briefly surveyed in the present section and details will be added only where the developments are relatively recent. It has been mentioned that, for direct determinations of dissociation pressures, the volatility of the components is of paramount importance. Since this applies particularly to the investigation of alloys, a few words may be added to assess the prospects for the study of alloys for uses at high temperatures. Such alloys will consist of the higher melting metals and it is therefore pertinent to compare the vapour pressures of these metals as a function of temperature. In particular, the temperatures at which their vapour pressures are in the range 10^{-3} to 10^{-5} atm are of interest, since this range is most readily accessible to measurement by current high-temperature experimental techniques. In the Table, selected vapour pressure values compiled by Hultgren *et al.*[36] are

Table *Temperatures T corresponding to selected vapour pressures p of the more refractory metals*

p/atm	10^{-5}	10^{-4}	10^{-3}		10^{-5}	10^{-4}	10^{-3}
Metal		T/K		*Metal*		T/K	
Mn	1229	1347	1495	Rh	2290	2496	2749
Ag	1283	1414	1577	Pt	2331	2551	2817
Cu	1510	1662	1851	Ru	2589	2813	3088
Cr	1657	1808	1992	Zr	2640	2893	3199
Au	1664	1834	2043	Hf	2690	2955	3273
Pd	1723	1898	2117	Ir	2714	2958	3253
Fe	1728	1890	2093	Mo	2742	2995	3314
Ni	1784	1951	2156	Nb	2944	3208	3525
Co	1790	1959	2167	Ta	3298	3600	3959
Si	1909	2103	2340	Re	3275	3584	3972
Ti	1980	2180	2435	W	3478	3775	4139
V	2100	2290	2525				

listed for the more refractory metals. The metals are tabulated in approximate order of decreasing volatility, which corresponds to increasing experimental difficulty of investigation.

At temperatures above about 2100 K, problems associated with specimen–container reaction and temperature measurement and control become severe and it can be seen by reference to the Table that accurate vapour-pressure measurements are thereby restricted roughly to metals of greater volatility than, say, vanadium. For investigation of metals of lower volatility than, say, vanadium, either experimental methods capable of measuring pressures much lower than 10^{-5} atm must be employed, or a reaction which is more

[36] R. Hultgren, R. L. Orr, P. D. Anderson, and K. K. Kelley, 'Selected Values of Thermodynamic Properties of Metals and Alloys', Wiley, New York, 1963 (and supplements).

favourable than direct dissociation must be studied (for an example, see below).

Much of the early vapour-pressure work was carried out using static methods of pressure measurement. These included various manometric techniques for materials of relatively high volatility, but for lower pressure ranges methods such as those described by Hargreaves [37] and Herasymenko [38] are more suitable. These involve the establishment of an equilibrium vapour pressure between an alloy specimen held at the high-temperature end of a sealed tube and the pure volatile component of the alloy held at the low-temperature end.

Whereas Hargreaves' method has nearly exhausted its usefulness, the method used by Herasymenko continues to find wide application under the name *isopiestic method*. In this technique the temperatures of the hot and cold ends of the tube are fixed and it is the equilibrium composition of the alloy that is determined. The activity of the volatile component in the alloy at the temperature of the hot end can then be calculated provided that the vapour pressure *versus* temperature relation for the volatile component is known. Komarek and Silver [39] have made interesting use of the method by combining it with a technique employed by Kubaschewski and Dench [40] for the determination of partial molar Gibbs energies of solution of oxygen in titanium. For these studies, specimens consisting of titanium and an alkaline earth (MgO) in close physical contact were placed within the temperature gradient of the sealed tube with an excess of the pure alkaline earth metal being held at the colder end. Reaction was allowed to proceed until equilibrium was reached where the partial pressure of oxygen over the titanium + oxygen alloy was equal to the partial pressure of oxygen in the reaction:

$$(\mathrm{Mg}) + \tfrac{1}{2}(\mathrm{O_2}) = \langle \mathrm{MgO} \rangle.$$

By varying the vapour pressure of the alkaline earth metal from one run to another, the chemical potential of oxygen in the Mg + MgO system was changed and hence, in turn, the chemical potential of oxygen in the titanium + oxygen alloy. The oxygen content of the alloys was calculated from the mass gain of the specimens after equilibration.

A further example of the static method of vapour pressure measurement which after teething troubles is finding increased favour is *absorption spectroscopy*. The method consists of heating a specimen in a closed vessel of known dimensions, usually a silica cell, maintained at a uniform temperature. A characteristic spectral source of high intensity is provided, for example, by the incorporation of the element of interest into a hollow cathode lamp. Radiation from the lamp is passed through the closed

[37] R. Hargreaves, *J. Inst. Metals*, 1939, **64**, 115.
[38] P. Herasymenko, *Acta Metallurgica*, 1956, **4**, 1.
[39] K. L. Komarek and M. Silver, Proceedings I.A.E.A. Symposium 'Thermodynamics of Nuclear Materials', Vienna, 1962, 749.
[40] O. Kubaschewski and W. A. Dench, *J. Inst. Metals*, 1953, **82**, 87.

vessel and the spectral lines characteristic of the element in the cathode are attenuated by the atoms of that element in the radiation path. The extent of absorption is dependent upon the vapour pressure of the species under investigation. The method shows extreme sensitivity and high selectivity, and there is seldom interference among resonant wavelengths of different species so that the vapour pressure of more than one volatile component can be determined in a single experiment. Thus Rapperport and Pemsler [41] have claimed a sensitivity of 10^{-12} atm in measurements of the vapour pressure of Ag and Cd over a silver + cadmium alloy, with a reproducibility of ± 20% in the vapour pressure values. Another application of the method has been described by Kibler *et al.*[42] who used both resonant line absorption and total absorption to determine metal and non-metal activities in ZrB_2 at temperatures between approximately 2000 and 2500 K.

The *transpiration technique* is the most important of the dynamic methods of vapour-pressure measurement for investigations of materials of low volatility. Pressures of about 10^{-1} to 10^{-6} atm can be determined using this technique. In transpiration experiments, vapour from the heated specimen is transported by means of a suitable carrier gas to a condensation area. The vapour pressure is calculated from a knowledge of the initial volume and pressure of the carrier gas, the molar mass of the vapour species, and the mass loss of the specimen and/or mass gain in the collector system. Phillips and Rand [43] have recently determined the vapour pressure of gold at temperatures up to 2100 K using this method, and in their report they discuss the scope and errors of the method in some detail. The most important condition to be observed is that the carrier gas be saturated with vapour from the specimen. It is known that at high flow rates saturation of the gas is not achieved and consequently the vapour density (mass loss per unit volume of carrier gas) is too low. On the other hand, very low flow rates lead to thermal diffusion of the vapour and increased mass losses of the specimen. Optimum working conditions obtain in the 'plateau' range, where the vapour density is independent of flow rate. An important advantage of the transpiration technique is that it offers the possibility for determination of activities of more than one volatile component in an alloy system when their vapour pressures are of a similar magnitude.

A recent and promising modification of the transpiration method incorporates levitation heating of the specimen as problems associated with specimen + container reactions are thereby avoided. Thus Svyzahin *et al.*[44] have combined levitation melting with transpiration measurements, using argon as carrier gas, in determinations of the vapour pressure of iron at temperatures around 2275 K, and Mills and Kinoshita [45] have used

[41] E. J. Rapperport and J. P. Pemsler, *Trans. Metallurg. Soc. A.I.M.E.*, 1968, **242**, 151.
[42] G. M. Kibler, T. F. Lyon, M. J. Linevsky, and V. J. DeSantio, *USAF Tech. Report* No. WADD-TR-60-646, Aug. 1964.
[43] B. A. Phillips and M. H. Rand, *AERE Harwell Report* R.5352, 1967.
[44] A. G. Svyazhin, A. F. Vishkarev, and V. I. Yavoyski, *Russ. Metallurgy*, 1968, **5**, 47.
[45] K. C. Mills and K. Kinoshita, *J. Metals*, 1964, **16**, 107.

levitation melting in thermodynamic studies of liquid iron + nickel alloys at 2275 to 2475 K. The main difficulties associated with levitation studies are those of accurate temperature measurement and control, but if these can be overcome, great advantages are offered for high-temperature vapour-pressure measurements.

The effusion methods are perhaps the most widely applied of the available vapour-pressure techniques at high temperatures. The most straightforward of these from the experimental viewpoint is the *Langmuir method*, in which mass losses due to evaporation from the free surface of a suitably shaped specimen heated in a vacuum are determined. Although the method avoids specimen + container reactions to a large extent and can be used to determine lower vapour pressures than is possible by most other methods, its application is generally limited to pure metals. This is due to the fact that for most materials the evaporation coefficient is not unity, as must be assumed for Langmuir calculations of vapour pressure, but is an unknown factor which can be temperature-dependent.

The *Knudsen effusion method* is the most popular method for determination of vapour pressures at high temperatures. In this a specimen is contained in a small cell which is heated in a vacuum. Vapour in equilibrium with the specimen effuses through an orifice in the lid of the cell. If m is the mass of material effusing in time t and at temperature T, from an orifice of area A, then the vapour pressure is given by the expression:

$$p = (m/KAt)(2\pi RT/M)^{1/2},$$

where K is a correction factor allowing for the finite length of the orifice, R is the gas constant, and M is the molar mass of the effusing species. The general range of pressures which can be measured by the Knudsen method is 10^{-3} to 10^{-7} atm and various techniques can be used to determine the m/t factor in the above equation. For example, the mass loss from the cell can be determined by weighing before and after an experiment, with suitable corrections being made for the loss incurred during the heating and cooling periods. A more accurate method incorporates continuous weighing of the cell during an experiment by use of a microbalance. Condensation and chemical analysis of the effusate, although sometimes a less reliable procedure, permits investigations of systems which contain more than one volatile species. The sensitivity of the method can be improved by the use of radioactive tracers, as for example in the work of Kubaschewski, Heymer, and Dench [46] in thermodynamic studies of chromium alloys. The apparatus of these investigators is illustrated in Figure 6. The main features to be noted are the methods used to provide uniform heating of the effusion cell and to collect and analyse the effusate. The furnace is resistance-heated, with tantalum and molybdenum radiation shields. A separately controlled top heater is used to compensate for the heat losses through the tube. The temperature of the effusion cell is kept constant to ± 0.5 K in the range

[46] O. Kubaschewski, G. Heymer, and W. A. Dench, *Z. Elektrochem.*, 1960, **64**, 801.

1275 to 1675 K. The effusate is collected partly on the inner surface of the molybdenum tube standing on the orifice-disk and partly on a tantalum target-disk held opposite the end of the tube. After solution of both the effusate and the molybdenum tube in acid, the radioactivity of the ^{51}Cr is determined with a Geiger counter and compared with the activity of a

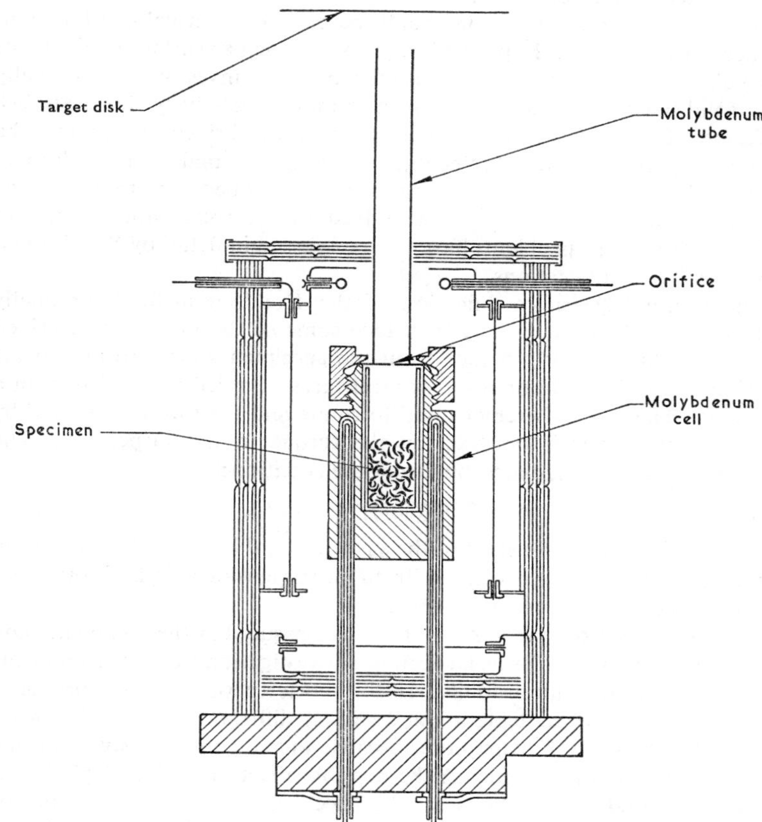

Figure 6 *Effusion cell and furnace for the determination of the vapour pressures of metals and alloys at* 1475 *to* 1675 K
(Reproduced by permission from *Z. Elektrochem.*, 1960, **64**, 801)

standard solution containing a known amount of Cr from the same irradiation batch as that from which the chromium alloy was prepared. The reaction

$$[\mathrm{Si}]_{\langle \mathrm{Me}+\mathrm{Si}\rangle} + \langle \mathrm{SiO_2}\rangle = 2(\mathrm{SiO}),$$

where Me stands for Cr, Mo, and W, has been studied in the range 1350 to 1700 K by T. G. Chart and Kubaschewski in as yet unpublished work, using the Knudsen effusion method. For the silicides of these metals, silicon monoxide pressures within the Knudsen method range are attained at temperatures about 700 K below those required for attaining the same pressures by direct dissociation of the silicides. Knudsen cells were constructed from silica and their mass losses were determined continuously, using a vacuum microbalance.

In all cases, careful attention should be paid to the usual conditions of Knudsen experiments. In particular, problems such as surface depletion of the volatile component in the specimen must be investigated, especially with solid materials where diffusion rates may be relatively slow. Surface diffusion of the effusate round the edges of the orifice, which may be particularly marked for knife-edge orifices, can make an additional contribution to the measured pressure, and non-obedience to the cosine law by the effusing vapour can also lead to incorrect vapour pressure values. All of these problems have been discussed in detail by Ward [47] in a recent series of publications.

The *torsion–effusion* modification of the Knudsen method, originally introduced by Volmer,[48] helps to reduce some of the problems described above. In this method, vapour from the specimen effuses from two (or possibly more) holes in opposite vertical faces of a cell suspended from a fine wire. The cell is thereby caused to rotate and the torque imparted by the effusing vapour is opposed by the torque in the suspension. The vapour pressure is calculated by use of the equation:

$$p = 2\tau\theta/(a_1q_1f_1 + a_2q_2f_2),$$

where τ is the torsion constant of the suspension, θ is the angle of rotation, and a, q, and f are the area, distance from the suspension axis, and correction factor respectively for orifices 1 and 2.

The range of vapour pressures that can be measured by the torsion method is basically the same as for usual Knudsen investigations, but measurements can be made much more rapidly. In the shorter period of time, problems associated with surface depletion and surface diffusion through the orifice are greatly reduced. Furthermore, because of the continuously recording nature of torsion measurements, the possible effect of surface depletion at the experimental temperatures can be checked. This has been done for example by Hultgren and Roy.[49] As can be seen from the torsion–effusion equation, no knowledge of the molar mass of the effusing species is required for calculation of vapour pressure values. The method has been used by Spencer and Pratt [50] to measure the vapour pressure of manganese both in the pure state and over its alloys, *e.g.* with copper [51] and gold[52].

[47] J. W. Ward, *J. Chem. Phys.*, 1967, **77**, 5, 10, 1710, 1718, 4030.
[48] M. Volmer, *Z. phys. Chem.*, *Bodenstein Festbd.*, 1931, 863.
[49] R. Hultgren and P. Roy, *Trans. Metallurg. Soc. A.I.M.E.*, 1965, **233**, 1811.
[50] P. J. Spencer and J. N. Pratt, *Brit. J. Appl. Phys.*, 1967, **18**, 1473.
[51] P. J. Spencer and J. N. Pratt, *Trans. Faraday Soc.*, 1968, **64**, 1470.
[52] P. J. Spencer and J. N. Pratt, *Rev. Internat. Hautes Tempér. et Réfract.*, 1968, **5**, 155.

The Knudsen and torsion–effusion methods have been combined in a single apparatus, for example by Lindscheid and Lange.[53] In this work the torsion cell was suspended from a microbalance, thus enabling simultaneous observations of mass loss and cell rotation to be made. The values of vapour pressure calculated from the two sets of results can be combined through the Knudsen and torsion–effusion equations to obtain the molar mass of the vapour species. Lindscheid and Lange carried out such measurements for Fe, Co, and Ni, and for Ni + Co alloys.

A further modification of the torsion method which incorporates evaporation from a free surface has been termed the *torsion–Langmuir* method by Alcock and Peleg[54] and has been used to investigate the vaporization kinetics of a number of ceramic oxides at temperatures around 2300 K. This particular investigation is of added interest because of the attempt made to determine evaporation coefficients directly by observing the torque resulting from the opposing effects of evaporation from a free surface (Langmuir evaporation) and of equilibrium effusion through an orifice.

Perhaps the most important recent addition to the techniques used for high-temperature vapour-pressure measurements is the mass spectrometer. The instrument can be used in conjunction with a Knudsen effusion cell to identify and investigate different gaseous species in a gas mixture, to measure dissociation pressures of compounds in the condensed state, or to obtain activities for more than one volatile component in an alloy system. The use of the mass spectrometer in high-temperature vaporization studies has been reviewed by Inghram and Drowart,[55] and recent examples of the work of the latter author include determinations of dissociation energies of gaseous rare-earth monoselenides and monotellurides[56] and of borides and silicides of some transition metals.[57] The sensitivity of the method is high and scope is offered for many different types of vaporization study. It should nevertheless be borne in mind that the usual errors of Knudsen experiments must be considered in addition to the variable calibration errors of the spectrometer itself. Particular attention should therefore be paid to the optimum conditions for operation of the Knudsen cell and detailed preliminary studies of orifice sizes, possible depletion or diffusion effects, and experimental temperature ranges are essential to avoid large experimental errors. Calibration problems have frequently hindered mass-spectrometric vapour-pressure studies, but in recent investigations of binary alloy systems in which both components have a significant vapour pressure,

[53] H. Lindscheid and W. Lange *Z. Metallkunde*, 1970, **61**, 193.
[54] C. B. Alcock and M. Peleg, *Trans. Brit. Ceram. Soc.*, 1967, **66**, 217.
[55] M. G. Inghram and J. Drowart, Proceedings of the International Symposium on High Temperature Technology, McGraw-Hill, New York, 1960, p. 219.
[56] C. Bergman, P. Coppens, J. Drowart, and S. Smoes, *Trans. Faraday Soc.*, 1970, **66**, 800.
[57] A. Vander Auwera-Mahieu, R. Peeters, N. S. McIntyre, and J. Drowart, *Trans. Faraday Soc.*, 1970, **66**, 809.

Belton and Fruehan[58] claim to have overcome problems associated with changes in instrument sensitivity or cell geometry by calculating activities from a series of measurements on the *ratio* of ion currents of the components. Their results for the Fe + Ni system[58] were found to be in excellent agreement with previous work and subsequent studies of liquid Fe + Al and Ag + Al[59] alloys have confirmed the reliability of their method.

As an example of a newly devised method for vapour pressure measurements, the isotope-exchange technique of De Dyk Man and Nesmeyanov[60] is described briefly. In this, disk-shaped specimens of active and inactive metal (or alloy) are placed in recesses on opposite sides of an exchange chamber and are separated by a molybdenum screen. When the chamber has been heated to the required temperature, the screen is removed to allow exchange of vapour between the active and inactive specimens. After a given time the screen is replaced and, after cooling, the activity of the initially inactive specimen is measured. Curves of activity against exposure time are obtained for different temperatures. The pressure is determined from the kinetic straight-line curves according to the formula:*

$$p/\text{Torr} = 17.14(I/I_0)(T/\text{K})^{1/2}/(M/\text{g mol}^{-1})^{1/2}(tA/\text{s cm}^2),$$

where I denotes target activity, I_0 specific activity, T thermodynamic temperature, M molar mass of vapour, t exposure time, and A specimen area.

In the experiments of De Dyk Man and Nesmeyanov on cobalt, nickel, and cobalt + nickel alloys, an uncertain Langmuir evaporation coefficient, α, had to be incorporated into the calculations of vapour pressure. This resulted in large inaccuracies in the final data. Nevertheless the method offers some advantages if it can be improved, particularly for studies of alloys whose components have similar vapour pressures.

7 Comparison of Gibbs Energy Methods

A discussion of vapour-pressure measurement techniques is perhaps best concluded by a survey of the relative merits and errors of the various methods available. This is particularly difficult when chemical errors vary with the system under investigation, when vapour pressures can change by very large amounts over small composition ranges, for example in alloy systems, and where the methods themselves may not be directly comparable in respect of the range of pressures to which they are applied. It is worth emphasizing, however, that vapour pressure is a very sensitive function of temperature and it is therefore important to be certain that recorded experimental temperatures are the true temperatures of measurement, no matter what experimental method is employed. As temperatures of

[58] G. R. Belton and R. J. Fruehan, *J. Phys. Chem.*, 1967, **71**, 1403.
[59] G. R. Belton and R. J. Fruehan, *Trans. Metallurg. Soc. A.I.M.E.*, 1969, **245**, 113.
[60] De Dyk Man and A. N. Nesmeyanov, *Russ. Metallurg. and Fuels*, 1960, (1), 50.

* Torr = atm/760.

investigation are increased and optical pyrometry is used for temperature measurement, the accuracy of vapour pressures decreases significantly. Errors of ± 20% in vapour pressure values are neither uncommon nor unreasonable, particularly when obtained from high-temperature studies. Phillips and Rand,[43] in a report on transpiration measurements of the vapour pressure of gold, concluded from an assessment of all available data that '. . . even for this material, which has been studied quite extensively, which is not difficult to contain and whose vapour is quite simple, the vapour pressure at its melting temperature of 1336 K cannot yet be given to better than ± 13%. The National Bureau of Standards have subsequently carried out a detailed statistical analysis of vapour pressure values obtained from eleven different laboratories for standard gold materials.[61] The accuracy of their published values corresponds to about ± 10%, but it should be noted that, in every case, vapour pressure measurements were made by an effusion method and thus the possibility remains of similar errors having been introduced into each set of measurements.

For high-temperature investigations static methods such as the Hargreaves or atomic absorption techniques are largely unsuitable. Both have been employed in nearly all cases with a transparent vessel as the container. Measurements are thereby restricted to the maximum working temperature of silica. Attempts have been made to detect the formation of the 'dew' on the walls by other than optical means in the Hargreaves method, and an opaque absorption cell with orifices at either end has been used in the absorption method, but the results obtained are perhaps not completely satisfying. The absorption method is very sensitive, however, so that it may be possible to measure the magnitudes of pressures obtained at lower temperatures, where other techniques would fail.

Problems associated with specimen + container reactions become serious as the temperature of vapour pressure measurement is increased towards 2000 K. For example, in effusion measurements it becomes difficult to find a cell material which is inert to the more refractory metals and which is not itself volatile. In transpiration and isopiestic studies, thermal diffusion errors are very difficult to avoid, although they can be considerably reduced by suitable precautions.

For high-temperature investigations involving more than one vaporizing species the choice of experimental method becomes limited, and although methods such as the transpiration or atomic absorption techniques can be used, it is for studies of this nature that the mass spectrometer is most suitable.

Although an error limit of ± 20% has been mentioned in connection with determinations of vapour pressure, it is frequently activity or partial molar Gibbs energy values that are obtained from vapour pressure studies. It is the ratio p'/p° (see page 338) which then becomes of importance in an

[61] R. C. Paule and J. Mandel, NBS Special Publication 260–19, January, 1970.

error analysis. Since the same sources of error in experimental measurement are normally repeated in determinations of both p' and p°, the errors associated with the ratio of these values will be reduced and a more reasonable general accuracy of activity and Gibbs energy data obtained from vapour pressure studies is ± 5% at best. In studies of alloy systems, however, the possible rapid change of vapour pressure of the volatile species with mole fraction of that component in the alloy can lead to wide variation in the accuracy of the results obtained across the system. Frequently it is the values for alloys with low mole fraction of the vaporizing species that are most inaccurate and this must be remembered if subsequent Gibbs–Duhem integrations are made to obtain values for other components in the system.

In discussions of the relative merits of various methods for obtaining Gibbs energy values, the question is often asked whether to give preference to e.m.f. or to vapour pressure methods for the investigation of compounds and alloys. There is, of course, no general answer to this question. Experimental conditions vary from one chemical system to another and many of the 'chemical errors', such as reaction of the substance investigated with its surroundings, interfere similarly with both types of method. For the few systems to which a variety of methods have been applied with equal thoroughness (*e.g.* Ag + Au, Cu + Zn, NiO, TiO_{2-x}), the e.m.f. method seems to produce slightly better results but, again, care must be taken to avoid sweeping statements. For a given system one method is always likely to be more suitable than another.

The best reproducibility in results obtained on alloys by means of the e.m.f. method appears to be that of Dench [24] on the Co + Cu system. He claims a reproducibility of ± 0.2 mV corresponding to an accuracy of ± 10 cal_{th} mol^{-1}. The available vapour pressure methods for the determination of Gibbs energies do not attain anything approaching this reproducibility although the example investigated by Dench was exceptionally favourable for the e.m.f. method. More normally, the best accuracies justifiably claimed for that method are between ± 1 and ± 5 mV, but even so, where the chemical properties of a system are such that both e.m.f. and vapour pressure methods could be employed, the present authors would choose the former method. Whereas the e.m.f. method may achieve accuracies between ± 10 and ± 250 cal_{th} mol^{-1}, the best accuracy one may expect to attain from a vapour pressure method is, say, ± 50 cal_{th} mol^{-1} but results with an error ± 300 cal_{th} mol^{-1} may still be considered good. Not too much attention should be paid to the statements of authors who quote much smaller errors, because their statements often relate to precision rather than accuracy.

Most of the factors that determine the applicability of a particular method to a particular chemical system have been mentioned in the preceding sections. In addition, the applicability depends on the absolute value of the Gibbs energy measured. This may be illustrated by the example of the metal oxides at an arbitrary temperature of 1273 K. At that temperature, the standard Gibbs energy of dissociation of metal oxides (to O_2) may vary from zero to more than 250 $kcal_{th}$ mol^{-1}. Figure 7 shows schematically the stability ranges within which the four most important methods may profitably be employed. The upper limit of the e.m.f. method using thoria-based electrolytes has not yet been established but is unlikely to reach far beyond the 150 $kcal_{th}$ mol^{-1} mark. There is thus a gap between about 150 and 180 $kcal_{th}$ mol^{-1} which cannot yet be covered experimentally. Other types of chemical system also show such gaps. There is thus ample scope for thermochemists to employ their ingenuity.

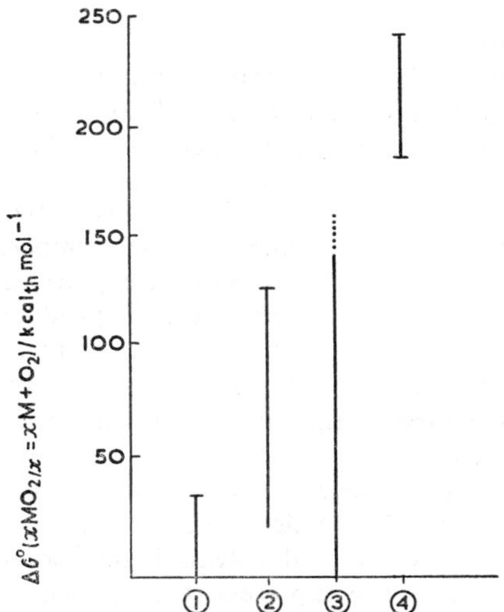

Figure 7 *Schematic representation of the range of application of various experimental methods to the investigation of the Gibbs energies of dissociation of metal oxides at* 1273 K. (1) *Direct dissociation;* (2) *equilibria with* $CO + CO_2$ *or* $H_2 + H_2O$ *mixtures;* (3) *e.m.f. methods employing solid oxygen-ion electrolytes;* (4) *equilibration method after Komarek and Kubaschewski*

Similar restrictions apply to experimental methods for the determination of enthalpies of reactions. The experimental accuracy to be expected from Gibbs energy methods has been delineated above. It must be clearly understood that these limits apply only to the actual Gibbs energy values at any one temperature. Values derived from the temperature coefficients, *e.g.* enthalpies, are likely to be much less reliable, as has been pointed out earlier when it was also mentioned that this shortcoming makes direct enthalpy measurements of paramount importance.

The best accuracy achieved by calorimetric methods, in particular in Calvet calorimetry, is about ± 10 cal_{th} mol^{-1}. That is about the same as the best accuracies obtained in e.m.f. work. Even where the accuracy is not better than, say, ± 100 cal_{th} mol^{-1}, it will still be superior to most enthalpy values obtained by second law evaluations.

Thus it follows that it is not possible to give general advice on the application of experimental methods. Every chemical system must be considered separately and its thermochemical investigation planned accordingly. The use of several independent methods would increase the reliability of the results. The most obvious advice is that once a suitable method has been selected it should be employed with the utmost regard for all the sources of error and the results considered with the exercise of adequate self-criticism by the observer.

8 Conclusion

Since scientific research is likely to remain under pressure for some time to move more closely to the interests of industry, future developments in metallurgical thermodynamics should be considered in the light of industrial demands.

Metallurgical thermodynamics can help industry by providing essential basic information in the development of high-temperature materials for use in power stations, gas turbines, and jet engines, as well as in giving advice on the improvement of processes for the production and refining of metals, mostly of high melting temperature. Compatibility of different materials for use at high temperatures is also a domain of thermodynamic study. For the metallurgical thermochemist, this means increasing use of high-temperature methods.

The most sensitive calorimeter, *i.e.* one of the Calvet type, is now available in a version suitable for application at 1275 K and there is no reason why temperatures a few hundred kelvins higher should not be studied. The adiabatic calorimeter has at present reached its limit at 1675 to 1775 K for the determination of good enthalpies of reaction, but the principle could be applied to temperatures up to, say, 2300 K for the determination of enthalpies of transition which can then be used to convert medium-temperature data to enthalpies of reaction at high temperature. Heat capacities can be measured at even higher temperatures (*e.g.* by pulse calorimetry). Where these methods fail to provide the desired enthalpies of formation, difference methods employing room-temperature calorimeters, such as fluorine combustion and acid solution calorimetry will, of course, continue to be used. The low-temperature calorimeter providing standard entropies has also not yet reached the limits of its range of application to chemical substances of interest to metallurgists.

The future of high-temperature e.m.f. methods is difficult to assess. Development depends mostly on the discovery of ionic conductors of extreme stability and the choice of substances is getting smaller and smaller. Vapour-pressure methods are likely to continue to provide the bulk of results for high-temperature Gibbs energies. The limitations of the

Langmuir method have already been pointed out, but in combination with the torsion–effusion method, that is in the form of the torsion–Langmuir method, its range of application has been considerably widened and temperatures well above 2300 K have been reached. Further variations of this very comprehensive method will probably be developed. Application of the transpiration method has been further extended by its combination with levitation melting thus simplifying the compatibility problem. When temperature measurement and control with levitation melting has been improved, this device is likely to play an increasingly important part in thermochemical research at high temperatures, while equilibration techniques show even greater potential. Mass spectrometry will, of course, continue to be employed frequently.

Further thermochemical work on low-melting-temperature alloys need not be encouraged. Data for high-melting-temperature alloys, on the other hand, are still in short supply. Iron alloys, it is true, have attracted much attention from thermochemists, for obvious reasons, but in view of the large number of elements found in alloyed steels more results are still required. Thermochemists should make sure, however, that the ferrous system they intend to study has not already been adequately investigated. More thermochemical data are needed for alloys of transition metals of technical interest, such as titanium, zirconium, niobium, tantalum, chromium, molybdenum, and tungsten.

The thermodynamic properties of metal chlorides are by now fairly well covered. Although the corresponding quantities for metal bromides and iodides are less well explored, only very limited interest attaches to them. In some cases, they may be useful to assess the possibilities of refining reactions involving the rarer metals. Good data for metal fluorides have recently become available in quantity and it is hoped that this effort will continue somewhat longer.

The thermodynamic properties of binary oxides are quite well covered but, for the high-affinity chalcogenides, the data are still lacking in desirable accuracy. The non-stoichiometric regions (*e.g.* titanium + oxygen) require further investigation, in particular the dilute solutions of oxygen in the metals. Quite a number of double oxides have been studied, but again investigators should pay much more attention to the, often pronounced, non-stoichiometric nature of oxide + oxide systems. Multicomponent oxides that occur in various slags, and not only ferrous slags, are of particular interest.

Similar considerations apply to the metal sulphides. Good thermochemical quantities for sulphur in dilute solution in transition metals are conspicuously absent. Recovery of small amounts of selenium and tellurium from sulphidic ores is gaining interest. Whether this position warrants an extensive study of the thermochemistry of compounds of selenium and tellurium is difficult to say. For the time being, the available information suffices.

Nitrogen is often an important impurity in metals. The non-stoichiometric nature of many systems of nitrogen with transition metals has been largely disregarded by thermochemists. Particular attention should again be paid to the dilute solutions. Interaction coefficients of nitrogen with alloying metals in solid iron have to some extent become available but further work will be welcome. The thermochemistry of the system iron + phosphorus and data on ternary solutions of phosphorus in iron could be improved.

The lack of good data for transition-metal + carbon systems, in particular those of high affinity, is also noteworthy. The reason is the non-stoichiometric nature of such systems, the high melting temperatures, and the difficulty of finding suitable experimental methods. Even the phase diagrams of high-melting-temperature carbide systems are often hopelessly inadequate so that good thermochemical research on a systematic basis would be very rewarding.

Interest in metal silicides stems from their potential use as high-temperature materials. Thermochemical investigation is somewhat simpler than with the carbides because they tend less to non-stoichiometry and experimental methods are more easily devised. Work on these substances on a modest scale is to be encouraged.

Interest in the metal borides is due mostly to the presence of boron in certain steels and its interaction with other alloying elements. An example of other industrial use of borides is the use of titanium borides for the grain refining of aluminium. Thermochemical data for metal borides are, however, virtually non-existent. The reason, again, is the difficulty of finding suitable experimental methods for enthalpy and Gibbs energy determinations.

When a sufficient number of thermochemical values have been obtained by one or more investigators for a particular system, the task of the thermochemist is still not finished. It is then necessary to convert the various experimental results into a consistent and complete set of data. The critical compilation of the thermodynamic properties of chemical substances has recently developed into a field of its own (see ref. 62). It requires highly skilled personnel who have not only to know the fundamental thermodynamic principles and rules but must also know the advantages and limitations of the experimental methods, and have in addition a flair for the handling of thermochemical values and the ability to recognize whether a value is likely to be correct. Because of these high demands, the number of skilled compilers is still rather small and individual investigators should be encouraged to try their expertise in critical compilation, at least within their more or less limited field of interest.

[62] M. H. Rand and O. Kubaschewski, 'Metallurgical Chemistry', Proceedings of a Symposium held at Brunel University and N.P.L., July 1971, HMSO, 1972, Paper 5.1.

Thus, there is a wide variety of problems that confront the metallurgical thermochemist and demand a profound knowledge of his chosen subject. It is a challenge that should attract the best of a young generation of physical chemists.

Author Index